$$\begin{array}{r} \overset{8}{\cancel{9}}\overset{9}{\cancel{0}}\overset{9}{\cancel{0}}\overset{10}{\cancel{0}} \\ -\quad 4 \\ \hline 8996 \end{array}$$

$$\begin{array}{r} \overset{}{\cancel{1}}\overset{9}{0}\overset{9}{0}\overset{10}{\cancel{0}} \\ \quad 9 \\ \hline \end{array}$$

2

$$\begin{array}{r} \overset{1}{\cancel{2}}000 \\ 2 \\ \hline \end{array}$$

FLOATING-POINT
COMPUTATION

Prentice-Hall
Series in Automatic Computation

George Forsythe, editor

MARTIN, *Teleprocessing Network Organization*

MARTIN AND NORMAN, *The Computerized Society*

MATHISON AND WALKER, *Computers and Telecommunications: Issues in Public Policy*

MCKEEMAN, et al., *A Compiler Generator*

MEYERS, *Time-Sharing Computation in the Social Sciences*

MINSKY, *Computation: Finite and Infinite Machines*

NIEVERGELT et al., *Computer Approaches to Mathematical Problems*

PLANE AND MCMILLAN, *Discrete Optimization: Integer Programming and Network Analysis for Management Decisions*

PRITSKER AND KIVIAT, *Simulation with GASP II: a FORTRAN-Based Simulation Language*

PYLYSHYN, editor, *Perspectives on the Computer Revolution*

RICH, *Internal Sorting Methods: Illustrated with PL/1 Program*

RUSTIN, editor, *Algorithm Specification*

RUSTIN, editor, *Computer Networks*

RUSTIN, editor, *Data Base Systems*

RUSTIN, editor, *Debugging Techniques in Large Systems*

RUSTIN, editor, *Design and Optimization of Compilers*

RUSTIN, editor, *Formal Semantics of Programming Languages*

SACKMAN AND CITRENBAUM, editors, *On-line Planning: Towards Creative Problem-Solving*

SALTON, editor, *The SMART Retrieval System: Experiments in Automatic Document Processing*

SAMMET, *Programming Languages: History and Fundamentals*

SCHAEFER, *A Mathematical Theory of Global Program Optimization*

SCHULTZ, *Spline Analysis*

SCHWARZ, et al., *Numerical Analysis of Symmetric Matrices*

SHERMAN, *Techniques in Computer Programming*

SIMON AND SIKLOSSY, *Representation and Meaning: Experiments with Information Processing Systems*

STERBENZ, *Floating-Point Computation*

STERLING AND POLLACK, *Introduction to Statistical Data Processing*

STOUTEMYER, *PL/1 Programming for Engineering and Science*

STRANG AND FIX, *An Analysis of the Finite Element Method*

STROUD, *Approximate Calculation of Multiple Integrals*

TAVISS, editor, *The Computer Impact*

TRAUB, *Iterative Methods for the Solution of Polynomial Equations*

UHR, *Pattern Recognition, Learning, and Thought*

VAN TASSEL, *Computer Security Management*

VARGA, *Matrix Iterative Analysis*

WAITE, *Implementing Software for Non-Numeric Application*

WILKINSON, *Rounding Errors in Algebraic Processes*

WIRTH, *Systematic Programming: An Introduction*

FLOATING-POINT COMPUTATION

PAT H. STERBENZ

IBM Systems Research Institute
New York, N.Y.

PRENTICE-HALL, INC.

ENGLEWOOD CLIFFS, N.J.

Library of Congress Cataloging in Publication Data

STERBENZ, PAT H.
   Floating-point computation.

   (Prentice-Hall series in automatic computation)
   Bibliography: p.
   1. IBM 360 (Computer)—Programming. 2. Floating-point arithmetic. 3. Compiling (Electronic computers)
I. Title.
QA76.8.I12S77 1974   519.4   73-6620
ISBN 0-13-322495-3

© 1974 by Prentice-Hall, Inc., Englewood Cliffs, N.J.

10 9 8 7 6 5 4 3 2 1

Printed in the United States of America.

PRENTICE-HALL INTERNATIONAL, INC., *London*
PRENTICE-HALL OF AUSTRALIA PTY. LTD., *Sydney*
PRENTICE-HALL OF CANADA, LTD., *Toronto*
PRENTICE-HALL OF INDIA PRIVATE LIMITED, *New Delhi*
PRENTICE-HALL OF JAPAN, INC., *Tokyo*

To My Wife

PREFACE

This book grew out of lecture notes for a course on floating-point computation given for several years at the IBM Systems Research Institute. It presents floating-point arithmetic in a somewhat generalized form which allows for variations in the radix and the word length. However, instead of striving for extreme generality, the book discusses the arithmetic of the IBM System/360 in detail and generalizes it where it is convenient to do so. The examples in the book refer primarily to the System/360 and to the FORTRAN and PL/I compilers currently available for it, but other machines and other compilers are discussed where appropriate. All the examples are presented in higher-level languages, so no knowledge of Assembler Language is necessary. However, it is assumed that the reader is familiar with either FORTRAN or PL/I. (It is not necessary for him to be familiar with both of these languages.)

The material presented here might constitute a second course in programming for someone interested in scientific computing. A first course in programming usually concentrates on a description of language features and the use of these features in writing programs. This book discusses the details of what actually happens when floating-point arithmetic is performed during the execution of the program, and the emphasis is on the quality of the answers produced. It is my hope that, by making the reader more aware of the arithmetic that will be performed as a result of the FORTRAN statement he writes, the book will contribute to the production of better programs.

This book is directed toward two different types of readers. First, it is addressed to the obvious audience of those who are interested in using higher-level languages to write programs which will perform floating-point computation. Second, it is also directed toward the compiler designers and machine designers who are concerned with floating-point operations. The material presented here has been found to be of interest to this group because, by illustrating the way floating-point arithmetic is used to solve problems, it leads to an understanding of the reasons for incorporating various features in the hardware and in the languages.

It is a pleasure to acknowledge the assistance I have received from many friends, colleagues, and students. Particularly important was my association with the SHARE Numerical Analysis Project, for it led to many helpful discussions with W. J. Cody, L. J. Harding, Jr., W. Kahan, H. Kuki, O. K. Smith, and L. R. Turner. I would especially like to thank W. J. Cody and D. W. Sweeney for reading the manuscript and making many helpful suggestions. Finally, I would like to thank Miss Katherine Chandri for carefully typing the manuscript.

PAT H. STERBENZ

CONTENTS

3 ERROR ANALYSIS   71

4 EXAMPLE   123

5 DOUBLE-PRECISION CALCULATION   154

FLOATING-POINT
COMPUTATION

1 FLOATING-POINT NUMBER SYSTEMS

1.1. FIXED-POINT CALCULATION

We shall begin with a brief look at fixed-point calculation in order to understand why one is led to use floating-point arithmetic. Fixed-point arithmetic is extensively used in computers, especially in business or commercial applications. Since many of the early stored program machines had only fixed-point arithmetic, at least insofar as the operation codes available in hardware were concerned, it has also been used for scientific computing.

Fixed-point arithmetic is the natural form of arithmetic when one is dealing with small integers. Here a "small" integer is one which is small enough so that we may record it and use it exactly—that is, without rounding. Usually, the limitation is either the word size of the machine or the maximum number of digits on which arithmetic can be performed in one step. This limit may be 10^{10}, 2^{35}, 10^{15}, etc., depending on the machine being used. On some variable word length machines, the bound may be so large that we are restricted only by the efficient use of storage. Of course, one can use more than one word to hold a number and use multiple-precision fixed-point arithmetic, but this becomes cumbersome, and it is seldom supported by higher-level languages. If all the data, intermediate results, and answers are small integers, all the arithmetic is exact, so no errors are introduced by the arithmetic operations. This is often (but not always) the situation in the calculations one finds in accounting and business applications of computers. (To keep all quantities in the realm of integers, one may have to express financial data in cents rather than dollars.) Consequently, machines designed for business or commercial applications of computers emphasize fixed-point arithmetic. For scientific computing, indexing provides a salient example of arithmetic involving only small integers.

By contrast, problems which are referred to as *scientific* frequently involve calculations in which the arithmetic produces only an approximate answer. If we want to divide 1 by 3 on a decimal machine, we would require an infinite number of places to represent the answer .333333 . . . exactly. Consequently, we are forced to *round* the result to a modest number of digits. Practical considerations lead to the same approach for multiplication. Although the product of several numbers, each having only a finite number of digits, could be computed and stored exactly, the number of digits required can grow quite rapidly. For example, we may require 50 digits to represent the product of 10 five-digit numbers. In this case we find it expedient to round the result to a reasonable number of digits, even though we could calculate the exact answer if we wanted to. Thus, we may distinguish between integer arithmetic, which is exact, and the fixed-point arithmetic of scientific computing, in which the computed answers are approximations for the true answers.

If we are using integer arithmetic on a variable word length machine, we may store each variable in a field just large enough to hold the number of digits required by the maximum value the variable may attain. We are then faced with the problem of estimating the maximum size of each value we develop. Underestimating the maximum size of any quantity can result in a catastrophic error, which, if undetected, can result in the program producing a ridiculous answer. However, since all the arithmetic is exact, we are not concerned with error analysis.

In contrast, in scientific computing we are continually faced with the problem of rounding numbers in order to reduce the number of digits required to a manageable size. This often leads to a fixed word length approach in which we select some reasonable word length for the number of digits which will be carried in each number. On a fixed word length machine, there is a compelling reason for selecting the word length of our numbers to be the word length of the machine, although there are cases in which one might pack two numbers in one word on a machine with long word length or use two words per number on a machine with short word length. On a variable word length machine, we may select the word length to be used in computation arbitrarily within some rather wide limits. Thus, we are led to treat each quantity as a signed p-digit number in the number base of the machine. Typical examples are a 10-digit decimal word, a 35-bit binary word, etc. The 10-digit decimal word length is quite common in the desk calculators designed for scientific computing, and many stored program machines have roughly this word length. These machines are usually capable of developing numbers twice as long in the registers. For example, we may be able to multiply two 10-digit numbers to produce a 20-digit result, and on a typical desk calculator we can add a 10-digit number into 10 consecutive positions of a 20-digit accumulator.

To illustrate the use of fixed-point arithmetic, we suppose that we are working with a 10-digit decimal machine. The decimal point is not actually stored as part of the number; instead, its position must be remembered by the programmer. Thus, instead of storing the number -12.34512345, we store the minus sign and the string of digits 1234512345. Suppose that we have decided to store a number x with three digits to the left of the decimal point and seven digits to the right. This means that we are convinced that $|x|$ will always be less than 1000. If, for some data, we have $x = .5432154321$, we have to store x as 0005432154. Now suppose that we want to compute $z = x + y$, where x is stored with three places to the left of the decimal point and y is stored with the decimal point at the left of the number. For example, we may have $x = 123.4512345$ and $y = .1111122222$. Before we can add x and y we must shift one of them to line up the decimal points. By shifting y three places to the right, we produce 0001111122, which has three places to the left of its decimal point. Then this value may be added to x to produce the value 123.5623467 for z.

A further complication is that there may be a high-order carry. Thus, even though $x < 1000$ and $y < 1$, we may have $z = x + y \geq 1000$, in which case z would require four places to the left of the decimal point. Unless we are convinced that $|z|$ will be less than 1000 for all runs of the program, we shall have to store z with four places to the left of the decimal point. Thus, for the data considered above we would store z as 0123562346, which sacrifices one digit of accuracy. Because we had to allow for the possibility that $|z|$ may be ≥ 1000, we have sacrificed accuracy whenever $|z| < 1000$. This is the fundamental problem that faces us whenever we use fixed-point arithmetic. We must estimate the maximum value for each quantity which is involved in the calculation, either as data, intermediate result, or final answer. If this estimate of the maximum number of digits required is exceeded, we lose high-order digits, which may cause us to produce ridiculous answers. But if we overestimate the maximum, we lose accuracy unnecessarily.

If we store x in a 10-digit word with three places to the left of the decimal point, we can represent x to within $\pm .5 \times 10^{-7}$ regardless of the size of x. Now if $x \approx 123.4512345$, we can represent x with small relative error, but if $x \approx .0000056$, we can save only two significant digits of x. Thus, in fixed-point computation we control absolute error rather than relative error or the number of significant digits. In some problems it is absolute error that we want to control, and fixed-point arithmetic can be used quite easily. In other problems, such as the solution of simultaneous equations, scale factors can be introduced so that the computation can be carried out using fixed-point arithmetic [see National Physical Laboratory (1961)]. But there are many problems which are quite difficult to handle in fixed-point. A particularly annoying aspect of fixed-point computing is that a slight change in the problem may change the bounds for various quantities in the program, so that extensive recoding becomes necessary.

As an illustration of the difficulty of programming in fixed-point, consider the problem of computing x^N for large N. Suppose that we have quite tight bounds for the range of x, say $.1 \leq x < 1$. If we want to compute x^{100}, we know only that $10^{-100} \leq x^{100} < 1$. If we are working with a 10-digit machine, we store x^{100} with the decimal point at the left. Then we shall store zero for x^{100} if x^{100} is less than 10^{-10}, that is, if x is less than about $.793$.

In writing a fixed-point program, we must decide which digits to save at the time we write the program, so we must make the decision without seeing the numbers involved. This is quite different from the situation in manual calculation. Whether we are working with pencil and paper or with a desk calculator, we record the decimal point with each number we write down and we look at the number before deciding which digits to keep. It is quite natural to try to follow this same approach in machine computation. For each number we shall store the first few significant digits and an indication of where the decimal point lies. We can see from the example of x^{100} that unless we carry a great many digits, we cannot guarantee that the decimal point will lie between the first and last digit we are carrying. Thus, instead of storing the decimal point as a character in the string of digits, it is convenient to store a count indicating how many digits of our number lie to the left of the decimal point. If this count is negative, it indicates the number of leading zeros that have been suppressed. Since we do not see the intermediate results, we must depend on the computer to select the proper digits for us. For each arithmetic operation we ask the computer to present us with the first few significant digits and the count telling us where the decimal point lies. These operations are referred to as floating-point arithmetic.

Floating-point arithmetic has proved to be very useful, and today most of what is thought of as scientific computing is performed in floating-point arithmetic. It is available as hardware operation codes on many machines, and it is accomplished by subroutines on others. It is widely exploited by higher-level languages through compilers and interpreters. In fact, the ability to write a program without keeping track of the decimal point adds a great deal to the ease of use of many higher-level languages.

1.2. FLOATING-DECIMAL REPRESENTATION OF NUMBERS

Because decimal numbers are much more familiar than binary or hexadecimal numbers, we shall begin by describing floating-decimal numbers and arithmetic. In Section 1.4 these will be generalized to an arbitrary radix, and throughout most of the book we shall deal with floating-point numbers with an arbitrary radix. However, many of the examples will use the decimal system.

As we have seen, our objective is to represent numbers by their first few significant digits and an indication of where the decimal point lies. The approach we shall follow is a slight modification of the familiar concept of *scientific notation.* To indicate which digits of a number are significant, it has long been the custom to write numbers such as the velocity of light as 1.86×10^5 miles per second instead of 186,000 miles per second. Thus, in scientific notation we write our number as a signed number x in the range $1 \leq |x| < 10$ times a power of 10. This could be implemented on a computer —and it sometimes has been. However, we shall modify this approach slightly and hold the significant digits with the decimal point at the left, so the velocity of light will be written as $.186 \times 10^6$ miles per second. That is, we write our numbers as $y \times 10^k$, where $.1 \leq |y| < 1$. Here the exponent on the 10 is the count we discussed in the last section. A further modification that we make to the idea of scientific notation is that instead of carrying only the significant digits in a number, we shall carry a fixed number of digits throughout the computation regardless of whether we can guarantee that the low-order digits are significant.

We now have to decide how many digits to carry in the floating-point numbers. It is natural to try to fit the floating-point representation of a number into one word, and this is the usual approach when floating-point arithmetic is to be performed by hardware operation codes. However, if the floating-point arithmetic is performed by subroutines, it is quite possible to use one word to hold the significant digits and another word to hold the power of 10. Suppose that we have a decimal machine in which each word holds a sign and 10 decimal digits. We shall illustrate a floating-decimal representation in which we use one word per number. The sign bit of the word holds the sign of the number, and we shall use eight decimal digits of the word to hold the high-order eight digits of the number (not all of which need be significant). It is assumed that there is a decimal point at the left of these eight digits and that the high-order digit is not zero. Thus, we represent the velocity of light as $.18600000 \times 10^6$.

We have two digits left to hold the power of 10. Now our number may be multiplied by either a positive or negative power of 10, so the exponent of the 10 is a signed integer. Since the sign bit of the word was used to hold the sign of the number, we have to hold a signed integer in two decimal digits. A common approach is to store the signed exponent plus 50 in these two digits. Then we can represent powers of 10 from 10^{-50} to 10^{49}, inclusive. We assume that the representation for the power of 10 is written ahead of the significant digits, so our representation for the velocity of light becomes $+5618600000$. Obviously other approaches are possible, and they will be discussed in Section 12.2.

From time to time we shall want to refer to the various parts of the floating-point representation, so it is desirable to introduce terminology for them. Unfortunately, there are several terms in common use. In this book we shall follow a commonly used terminology borrowed from logarithms. We shall refer to the string of significant digits with its sign and the decimal point on the left as the *mantissa* of the floating-point number. The mantissa will have a fixed number of digits, so the low-order digits may not be significant.

The power of 10 will be called the *exponent* of the number and the exponent plus 50 will be called the *characteristic*. Thus, for our representation of the velocity of light as $+5618600000$, the mantissa is $+.18600000$, the exponent is 6, and the characteristic is 56. This terminology becomes awkward only when we talk about logarithms of floating-point numbers. Since other names that are sometimes used to describe the parts of the floating-point representation are also used elsewhere in mathematics, changing the nomenclature merely changes the ambiguity to some other area. Thus, the mantissa is called the *fraction* in Campbell (1962) and in Cody (1971a), it is called the *fractional part* in Knuth (1969), and it is called the *coefficient* in Ashenhurst (1965a, 1965b). Forsythe and Moler (1967) avoid ambiguity by referring to the mantissa as the *significand*, but this name has not yet achieved wide use. The characteristic is often called the *biased exponent*.

Another aspect of number representation is the distinction between *normalized* and *unnormalized* numbers. A nonzero number is said to be normalized if the leading digit of its mantissa is not zero. Since we associate the sign with the mantissa m, this means that $.1 \leq |m| < 1$. Now if the mantissa is zero, the number is zero regardless of what the characteristic is. The representation of zero is said to be normalized if its sign is plus and its characteristic is zero. Thus, a *normal zero* is $+0000000000$. A number which is not normalized is said to be unnormalized. In most of our work we shall assume that all the numbers we are dealing with are normalized and the floating-point arithmetic always produces normalized answers. In Section 12.4 we shall discuss unnormalized operands and arithmetic operations which are allowed to produce unnormalized results.

We have tacitly assumed that the mantissa m is represented by a sign followed by a positive number which represents $|m|$. The representation of negative numbers by complements is discussed in Section 12.2.

When we are writing programs in a higher-level language, such as FORTRAN or PL/I, we can usually think of our numbers as they are written analytically, rather than as they are represented in the machine. Thus, we think of the velocity of light as $.18600000 \times 10^6$ instead of as $+5618600000$. There are some situations, such as the dismantling of floating-point numbers (discussed in Section 4.4), in which we must know exactly how the numbers are represented in the machine. But these cases are atypical, and for most of this book we shall be able to deal with numbers as they are written analytically.

1.3. FLOATING-DECIMAL ARITHMETIC

Before proceeding to a more general setting, it is desirable to see how to perform the standard arithmetic operations of addition, subtraction, multiplication, and division for floating-decimal numbers with the representation described in the previous section. For each operation, our objective is to produce the first eight digits of the result as a normalized floating-point number. We shall ignore the limitation on the size of the exponent until the discussion of overflow and underflow in Chapter 2.

We shall perform these operations on the normalized floating-point numbers x and y, where $x = 10^e m$ and $y = 10^f n$. Here m and n are eight-digit decimal numbers, and if they are not zero, we have $.1 \leq |m| < 1$ and $.1 \leq |n| < 1$.

First, consider multiplication. If either factor is zero, we produce a normal zero as the answer. If both factors are nonzero, we may easily determine the sign of the answer by checking whether x and y have like or unlike signs. To see how the absolute value of the answer is computed, we may assume that x and y are both positive, so $.1 \leq m, n < 1$. We want the first eight significant digits of

$$xy = 10^{e+f} mn.$$

We first compute the 16-digit product mn. If $mn \geq .1$, the mantissa of the answer is the first eight digits to the right of the decimal point in mn and the characteristic of the answer is

$$e + f + 50 = \text{characteristic } (x) + \text{characteristic } (y) - 50.$$

On the other hand, suppose that $mn < .1$. If we took the answer to be the first eight digits to the right of the decimal point in mn, we would produce an unnormalized result. Since both m and n are $\geq .1$, we have

$$.01 \leq mn < .1,$$

so mn has exactly one leading zero. We shift mn one place to the left, which is equivalent to multiplying it by 10, and we subtract 1 from $e + f$. Since

$$xy = 10^{e+f-1}(10mn)$$

and $.1 \leq 10mn < 1$, the mantissa of the answer is the first eight digits to the right of the decimal point in $10mn$, and the characteristic of the answer is

$$e + f - 1 + 50 = \text{characteristic } (x) + \text{characteristic } (y) - 50 - 1.$$

Here the normalization which we had to perform after multiplication is called *postnormalization*.

Next, consider division. If x is zero, the answer is zero, and if y is zero, the answer is undefined. Since we determine the sign of the answer by checking whether x and y have like or unlike signs, we may assume that x and y are positive. Now

$$\frac{x}{y} = 10^{e-f} \frac{m}{n},$$

and, since $.1 \leq m, n < 1$, we have

$$.1 < \frac{m}{n} < 10.$$

If $m < n$, we have $.1 < m/n < 1$. In this case we take the mantissa of the answer to be the first eight digits to the right of the decimal point in m/n and the characteristic of the answer to be

$$e - f + 50 = \text{characteristic } (x) - \text{characteristic } (y) + 50.$$

On the other hand, if $m \geq n$, we have

$$1 \leq \frac{m}{n} < 10.$$

In this case we shift m one place to the right, which is equivalent to dividing it by 10, and add 1 to e. Since

$$\frac{x}{y} = 10^{e-f+1} \frac{.1m}{n},$$

the mantissa of the answer is the first eight digits to the right of the decimal point in $(.1m)/n$, and the characteristic of the answer is

$$e - f + 1 + 50 = \text{characteristic } (x) - \text{characteristic } (y) + 50 + 1.$$

Finally, consider addition and subtraction. Since x and y are signed numbers, to subtract y from x we simply change the sign of y and add it to x. Therefore, it suffices to consider addition. As in manual computation, we must first line up the decimal points of x and y. We begin by comparing the characteristics of x and y and interchanging x and y, if necessary, to make x the number with the larger characteristic. Thus, we obtain $e \geq f$. We then write y as $10^e n'$, where $n' = 10^{-(e-f)}n$ is obtained by shifting n to the right $e - f$ places. If $e \neq f$, y will now be unnormalized and n' will have $8 + e - f$ places to the right of the decimal point. Then

$$x + y = 10^e(m + n'),$$

so if $m + n'$ is in the range $.1 \leq |m + n'| < 1$, the characteristic of the answer is $e + 50$, and the mantissa of the answer is the first eight digits to the right of the decimal point in $m + n'$. Suppose that $.1 \leq |m + n'| < 1$ fails to hold. First, consider the case in which x and y have the same sign. Then $|m + n'| \geq |m| \geq .1$, so we have $|m + n'| \geq 1$. Since

$$|m + n'| = |m| + |n'| < 2,$$

there is exactly one digit to the left of the decimal point. We shift $m + n'$ one place to the right and add 1 to e. Since

$$x + y = 10^{e+1}[.1(m + n')],$$

the characteristic of the answer is $e + 1 + 50$, and the mantissa of the answer is the first eight digits to the right of the decimal point in $.1(m + n')$.

Finally, suppose that x and y have opposite signs. Then

$$|m + n'| \leq \max(|m|, |n'|) < 1,$$

so we have $|m + n'| < .1$. If $m + n'$ is zero, we produce a normal zero as the answer. Otherwise, we have $0 < |m + n'| < .1$, so we now normalize the answer and refer to this procedure as postnormalization. Let k be the number of leading zeros in $m + n'$, so $.1 \leq 10^k |m + n'| < 1$. Then we shift $m + n'$ to the left k places and subtract k from e. Since

$$x + y = 10^{e-k}[10^k(m + n')],$$

the characteristic of the answer is

$$e - k + 50 = \text{characteristic } (x) - k,$$

and the mantissa of the answer is the first eight digits to the right of the decimal point in $10^k(m + n')$.

1.4. FLOATING-POINT NUMBER SYSTEMS

We shall now generalize the idea of floating-decimal arithmetic discussed in the last two sections to include many of the systems actually in use on computers. We want to do this in such a way as to include decimal, binary, octal, and hexadecimal representations of numbers, to allow for variations in word length from one machine to another, and to allow for variations in the details of how the arithmetic is performed. We shall designate the floating-point number system by FP(r, p, a). Here r is the *radix* or base of the number system. Thus, r is 10 for a decimal machine, 2 for a binary machine, 8 for an octal machine, and 16 for a hexadecimal machine. Although these are the commonly used values of r, our approach will allow r to be any integer ≥ 2. Since we shall occasionally want to use examples from systems other than decimal, we shall adopt the following convention about writing numbers: The radix may be specified by writing a letter as a subscript following the string of digits. The subscript will be D for decimal, B for binary, O for octal, and H for hexadecimal. If no radix is specified, the number is decimal. Thus,

(1.4.1) $$25 = 25_D = 19_H = 31_O = 11001_B.$$

For hexadecimal numbers, the digits "ten" through "fifteen" will be designated by A through F, respectively. Thus, $12_D = C_H$.

In the symbol FP(r, p, a), p stands for *precision*, and it designates the number of base r digits contained in the mantissa. For a we shall substitute various symbols specifying the details of how the *arithmetic* is to be performed. Thus, c will stand for *chopped* arithmetic, R for *rounded* arithmetic, etc. (The precise meaning of rounding and chopping is discussed in Section 1.5.) The system discussed in Sections 1.2 and 1.3 is designated by FP(10, 8, c), indicating that it used eight-digit decimal numbers and that it employed chopped arithmetic. Similarly, a 27-bit binary machine using chopped arithmetic would be designated by FP(2, 27, c).

Now FP(r, p, a) will denote a system comprised of a set S of numbers which we shall call *floating-point numbers* and a definition of the four arithmetic operations of *floating-point addition, floating-point subtraction, floating-point multiplication*, and *floating-point division*, defined for elements of S. The set S of floating-point numbers depends on r and p but not on a. When r and p are not fixed by context, we shall write $S(r, p)$ instead of S. The set $S(r, p)$ contains zero and all numbers of the form

$$(1.4.2) \qquad\qquad x = r^e m,$$

where e is any integer (positive, negative, or zero) and m is any positive or negative fraction satisfying

$$(1.4.3) \qquad\qquad r^{-1} \leq |m| < 1$$

whose absolute value can be expressed in the base r using at most p digits. That is,

$$|m| = r^{-p}M,$$

where M is an integer in the range $r^{p-1} \leq M < r^p$. In (1.4.2), the signed number m is called the *mantissa* of x and e is called the *exponent* of x.

Since every floating-point number is a real number, we can perform the standard arithmetic operations of addition, subtraction, multiplication, and division upon the elements of S viewed as real numbers. Thus, for x and y in S, we may form $x \pm y$, xy, and x/y. However, it is quite possible that these operations will produce numbers not in S. For example, we may need $2p$ digits to represent the product of two p-digit numbers, and division may produce a result requiring an infinite number of digits. Since the results of floating-point arithmetic always lie in S, the floating-point operations produce results which may differ from the results produced by the arithmetic operations in the field of real numbers. Thus, we define four new operations, called *floating-point addition, floating-point subtraction, floating-point multiplication*, and *floating-point division* for which we use the symbols $\oplus$, $\ominus$, $*$, and $\div$, respectively. The use of $*$ to denote floating-point multiplication is so familiar from FORTRAN that it seems very natural. On the other hand, the use of $\div$ for floating-point division may require some care. If x is to be divided by y in the field of real numbers, we shall write x/y or $\dfrac{x}{y}$, never $x \div y$.

In general, we expect the operations $\oplus$, $\ominus$, $*$, and $\div$ to produce results which are close to the results produced by $+$, $-$, $\cdot$, and $/$. That is, we expect to have $x * y \approx xy$, etc. The symbol substituted for a in FP(r, p, a) will specify the details of exactly how the floating-point operations are defined.

Our definition of FP(r, p, a) omits several details of the floating-point number representation which would have to be specified in order to give a description of the system that is complete enough to allow an engineer to design a computer to handle it. In fact, some of the details we have omitted may affect the programmer who is using the machine. The most striking omission is that we have not specified any bounds on the range of the exponent. In practice, such bounds always exist. For the system described in Section 1.2, the exponent had to lie in the range $-50 \le e \le 49$, and for the IBM System/360 the exponent must lie in the range $-64 \le e \le 63$. Attempting to produce a number whose exponent lies outside this range results in overflow or underflow, so the bounds for the exponent are of interest to the programmer. But in analyzing a program, problems related to overflow and underflow are often studied separately from problems related to the errors introduced by the floating-point arithmetic. This will be our approach here. We shall study the effects of performing arithmetic in a hypothetical system FP(r, p, a) which places no bounds on the exponent, and we shall relegate the study of overflow and underflow to a separate chapter (Chapter 2).

Even though we do not place bounds on the exponent of the numbers in $S(r, p)$, from time to time we shall want to refer to the characteristic of a floating-point number. We shall assume that there is a number γ such that the signed exponent e is actually stored as a *characteristic* which is defined to be $e + \gamma$. Then γ is 50 for the system described in Section 1.2, 64 for the IBM System/360, and 128 for the IBM 7090. A machine which stores the exponent as a signed number would have $\gamma = 0$. We shall assume that a normalized zero is stored with the smallest allowable characteristic.

Another omission is that we have not allowed for variations in the form in which negative numbers are stored. We shall assume that negative numbers are stored with a minus sign and true value of the magnitude of the mantissa. Of course other approaches are possible. Machines have been built which use either r's complements or $(r - 1)$'s complements. The use of complements will be discussed in Section 12.2, but throughout the rest of this book we shall assume that negative numbers are stored with *sign and true magnitude*.

Still another omission is that we have not allowed for variations in where the radix point falls in the mantissa of a floating-point number. Throughout this book we shall assume that the radix point lies at the left of the mantissa, so the mantissa of a normalized, nonzero number satisfies (1.4.3). However, other schemes have been implemented. For example, on the CDC 6600, floating-point numbers are represented in the form $2^e c$, where c is called the *coefficient* and is an integer. In this case, the binary point is at the right of the mantissa. Since results from FP(r, p, a) are readily translated to such a system, we shall assume throughout that the radix point lies at the left.

We have specified that all floating-point numbers are normalized, and we shall assume that the floating-point operations $\oplus$, $\ominus$, $*$, and $\div$ produce normalized results. Many machines offer the programmer the option of producing unnormalized results by suppressing the postnormalization which may occur in $\oplus$, $\ominus$, and $*$. This is true for $\oplus$ and $\ominus$ on the IBM System/360, but the FORTRAN and PL/I compilers for that machine translate the arithmetic operations which appear in an arithmetic expression into the normalized operation codes. Thus, we are dealing with the system the programmer sees when he writes programs in FORTRAN or PL/I for the IBM System/360. In Section 12.4, we shall discuss both unnormalized operation codes and unnormalized operands.

Finally, some machines have special numbers which are treated differently. For example, the CDC 6600 has an ∞ and an indeterminant form. The IBM 7030 had flag bits in the floating-point word which could cause interrupts. These features will be discussed in Section 12.2.

1.5.   FP(r, p, c) AND FP(r, p, R)

We shall now specify two different ways in which the floating-point operations $\oplus$, $\ominus$, $*$, and $\div$ may be defined, yielding two different systems which will be designated by FP(r, p, c) and FP(r, p, R). The letters c and R will denote *chopped* and *rounded* arithmetic, respectively. In chopped arithmetic, the result is first normalized, and then its low-order digits are discarded and its high-order p digits are retained unchanged. One often sees this approach referred to as *truncation* instead of chopping. However, this use of the word truncation may lead to confusion with the term *truncation error*, which is a poorly defined and overworked term in numerical analysis. It is used to refer to the error introduced by replacing an infinite series by a finite number of terms of the series, and it is sometimes used in a much wider context to refer to the error introduced by replacing a continuous problem by a discrete problem. To avoid the possibility of confusion between this type of error and the error introduced by the floating-point arithmetic, we use the term chopping instead of the more commonly used term truncation. The term chopping has been used in this context by other authors. [See, for example, McCracken and Dorn (1964).]

If x is any real number, let $\bar{x}$ denote x chopped to p digits in the base r. More specifically, for any real number x let T be the set of all numbers y in $S(r, p)$ with $|y| \leq |x|$. Then $\bar{x}$ is the element in T which is closest to x. Thus, in FP(10, 8, c), $\overline{.123456789} = .12345678$ and $-\overline{.123456789} = -.12345678$. To perform arithmetic in FP(r, p, c), we first perform the corresponding operation in the real number system, and then we chop the result to p digits in the base r. Thus

$$x \oplus y = \overline{x + y}$$
$$x \ominus y = \overline{x - y}$$
$$x * y = \overline{xy}$$
$$x \div y = \overline{x/y}$$

Similarly, we introduce the concept of rounded arithmetic. Here rounding means that we round to the closest p-digit number in the base r. If two such numbers are equally close, we round the magnitude upward. This is sometimes referred to as *symmetric rounding*. Other rules for rounding are possible, but they are more complicated, and they are seldom implemented on computers. For any real number x, let $\overset{\circ}{\overline{x}}$ be x rounded to p digits in the base r. The is, $\overset{\circ}{\overline{x}}$ in the number in $S(r, p)$ which is closest to x. If two numbers in $S(r, p)$ are equally close to x, $\overset{\circ}{\overline{x}}$ will denote the one with larger magnitude. Thus, in FP(10, 8, R), $\overline{.123456789}^{\circ} = .12345679$, $\overline{.123456785}^{\circ} = .12345679$, $\overline{.123456783}^{\circ} = .12345678$, and $-\overline{.123456785}^{\circ} = .12345679$. To perform arithmetic in FP(r, p, R), we first perform the operation in the real number system, and then we round the results to p digits in the base r. Thus, in FP(r, p, R):

$$x \oplus y = \overline{x + y}^{\circ}$$
$$x \ominus y = \overline{x - y}^{\circ}$$
$$x * y = \overline{xy}^{\circ}$$
$$x \div y = \overline{x/y}^{\circ}$$

Most implementations of floating-point arithmetic have tried to produce results which were either approximately the correctly chopped results or approximately the correctly rounded results. However, computers have seldom, if ever, produced exactly the results which would be produced in FP(r, p, c) or FP(r, p, R). (In Section 12.3 we shall discuss how a machine could be designed to produce exactly these results.) Thus, both FP(r, p, c) and FP(r, p, R) are idealized systems which probably do not describe exactly the arithmetic the programmer is using. But since the ideas of chopping and rounding are easy to work with, it is often convenient to study the results produced in these systems, without considering the modifications which have been introduced by the machine designers to make the arithmetic easier to perform. In Section 1.8 we shall discuss the system FP(r, p, clq), which has been implemented on many machines, including the IBM System/360. In some cases there is no difference between the results in FP(r, p, c) and FP(r, p, clq), and in other cases the difference is not important, so we may draw our examples from FP(r, p, c). In still other cases the distinction between FP(r, p, c) and FP(r, p, clq) is important, so we shall deal with FP(r, p, clq).

Since we shall place special emphasis on the arithmetic of the IBM System/360, we shall be most concerned with FP(r, p, clq). Because this system is closely related to FP(r, p, c), we shall devote much more time to the study of FP(r, p, c) than we do to the study of FP(r, p, R). However, FP(r, p, R) will be discussed where appropriate, and many of the results obtained for FP(r, p, c) will be carried to FP(r, p, R) in the exercises.

1.6. LAWS OF ALGEBRA

Algebraic manipulation of formulas is based on the validity of a few fundamental laws. Specifically, we appeal to the fact that the real numbers form a field. This means that the sum and product of real numbers are defined and that the following six axioms hold for any real numbers a, b, c:

1. *Closure*: The product ab and sum $a + b$ of the real numbers a and b are real numbers.

2. *Commutative laws*:

(1.6.1)
$$a + b = b + a$$
$$ab = ba$$

3. *Associative laws*:

(1.6.2)
$$(a + b) + c = a + (b + c)$$
$$(ab)c = a(bc)$$

4. *Distributive law*:

(1.6.3)
$$a(b + c) = ab + ac$$

5. There are real numbers 0 and 1 such that

(1.6.4)
$$a + 0 = 0 + a = a$$

and

(1.6.5)
$$a \cdot 1 = 1 \cdot a = a$$

hold for all a.

6. For any real number a there is a real number $-a$ such that

$$(1.6.6) \qquad a + (-a) = (-a) + a = 0,$$

and if $a \neq 0$, there is a real number a^{-1} such that

$$(1.6.7) \qquad aa^{-1} = a^{-1}a = 1.$$

A consequence of these axioms is that there are no *divisors of zero*, that is, if $ab = 0$, then at least one of the factors a, b must vanish. Another consequence of these axioms is the cancellation law:

$$(1.6.8) \qquad \text{If } ab = ac \text{ and } a \neq 0, \text{ then } b = c.$$

We now define subtraction by

$$a - b = a + (-b),$$

and if $b \neq 0$, we define division by

$$\frac{a}{b} = ab^{-1}.$$

Two immediate consequences of these definitions, along with (1.6.1) and (1.6.2), are

$$(1.6.9) \qquad (a + b) - b = a$$

and

$$(1.6.10) \qquad a\left(\frac{b}{a}\right) = b.$$

We now ask whether these laws are valid in our floating-point number system. Since their validity may depend on the details of how the arithmetic is performed, we study the specific system $FP(r, p, c)$. [The question of whether these laws are valid in $FP(r, p, R)$ forms Exercises 3–7.] We shall see that several of these laws fail to hold, although some of the ones which fail do hold "approximately" in the sense that the two expressions are approximately equal. The investigation of which of the laws hold approximately will be postponed until Section 3.4. The fact that some of these laws fail to hold is more than an oddity of the floating-point number system. Because they are the basis of the algebraic manipulation of formulas, the failure of any of them means that the programmer must think of his computation as being performed in $FP(r, p, c)$ instead of in the real number system. It may affect the best way to write a formula in FORTRAN or PL/I.

The following theorem is an immediate consequence of the definitions and the validy of the commutative laws in the real number system.

THEOREM

In FP(r, p, c), the floating-point sum $a \oplus b$ and the floating-point product $a * b$ of two floating-point numbers a and b are floating-point numbers. Also, for any a and b in $S(r, p)$,

$$a \oplus b = b \oplus a$$
$$a * b = b * a$$
$$a \oplus 0 = 0 \oplus a = a$$
$$a * 1 = 1 * a = a$$
$$a \oplus (-a) = (-a) \oplus a = 0.$$

It will be shown in Section 1.9 that for any $a \neq 0$ in $S(r, p)$ there is a number b in $S(r, p)$ with $a * b = 1$. That is, every nonzero element in FP(r, p, c) has an inverse. However, this is not as helpful as it seems. In the real number system, the existence of the inverse is used to define division, and it enables us to solve the equation $ax = b$. But we have defined division directly. Moreover, we shall see in Section 1.9 that, because of the failure of the associative law of multiplication in FP(r, p, c), the existence of a number c with $a * c = 1$ is not helpful in solving the equation $a * x = b$.

We shall now show that the remaining laws, namely (1.6.2), (1.6.3), (1.6.8), (1.6.9), and (1.6.10), fail to hold in FP(r, p, c) for nontrivial combinations of r and p. [They may hold, for example, in FP(2, 1, c).] By this we mean that it is not true that they hold for all a, b, c in $S(r, p)$. There may be some values of a, b, c for which they hold, but for each law we shall display an example for which the law fails.

Failure of the Associative Law of Addition

To show that

$$(a \oplus b) \oplus c = a \oplus (b \oplus c)$$

fails to hold in FP(r, p, c), let $a = r^{-2p}$, $b = 1$, and $c = -1$. Then

$$a \oplus b = \overline{1 + r^{-2p}} = 1,$$

so

$$(a \oplus b) \oplus c = 1 \oplus (-1) = 0.$$

But

$$a \oplus (b \oplus c) = r^{-2p} \oplus 0 = r^{-2p},$$

so the law fails.

If the associative law of addition were valid in FP(r, p, c), then (1.6.9) would also hold. That is, we would have

$$(a \oplus b) \ominus b = a.$$

But the example above shows that this fails to hold for $a = r^{-2p}$ and $b = 1$.

Sometimes we can exploit the fact that these laws fail. Suppose that we want the integer part of a number x which is known to be positive and less than r^{p-1}. Let $y = r^{p-1}$. Then the integer part of x is given by $(x \oplus y) \ominus y$. For example, suppose that we are working in FP($10, 8, c$) and that $x = 12.345678$. Then $x + y = 10000012$, so $(x + y) - y$ is 12. (Another way to produce the integer part of x is given in Section 12.4.)

In other cases, the failure of (1.6.9) may be more annoying. For example, in FP($10, 8, c$) let $a = 3.3333333$ and $b = .22222222$. Then $a \oplus b = 3.5555555$ and $(a \oplus b) - b$ is 3.33333328, so $(a \oplus b) \ominus b$ is 3.333332, which is different from a.

From the examples we have considered, it might appear that the failure of the associative law was due to the fact that subtractions were involved. We now give an example which shows that the associative law of addition may fail to hold even if a, b, and c all have the same sign. Let $a = 1, b = (r - 1)r^{-p}$, and $c = r^{-p}$. Then

$$(a \oplus b) \oplus c = \overline{1 + (r - 1)r^{-p}} \oplus r^{-p} = 1 \oplus r^{-p} = 1,$$

but

$$a \oplus (b \oplus c) = 1 \oplus r^{-(p-1)} = 1 + r^{-(p-1)},$$

so the associative law fails to hold. However, in Section 3.4 we shall show that the associative law holds approximately if a, b, and c all have the same sign.

Failure of the Associative Law of Multiplication

We shall show that the associative law of multiplication

$$(a * b) * c = a * (b * c)$$

fails to hold in FP(r, p, c), except for trivial combinations of r and p. To simplify notation, let $s = p - 1$. We shall assume that

$$4r^{-2s} \leq r^{-p},$$

which is true for all interesting combinations of r and p. Let $a = b = 1 + r^{-s}$ and $c = 1 - 2r^{-s}$. Then

$$(a * b) * c = \overline{1 + 2r^{-s} + r^{-2s}} * c = (1 + 2r^{-s}) * (1 - 2r^{-s})$$
$$= \overline{1 - 4r^{-2s}} = 1 - r^{-p}.$$

But

$$a * (b * c) = a * \overline{1 - r^{-s} - 2r^{-2s}} = (1 + r^{-s}) * (1 - r^{-s} - r^{-p})$$
$$= \overline{1 - r^{-p} - r^{-2s} - r^{-(s+p)}}$$
$$= 1 - 2r^{-p}.$$

Thus, the associative law of multiplication fails to hold in $FP(r, p, c)$ if $4r^{-2s} \le r^{-p}$, that is, if $4 \le r^{p-2}$. By considering various special cases, it can be shown that the associative law of multiplication fails to hold in $FP(r, p, c)$ except for the four trivial systems $FP(3, 1, c)$, $FP(2, 1, c)$, $FP(2, 2, c)$, and $FP(2, 3, c)$.

The failure of the two associative laws has an annoying logical consequence. In the real number system, we define the product of two real numbers; then, since the associative law of multiplication holds, we may simply write the product of three real numbers as abc and let the reader group them in any way he wants to. But, since the associative law of multiplication fails to hold in $FP(r, p, c)$ for interesting values of r and p, it is not legitimate to use this approach for floating-point multiplication. Technically, we should be required to specify whether we want $(a * b) * c$ or $a * (b * c)$; we should not be allowed to write $a * b * c$. However, the FORTRAN and PL/I compilers allow us to write $a * b * c$ without inserting parentheses. A justification for this lies in the fact that the associative law of multiplication holds approximately in $FP(r, p, c)$, as will be shown in Section 3.4. With somewhat less justification, the compilers also allow us to write a sum, such as $a \oplus b \oplus c \oplus d$, without parentheses. Some compilers, for example the FORTRAN and PL/I compilers for the IBM System/360, specify that these operations be performed from left to right. That is,

(1.6.11)
$$a * b * c * d = ((a * b) * c) * d$$
$$a \oplus b \oplus c \oplus d = ((a \oplus b) \oplus c) \oplus d.$$

On the other hand, there have been some compilers which did not guarantee the order in which these operations would be performed. Since we are emphasizing the IBM System/360 and its compilers, we shall use (1.6.11) as the definition of repetitive multiplication and repetitive addition in our floating-point number systems $FP(r, p, a)$.

Consider the two expressions $a * b * c$ and $a * c * b$ in $FP(r, p, c)$. Now

$$a * b * c = (a * b) * c = c * (a * b)$$

and

$$a * c * b = (a * c) * b = (c * a) * b.$$

Then these two expressions may fail to be equal because the associative law does not always hold. As an example, consider the expression $2xy$, which we may code as $2 * x * y$. In FP(10, 8, c), suppose that $x = .88111117$ and that $y = .44444444$. Then

$$2 * x = \overline{1.76222234} = 1.7622223,$$

while

$$2 * y = .88888888 = 2y.$$

Thus, $2 * y$ is exact, but $2 * x$ is not. Then $2 * y * x = \overline{2xy}$, but $2 * x * y$ may be smaller than this. Indeed, by carrying out the computations we find that $2 * y * x = .78320992$ but that $2 * x * y = .78320990$. Similarly, in FP(16, 8, c), if $x = .88111117_H$ and $y = .44444444_H$, then

$$2 * x = \overline{1.1022222E_H} = 1.1022222_H$$

and

$$2 * y = .88888888_H = 2y.$$

Direct computation yields $2 * y * x = .4891A2B6_H$, while $2 * x * y = .4891A2B2_H$. Although multiplication by 2 is always exact on a binary machine, on a machine with any other radix it can introduce error if the absolute value of the mantissa is greater than one-half. Thus, the order of the factors can be important even in so simple an expression as $2xy$. If we know something about the size of x and y, we may have a preference for one or the other of the forms $2 * x * y$ or $2 * y * x$.

Failure of the Cancellation Law

To show that the cancellation law (1.6.8) fails to hold in FP(r, p, c), except for trivial combinations of r and p, we find values a, b, and c in S(r, p) such that $a \neq 0$, $b \neq c$, and $a * b = a * c$. To this end, we first consider the case in which $r > 2$ and $p \geq 2$. Let $a = 2$, $b = r - 1$, and $c = r - 1 + r^{-(p-1)}$. Then $b \neq c$, but

$$a * b = r + r - 2$$

and

$$a * c = \overline{r + r - 2 + 2r^{-(p-1)}} = r + r - 2.$$

Thus, $a * b = a * c$, so the cancellation law fails. It is particularly annoying that we cannot cancel so simple a multiplier as 2.

For FP(2, p, c), we suppose that $p \geq 4$. Let $a = b = 1.1_B$ and $c = 1.1_B + 2^{-(p-1)}$. Then $b \neq c$, but

$$a * b = a * c = 10.01_B,$$

so the cancellation law fails.

By considering the remaining cases, it can be shown that the cancellation law fails to hold in $FP(r, p, c)$ except for the four systems $FP(3, 1, c)$, $FP(2, 1, c)$, $FP(2, 2, c)$, and $FP(2, 3, c)$.

Failure of the Distributive Law

We shall show that the distributive law

$$a * (b \oplus c) = (a * b) \oplus (a * c)$$

fails to hold in $FP(r, p, c)$ for interesting values of r and p. First, suppose that $r > 2$ and $p \geq 2$. Let $a = r - 1$, $b = r - 1 + r^{-(p-1)}$, and $c = (r - 1)r^{-(p-1)}$. Then

$$a * b = (r - 2)r + 1$$

and

$$a * c = [(r - 2)r + 1]r^{-(p-1)},$$

so

$$(a * b) \oplus (a * c) = \overline{(r - 2)r + 1 + (r - 2)r^{-(p-2)} + r^{-(p-1)}}$$
$$= (r - 2)r + 1 + (r - 2)r^{-(p-2)}.$$

But

$$a * (b \oplus c) = (r - 1) * (r - 1 + r^{-(p-2)})$$
$$= (r - 2)r + 1 + (r - 1)r^{-(p-2)},$$

so the distributive law fails to hold in $FP(r, p, c)$ for $r > 2$ and $p \geq 2$.

For $FP(2, p, c)$, suppose that $p \geq 4$. Let $a = 1.1_B$, $b = 1.1_B + 2^{-(p-1)}$, and $c = 2^{-(p-1)}$. Then $a * b = 10.01_B$ and $a * c = (1.1_B)2^{-(p-1)}$, so

$$(a * b) \oplus (a * c) = \overline{(10.01)_B + 2^{-(p-1)} + 2^{-p}} = 10.01_B.$$

But

$$a * (b + c) = 1.1_B * (1.1_B + 2^{-(p-2)}) = 10.01_B + 2^{-(p-2)},$$

so the distributive law fails. By considering the remaining cases, it can be shown that the distributive law fails to hold in $FP(r, p, c)$ except for the single case of $FP(2, 1, c)$.

Failure of the Relation a * (b ÷ a) = b

In $FP(r, p, c)$, let a and b be positive and $c = b \div a$. Then $c = \overline{b/a}$, so unless b/a can be expressed in p digits in the base r, we have $c < b/a$. Then $\overline{ac} < b$, so

$$a * (b \div a) = \overline{ac} < b.$$

Thus, the relation

(1.6.12) $a * (b \div a) = b$

holds if and only if $b \div a = b/a$.

Except for the trivial case of FP(2, 1, c), we may select $b = 1$ and let a be an integer in the range $1 < a < r^p$ which is relatively prime to r. Then b/a cannot be represented in a finite number of digits in the base r, so (1.6.12) fails. Thus, (1.6.12) fails to hold in FP(r, p, c) except for the trivial case of FP(2, 1, c).

1.7. INEQUALITIES IN FP(r, p, c)

We shall now investigate the extent to which FP(r, p, c) preserves the order relationships which we are accustomed to for the real number system. Since every number in $S(r, p)$ is a real number, the relations $x < y$, $x \leq y$, etc., are defined for them. The following laws are fundamental for the manipulation of inequalities in the real number system:

1. If $a < b$, then for all c

$$a + c < b + c.$$

2. If $a < b$ and $c < d$, then

$$a + c < b + d.$$

3. If $b < c$ and a is positive, then

$$ab < ac.$$

We would like the corresponding laws to hold in FP(r, p, c).

First, we observe that if x and y are real numbers with $x < y$, then $\bar{x} \leq \bar{y}$. Of course we may have $\bar{x} = \bar{y}$ even though $x < y$. This occurs whenever the first p digits of x and y are the same. The following theorem follows immediately from this observation and the definitions of $\oplus$ and $*$.

THEOREM 1.7.1

In FP(r, p, c),

1. If $a < b$, then $a \oplus c \leq b \oplus c$ holds for all c.
2. If $a < b$ and $c < d$, then $a \oplus c \leq b \oplus d$.
3. If $b < c$ and a is positive, then $a * b \leq a * c$.

Unfortunately, these relationships, which were strict inequalities in the real number system, have been weakened to $\leq$ in FP(r, p, c). We shall show below that the strict inequalities fail to hold in FP(r, p, c), so this theorem is the strongest statement that can be made.

For (1), let $a = r^{-2p}$, $b = 2r^{-2p}$, and $c = 1$. Then $a < b$, but $a \oplus c = b \oplus c = 1$.

For (2), we assume that $p \geq 2$. Let

$$a = 1 - r^{-p}$$
$$b = 1$$
$$c = r^{-p}$$
$$d = r^{-p} + r^{-(p+1)}.$$

Then $a < b$ and $c < d$, but $a \oplus c = b \oplus d = 1$. By considering the remaining case of $p = 1$, it can be shown that the strict inequality holds in (2) only for FP(2, 1, c).

For (3), we note that in any system FP(r, p, c) in which the cancellation law fails to hold, we have positive numbers a, b, c with $b < c$ and $a * b = a * c$. Thus, the strict inequality holds in (3) only for the systems FP(3, 1, c), FP(2, 1, c), FP(2, 2, c) and FP(2, 3, c).

The importance of these results lies in the fact that once we have established that an inequality holds, we expect to be able to deduce other relationships from it. Suppose that we have compared x and y in a program and found that $x < y$. If we now decide to change the units in which they are expressed by multiplying both of them by a positive constant, we expect the resulting values to satisfy the same inequality. However, we have seen that they may become equal. Indeed, if $r \neq 2$, even multiplication by 2 may convert unequal numbers into equal numbers. This may have annoying consequences. For example, if the denominator of a fraction is $a * b - a * c$, then determining that $b \neq c$ is not sufficient protection against division by zero. Another annoying consequence may arise in trying to debug a program which is misbehaving. If our output shows us that $a * b = a * c$, we cannot be absolutely certain which branch we took if the program branches on the condition $b = c$. This might lead us to search for the bug in the wrong part of the program.

1.8. FP(r, p, c/q)

When we introduced the system FP(r, p, c), we mentioned that many machines use approximately, but not exactly, this system. It may be viewed as an ideal system which is not quite attained in practice. We shall now describe a system FP($r, p, c/q$), which is a slight modification of FP(r, p, c) and which describes exactly the arithmetic that has been implemented on many machines. It includes both the single- and double-precision arithmetic of the IBM System/360 and the single-precision arithmetic of the IBM 709, 7090, and 7094 (but not the 704). It does not describe the double-precision arithmetic of the IBM 7094, which is more closely related to the programmed double-precision arithmetic described in Chapter 5. We are still ignoring the bounds on the range of the exponent.

The symbol *clq* means that we shall perform *chopped* arithmetic using a *low-* order register which is q digits long, where q may be any integer ≥ 0. This low-order register will be used in the operations $\oplus$, $\ominus$, and $*$ to hold low-order digits of intermediate results which have more than p digits. More specifically, in the operation $*$ it will hold the next q digits of the product, and in the operations $\oplus$ and $\ominus$ it will hold the next q digits of the operand which is shifted. The 7090 has a 27-bit low-order register called the MQ, and the IBM System/360 uses a one-digit low-order register called the *guard digit*. Thus, the arithmetic on the IBM 7090 is performed in the system FP(2, 27, *cl*27), and the single-precision arithmetic on the IBM System/360 is performed in the system FP(16, 6, *cl*1). When the early copies of the IBM System/360 were delivered, there was no guard digit for double-precision arithmetic, so the double-precision arithmetic was performed in the system FP(16, 14, *cl*0). Later, during 1968, the architecture of the IBM System/360 was changed to incorporate a guard digit in double-precision arithmetic. This change was also made in the machines already installed, so double-precision arithmetic on the IBM System/360 is now performed in the system FP(16, 14, *cl*1). Since the length of the low-order register determines the length of the intermediate results which may be held, we may think of FP(*r*, *p*, *c*) as FP(*r*, *p*, *cl*∞).

We still assume that our floating-point arithmetic takes p-digit normalized operands and produces a p-digit normalized result. We are interested in the low-order register only as it affects the high-order digits which are returned as the answer. Of course, it would be desirable to make the low-order digits available to the programmer—at least to the assembly language programmer —since they are useful for programming rounding or programming higher-precision arithmetic. However, this is not a requirement for the system FP(*r*, *p*, *clq*). Thus, in the IBM System/360 the guard digit, which is used while the arithmetic is being performed, is never saved in a register, so there is no way the programmer can get access to it—even in assembly language. Nevertheless, the guard digit meets our requirement for a one-digit low-order register.

Floating-Point Division

In the machines we are modeling here, the IBM 7090 and the IBM System/360,† the floating divide operation produces the correctly chopped result. Thus, in FP(*r*, *p*, *clq*) we define $a \div b$ to be $\overline{a/b}$.

---

†Exceptions to this rule are the IBM System/360 models 91, 95, and 195. On these models, the floating divide operation may produce a result which differs from the result produced by other models of the IBM System/360. [See International Business Machines (1966 and 1969).]

Floating-Point Multiplication

We define the product $a * b$ to be zero if either factor is zero. If $ab \neq 0$, the sign of the product is $+$ or $-$ depending on whether a and b have like or unlike signs. Then we may assume that a and b are positive. Let

$$a = r^e m, \qquad r^{-1} \leq m < 1,$$

and

$$b = r^f n, \qquad r^{-1} \leq n < 1.$$

Let $\mu' = mn$, so

$$ab = r^{e+f} \mu'.$$

Since

$$r^{-2} \leq \mu' < 1,$$

μ' is a $2p$-digit number with the radix point at the left and at most one leading zero. We assume that we can hold only $p + q$ digits of the result, so we let μ'' be the first $p + q$ digits to the right of the radix point in μ'. Thus, if $q \geq p$, we have $\mu'' = \mu'$, but if $q < p$, μ'' is obtained by discarding the low-order $p - q$ digits of the $2p$-digit number μ'. Let

$$a * b = r^g \mu,$$

where g and μ are defined as follows: If $r^{-1} \leq \mu < 1$, then $g = e + f$ and $\mu = \overline{\mu''}$. On the other hand, if $\mu'' < r^{-1}$, then we shift μ'' one place to the left to normalize it and compensate by decreasing the exponent by 1. (This shift is referred to as postnormalization.) In this case we have $g = e + f - 1$ and $\mu = \overline{r\mu''}$. We may summarize these two cases by writing

$$g = e + f - k$$
$$\mu = \overline{r^k \mu''},$$

where k is 1 or 0 depending on whether or not postnormalization is required.

We shall now compare the results obtained for $a * b$ in FP(r, p, clq) and FP(r, p, c). If no postnormalization is required, in each case the mantissa of $a * b$ is the high-order p digits of the $2p$-digit number μ', so the results are the same. On the other hand, if postnormalization is required, the mantissa of $a * b$ in FP(r, p, clq) is digits 2 through $p + 1$ of μ''. Now if $q \geq 1$, these are the same as the digits of μ', so $a * b$ produces the same result in FP(r, p, clq) as it does in FP(r, p, c). But if $q = 0$, the $(p + 1)$st digit of μ'' is zero, so, after a left shift of one place for postnormalization, the pth digit of the product is zero. Thus, we have proved the following theorem:

THEOREM 1.8.1

The floating-point product $*$ produces the same result in FP(r, p, clq) for $q \geq 1$ as it does in FP(r, p, c). For $q = 0$ the product $*$ produces the same result in FP(r, p, $cl0$) as it does in FP(r, p, c) whenever no postnormalization is required. If postnormalization is required, the product $a * b$ in FP(r, p, $cl0$) is obtained from the product $a * b$ in FP(r, p, c) by replacing the pth digit of the mantissa by zero.

A particularly annoying consequence of the result for $q = 0$ (no guard digit) concerns multiplication by a power of r. If a is a power of r, say r^e, then we write a as $r^{e+1} \cdot r^{-1}$, so the mantissa of a is r^{-1}. (This is written in the base r as .1.) Then $\mu' = r^{-1}n < r^{-1}$, so postnormalization is required. Thus, if $q = 0$, multiplication by a power of r replaces the low-order digit of the mantissa by zero. In many scientific calculations, scale factors are chosen to be a power of r so that scaling will not introduce rounding errors. [For example, see Forsythe and Moler (1967).] We see that this fails to be true in FP(r, p, $cl0$). Even more annoying, since $r^0 = 1$, we find that multiplication by 1 may change a number. Indeed, multiplication by 1 in FP(r, p, $cl0$) has exactly the effect of replacing the low-order digit by zero, so FP(r, p, $cl0$) does not have a unit element satisfying (1.6.4). The engineering change on the IBM System/360 mentioned above, which added a guard digit to the double-precision arithmetic, removed these problems by changing the system from FP(16, 14, $cl0$) to FP(16, 14, $cl1$).

Floating-Point Addition and Subtraction

We define

(1.8.1) $$a \ominus b = a \oplus (-b),$$

so we may restrict our attention to the floating-point addition of signed numbers. As above, we write $a = r^e m$ and $b = r^f n$. We shall assume that a and b are normalized and that $e \geq f$. (If $e < f$, we interchange a and b to produce the situation described above.) Then we write

$$b = r^e n',$$

where $n' = r^{-(e-f)}n$ is obtained by shifting n to the right $e - f$ places. Of course n' is not normalized unless $e = f$. We are assuming that we have only a q-digit register to hold the low-order digits shifted out of the p-digit register holding b, so we let n'' be the high-order $p + q$ digits of the $[p + (e - f)]$-digit number n'. If $e - f \leq q$, we have $n'' = n'$, but if $e - f > q$, then n'' is obtained from n' by discarding the low-order $e - f - q$ digits. Any digits of n' which do not appear in n'' are lost and cannot enter the calculation. (The shift of $e - f$ places to the right, retaining only the high-order $p + q$ digits, is called the *preshift*.) We then form $\mu' = m + n''$ and set

$$a \oplus b = \overline{r^e \mu'}$$

We note that if one of the operands is zero (in normalized form) it has the smallest allowable characteristic, so its exponent is not greater than the exponent of the other operand. Thus, if $e \neq f$, it is the zero which is shifted to the right, so $n'' = n' = n$. Then our definition produces

(1.8.2)
$$a \oplus 0 = a \ominus 0 = a$$
$$0 \oplus b = b$$
$$0 \ominus b = -b,$$

as expected.

To analyze the effects of this definition in more detail, it is convenient to separate the discussion into two cases, depending on whether the addition of signed numbers results in the addition or subtraction of their magnitudes. In each case we shall write the results as

(1.8.3)
$$a \oplus b = r^g \mu.$$

Add Magnitude Case

This case arises if we add numbers having the same sign or subtract numbers having opposite signs. A consequence of our definition is that

(1.8.4)
$$(-a) \oplus (-b) = -(a \oplus b).$$

Using (1.8.1), (1.8.2), and (1.8.4), we may reduce the discussion of the add magnitude case to the discussion of $a \oplus b$ where a and b are positive. Since m and n' are less than 1,

$$\mu' = m + n'' < 2$$

and

$$\mu' \geq m \geq r^{-1}.$$

Then μ' can be represented as a $(p + q + 1)$-digit number with the radix point after the first digit (which may be zero). If $\mu' < 1$, we write $g = e$ and $\mu = \overline{\mu'}$. On the other hand, if $\mu' \geq 1$, we write

$$r^e \mu' = r^{e+1}(r^{-1}\mu'),$$

so $g = e + 1$ and $\mu = \overline{r^{-1}\mu'}$. In either case, we retain the high-order p digits of $m + n''$, and these are the same as the high-order p digits of $m + n$. Therefore, in the add magnitude case, $a \oplus b$ and $a \ominus b$ produce the same results in FP(r, p, clq) for all $q \geq 0$ as they do in FP(r, p, c). Any digits of n' which were discarded to produce n'' would have been discarded later when μ' was chopped to produce μ.

Subtract Magnitude Case

This case arises if we add numbers having opposite signs or subtract numbers having the same sign. Using (1.8.1), (1.8.2), and (1.8.4), we may reduce this case to the problem of computing $a \oplus b$ where

$$a > 0 > b$$

and

$$a \geq |b|.$$

Then

$$\mu' = m + n'' \leq m < 1.$$

If $\mu' = 0$, we set $a \oplus b$ equal to a normalized zero. If μ' is not zero, we need to ask only whether it is normalized. If it is not, we normalize it and refer to this operation as postnormalization. Let k be the number of leading zeros in μ'. Since

$$r^e \mu' = r^{e-k}(r^k \mu'),$$

we set $g = e - k$ and $\mu = \overline{r^k \mu'}$.

We first suppose that $q > 0$. Before we computed $m + n''$ by subtracting $|n''|$ from m, we shifted n to the right $e - f$ places. Now if $e - f \geq 2$, then $|n''| < r^{-2}$, so

$$\mu' = m - |n''| > r^{-1} - r^{-2} \geq r^{-2}.$$

Thus, when $e - f \geq 2$ we never have to shift μ' more than one place to the left to postnormalize it, so k is either 0 or 1. Therefore, if $k \geq 2$, then $e - f$ must be 0 or 1. This means that whenever $k \geq 2$ we have $n'' = n'$, so $\mu' = m + n'$ and it may be represented with at most $p + 1$ digits to the right of the radix point. Then, when we shift μ' to the left $k \geq 2$ places, we are able to hold all its digits, so $\mu = r^k(m + n')$. That is, if $q > 0$ and $k \geq 2$, the operation $\oplus$ introduces no error, so

$$a \oplus b = a + b.$$

This is a rather surprising result, since the postshift of two or more places indicates that the subtraction $a - |b|$ has produced leading zeros and therefore resulted in the loss of significance. The secret lies in the fact that although the operation $a \oplus b$ produces exactly the correct result for the operands a and b, the result is sensitive to errors in a and b.

Next, we note that if $e - f \leq q$, then $n'' = n'$, so $\mu' = m + n'$. In this case the operation $a \oplus b$ produces the correctly chopped result—that is, it produces the same result as it would in FP(r, p, c).

Finally, suppose that $e - f > q > 0$. Then some of the low-order digits of n' were chopped during the preshift. Unless these digits were zero, we have subtracted too little from m, so

$$\mu' > m - |n'|.$$

But we may have to chop nonzero digits of μ' in order to shorten μ' to p digits, and this would make the answer smaller. Thus, we have two effects which tend to compensate. Chopping n' tends to make μ too large, while chopping μ' tends to make μ too small. Since $q \geq 1$, we must have $e - f \geq 2$, so k is either 0 or 1. Then μ will be either digits 1 through p or else digits 2 through $p + 1$ of μ'. Now we have retained $p + q$ digits of n', so

$$(1.8.5) \qquad |n'| - |n''| < r^{-(p+q)}.$$

(For the IBM 7090, $q = 27$ so the difference between n' and n'' is seldom important. But for the IBM System/360, we have $q = 1$ and the difference may be noticeable.) From (1.8.5) we see that if μ' is too large, it is in error by less than 1 in the $(p + q)$th position to the right of the radix point. Then if μ is too large, it is too large by less than 1 in the $(p + q - 1)$st position. In particular, this means that if $a \oplus b \neq \overline{a + b}$, than $|a \oplus b|$ is greater than $\overline{|a + b|}$ by 1 in the last place. We summarize these results in the following theorem.

THEOREM 1.8.2

For the subtract magnitude case with $q > 0$,

1. If the postshift is two or more places, $a \oplus b = a + b$.

2. If the preshift is q or fewer places, the operation $a \oplus b$ produces the same result in FP(r, p, clq) as it does in FP(r, p, c).

3. If the preshift is more than q places, the operation $a + b$ either produces the same result in FP(r, p, clq) as it does in FP(r, p, c), or else the result in FP(r, p, clq) may be obtained by increasing the absolute value of the result in FP(r, p, c) by 1 in the last place. In this case, $\overline{|a + b|} < |a + b| < |a \oplus b|$, and $|a \oplus b|$ is greater than $|a + b|$ by less than 1 in the $(p + q - 1)$st position.

Finally, suppose that $q = 0$. In this case we may produce a result which is quite bad. For example, suppose that $a = 1$ and $b = -(1 - r^{-p})$. Here $m = r^{-1}$, $n = -(1 - r^{-p})$, $n' = -[r^{-1} - r^{-(p+1)}]$, and $n'' = -(r^{-1} - r^{-p})$. Then in FP($r, p, cl0$),

$$a \oplus b = r^{-p}$$

but

$$a + b = r^{-(p+1)}.$$

Thus, the result in FP($r, p, cl0$) is r times as large as the result in FP(r, p, c). This was one of the reasons for adding a guard digit to the double-precision arithmetic on the IBM System/360.

In summary, we see that for $q > 0$ the four operations $\oplus$, $\ominus$, $*$, and $\div$ produce results in FP(r, p, clq) which are close to the results produced in FP(r, p, c). In fact, the only difference arises in the subtract magnitude case for the operations $\oplus$ and $\ominus$. Since we are primarily interested in the case $q > 0$, we may often ignore the distinction between these systems and deal with FP(r, p, c). Indeed, in all the examples discussed in Sections 1.6 and 1.7

in the study of the laws of algebra and inequalities, the same results would be produced in FP(r, p, clq) for $q > 0$ as in FP(r, p, c). Thus, we have demonstrated the failure of these laws in FP(r, p, clq) for $q > 0$ and nontrivial combinations of r and p.

1.9. THE SOLUTION OF $a * x = b$ IN FP(r, p, c)

In this section we shall consider the question of whether or not an equation of the form

$$(1.9.1) \qquad\qquad a * x = b$$

has a solution in FP(r, p, c). [Our analysis will also apply to FP(r, p, clq) for all $q > 0$, since the operation $*$ produces the same result in this system as it does in FP(r, p, c).] Since the solution of $ax = b$ is b/a, it is natural to ask whether $b \div a$ satisfies (1.9.1). But if it did, we would have

$$a * (b \div a) = b,$$

and we saw in Section 1.6 that this holds if and only if the division $b \div a$ is exact. Thus, $b \div a$ seldom satisfies (1.9.1). For any a and c in $S(r, p)$, we may set $b = a * c$ and obtain an equation of the form (1.9.1) which does have a solution, although this solution may be different from $b \div a$. In this section we shall show that for nontrivial combinations of r and p there are always nonzero a and b in $S(r, p)$ for which (1.9.1) does not possess a solution.

Assume that a and b are given, and write $a = r^e m$. Since we do not expect to solve (1.9.1) if $a = 0$, and we clearly can solve it if $b = 0$, we may assume that neither a nor b is zero. Changing the sign of x changes the sign of $a * x$, so we need consider only the case in which a and b are positive. Then $r^{-1} \leq m < 1$, and we may write $x = r^f n$, $r^{-1} \leq n < 1$. Now if a or x is multiplied by a power of r, $a * x$ is multiplied by the same power of r, so the problem of solving (1.9.1) reduces to the question of whether or not we can find a number x such that $a * x$ has the same mantissa as b. Clearly this depends only on the mantissas of a and b. Then (1.9.1) is solvable for all b if and only if the mantissa of $a * x$ takes on all the $(r - 1)r^{p-1}$ possible nonzero values as the mantissa of x varies from r^{-1} to $1 - r^{-p}$. But the failure of the cancellation law for nontrivial systems FP(r, p, c) implies that there are numbers a, x, and y in $S(r, p)$ such that $a \neq 0$, x and y have different mantissas, and $a * x = a * y$. For this value of a there are duplications among the mantissas of $a * x$ as the mantissa of x varies from r^{-1} to $1 - r^{-p}$, so there must also be omissions. That is, there are values of b for which $a * x$ is never b. The failure of the cancellation law also shows that there are values of a and b for which the solution of (1.9.1) is not unique.

We shall now examine the computation $a * x$ in more detail. Write

$$a * x = r^{e+f-k}\mu,$$

where $\mu = \overline{r^k mn}$ and k is 1 or 0 depending on whether or not postnormalization is required. Now

$$r^{-1}m \leq mn < m,$$

so we have either $r^{-1}m \leq mn < r^{-1}$ and $k = 1$ or else $r^{-1} \leq mn < m$ and $k = 0$. Similarly, if $k = 1$, we have $r^{-1}n \leq mn < r^{-1}$, and if $k = 0$, we have $r^{-1} \leq mn < n$. Thus, if postnormalization is required, the mantissa μ of $a * x$ is $\geq$ both m and n. On the other hand, if no postnormalization is required, μ is less than both m and n. Surprisingly, μ can never lie between m and n.

We note that if $m = r^{-1}$, then $\mu = n$, so Eq. (1.9.1) can be solved for all b. Thus, we may assume that $m > r^{-1}$. If $n = 1 - r^{-p}$, we have

$$m > mn = m - r^{-p}m > m - r^{-p}$$

so $\mu = \overline{mn} = m - r^{-p}$. In this case the mantissa of $a * x$ is less than m by 1 in the last place.

Now suppose that $n < 1 - r^{-p}$ and consider the number y obtained by increasing x by 1 in the last place. Then $y = r^f l$, where $l = n + r^{-p}$. If $k = 0$,

$$(1.9.2) \qquad ml = mn + mr^{-p} < mn + r^{-p}$$

yields $\overline{ml} \leq \overline{mn} + r^{-p}$. On the other hand, if $k = 1$, we have

$$(1.9.3) \qquad ml = mn + r^{-p}m \geq mn + r^{-(p+1)},$$

so $\overline{rml} \geq \mu + r^{-p}$. Then, if $ml < r^{-1}$, we find that increasing x by 1 in the last place increases $a * x$ by at least 1 in the last place. For the special case in which $a * x$ requires postnormalization but $a * y$ does not, we have

$$(1.9.4) \qquad mn < ml = mn + r^{-p}m < mn + r^{-p} < r^{-1} + r^{-p},$$

so $\overline{ml} = r^{-1}$. Clearly $a * y$ is greater than $a * x$ in this case. Finally, if $n = 1 - r^{-p}$, then increasing x by 1 in the last place produces $y = r^f$, so $a * y = r^f a$, which is greater than $a * x$ by 1 in the last place. We have proved:

THEOREM 1.9.1

In FP(r, p, c), let a and x be positive. If $a * x$ requires postnormalization, increasing x by 1 in the last place increases $a * x$ by at least 1 in the last place. If $a * x$ does not require postnormalization, increasing x by 1 in the last place either leaves $a * x$ unchanged or increases it by 1 in the last place.

We shall now consider the changes in $a * x$ as n varies from r^{-1} to $1 - r^{-p}$. We shall still assume that $m > r^{-1}$. If $n = r^{-1}$, $\mu = m$ and postnormalization is required. If $n = 1 - r^{-p}$, $\mu = m - r^{-p}$ and no postnormalization is required. Let v be the smallest mantissa n for which no postnormalization is required. Then $r^{-1} < v \leq 1 - r^{-p}$, and from (1.9.4) we see that $\overline{mv} = r^{-1}$. An immediate consequence of this is that $z = r^{1-e}v$ is a solution of

$$(1.9.5) \qquad\qquad\qquad a * z = 1.$$

That is, every nonzero element of FP(r, p, c) has an inverse under the operation $*$. Now if the associative law were valid in FP(r, p, c), we could solve (1.9.1) by letting z be a solution of (1.9.5) and setting $x = z * b$. Then $a * x$ would be equal to $(a * z) * b$, which is b. Unfortunately, since the associative law of multiplication fails to hold in FP(r, p, c) for nontrivial systems, the existence of an inverse does not allow us to solve (1.9.1).

Now as n varies from v to $1 - r^{-p}$, μ varies from r^{-1} to $m - r^{-p}$, and in this range increasing n by 1 in the last place increases μ by at most 1 in the last place. Then μ takes on every value less than m, so (1.9.1) can always be solved if the mantissa of b is smaller than the mantissa of a. It can also be solved if a and b have the same mantissa, since $n = r^{-1}$ yields $\mu = m$. Thus, if (1.9.1) fails to have a solution, the mantissa of b must be larger than the mantissa of a.

We shall now ask whether we can solve (1.9.1) for all $b > 0$. We surely can if $m = 1 - r^{-p}$, because there are no mantissas which are larger. Then we may assume that $r^{-1} < m < 1 - r^{-p}$. There are $(1 - m)r^p - 1$ mantissas greater than m. If (1.9.1) has a solution for all b, $a * x$ cannot skip any of these mantissas as n varies from $r^{-1} + r^{-p}$ to $v - r^{-p}$. Since there are no repetitions among the values of μ corresponding to n in this range, a necessary and sufficient condition for (1.9.1) to have a solution for all b is that

$$v - r^{-p} = r^{-1} + [(1 - m)r^p - 1]r^{-p},$$

that is,

$$(1.9.6) \qquad\qquad\qquad v = r^{-1} + 1 - m.$$

Since the right-hand side of (1.9.6) is obtained by increasing r^{-1} by $(1 - m)r^p$ units in the last place, it is surely large enough so that $m(r^{-1} + 1 - m) \geq r^{-1}$. Thus, v can never exceed the right-hand side of (1.9.6). Therefore, a necessary and sufficient condition for (1.9.1) to have a solution for all b is that

$$m(r^{-1} + 1 - m - r^{-p}) < r^{-1},$$

which reduces to

$$m^2 - (1 + r^{-1} - r^{-p})m + r^{-1} > 0$$

Let

$$g(t) = t^2 - (1 + r^{-1} - r^{-p})t + r^{-1}.$$

Then (1.9.1) has a solution for all b if and only if $g(m) > 0$. By direct substitution we find

$$g(r^{-1}) = r^{-(p+1)}$$
$$g(r^{-1} + jr^{-p}) = -jr^{-p} + (j + 1)r^{-(p+1)} + j(j + 1)r^{-2p}$$
$$g(1 - r^{-p}) = r^{-(p+1)}$$
$$g(1 - jr^{-p}) = -(j - 1)r^{-p} + jr^{-(p+1)} + j(j - 1)r^{-2p}.$$

Now $g(t)$ is a quadratic expression which is positive for large $|t|$. If there are two values $t_1 < t_2$ for which $g(t)$ is negative, then $g(t)$ is negative for all t in the range $t_1 \leq t \leq t_2$. Clearly $g(r^{-1})$ and $g(1 - r^{-p})$ are always positive. If $r > 2$ and $p \geq 2$, we find that $g(r^{-1} + r^{-p})$ and $g(1 - 2r^{-p})$ are negative, so (1.9.1) has a solution for all b if and only if m is r^{-1} or $1 - r^{-p}$. For $r = 2$ and $p \geq 4$, we find that $g(2^{-1} + 2^{-p})$ and $g(1 - 2^{-(p-1)})$ are positive, but $g(2^{-1} + 2^{-(p-1)})$ and $g(1 - 3 \cdot 2^{-p})$ are negative, so (1.9.1) has a solution for all b if and only if m has one of the four values 2^{-1}, $2^{-1} + 2^{-p}$, $1 - 2^{-p}$, or $1 - 2^{-(p-1)}$. By considering the remaining cases we may prove the following theorem:

THEOREM 1.9.2

In FP(r, p, c), Eq. (1.9.1) with $a \neq 0$ has a solution if $b = 0$ or if the absolute value of the mantissa of b is not larger than the absolute value of the mantissa of a. In the four systems FP$(3, 1, c)$, FP$(2, 1, c)$, FP$(2, 2, c)$, and FP$(2, 3, c)$, (1.9.1) always has a solution if $a \neq 0$. In any other system FP(r, p, c) with $r \geq 3$, (1.9.1) has a solution for all b if and only if the absolute value of the mantissa of a is r^{-1} or $1 - r^{-p}$. In FP$(2, p, c)$ with $p \geq 4$, (1.9.1) has a solution for all b if and only if the absolute value of the mantissa of a is one of the four numbers 2^{-1}, $2^{-1} + 2^{-p}$, $1 - 2^{-p}$, or $1 - 2^{-(p-1)}$.

Our study of $a * x$ also leads to a better understanding of the relationship between $(b \div a) * a$ and b. We suppose that a and b are positive, and let $x = b \div a$. We saw in Section 1.6 that $x * a$ will be less than b unless the division $b \div a$ is exact, that is, unless b/a is in $S(r, p)$. Suppose that $x < b/a$, so $a * x < b$. Let $x = r^e m$, $r^{-1} \leq m < 1$, and let $y = r^e(m + r^{-p})$. Then $ax < b < ay$, so

$$(1.9.7) \qquad\qquad a * x < b \leq a * y.$$

Now (1.9.2) and (1.9.3) show that the mantissa of $a * y$ cannot exceed the mantissa of $a * x$ by more than r units in the last place. Then (1.9.7) yields the following theorem:

THEOREM 1.9.3

In FP(r, p, c) or in FP(r, p, clq) with $q \geq 1$, let $a \neq 0$ and $c = (b \div a) * a$. Then $|b|$ cannot exceed $|c|$ by more than r units in the last place of c.

1.10. DIVISION

We have defined division in both FP(r, p, c) and FP(r, p, clq) by

$$a \div b = \overline{a/b}.$$

We now look more closely at the details of this computation. If b is zero, the quotient is undefined, and for any $b \neq 0$, we have $0 \div b = 0$. Then we may assume that a and b are nonzero and normalized. Let

$$a = r^e m, \qquad r^{-1} \leq |m| < 1,$$
$$b = r^f n, \qquad r^{-1} \leq |n| < 1.$$

Then

$$\frac{a}{b} = r^{e-f} \frac{m}{n},$$

and

$$r^{-1} < \left| \frac{m}{n} \right| < r.$$

Write $a \div b = r^g \mu$. If $|m/n| < 1$, we set $g = e - f$ and $\mu = \overline{m/n}$. On the other hand, if $|m/n| \geq 1$, we have

$$\frac{a}{b} = r^{e-f+1} \left(r^{-1} \frac{m}{n} \right)$$

and

$$r^{-1} \leq \left| r^{-1} \frac{m}{n} \right| < 1,$$

so we set $g = e - f + 1$ and $\mu = \overline{r^{-1}m/n}$. Let k be 0 if $|m| < |n|$ and 1 if $|m| \geq |n|$. Then

$$g = e - f + k$$
$$\mu = \overline{r^{-k}m/n}.$$

We note that in forming $r^{-k}m$ we may have to shift m one place to the right, so we must be able to handle a $(p + 1)$-digit dividend.

Now on some machines—for example, the IBM 7090—the floating-point divide operation produces both a quotient and a remainder. To see how the remainder is defined, we recall that for positive integers A and B we may divide A by B to produce quotient Q and a remainder R less than B. That is, there are unique integers Q and R such that

$$A = BQ + R$$
$$0 \leq R < B.$$

This is readily extended to any integers A and B with $B \neq 0$, so we have the following theorem:

THEOREM 1.10.1

If A and B are any integers with $B \neq 0$, there are unique integers Q and R such that

1. $A = BQ + R$.
2. $0 \leq |R| < |B|$.
3. If Q is not zero, it has the same sign as A/B.
4. If R is not zero, it has the same sign as A.

Theorem 1.10.1 is often the basis of the fixed-point divide operation on computers.

We wish to divide m' by n, where $m' = r^{-k}m$. Set $A = r^{2p}m'$ and $B = r^p n$. Then the Q and R of Theorem 1.10.1 satisfy

$$r^{2p}m' = r^p nQ + R,$$

so

(1.10.1) $$m' = nr^{-p}Q + r^{-2p}R.$$

We set $q = r^{-p}Q$ and $s = r^{-2p}R$, so

(1.10.2) $$m' = nq + s.$$

Since $|r^{2p}n| > |r^{2p}m'|$, we have $|Q| < r^p$ and hence $|q| < 1$. Also, since $|R| < |r^p n|$, we have

(1.10.3) $$|s| < |r^{-p}n|.$$

In both (1.10.1) and (1.10.2), the two terms on the right-hand side have the same sign, so

$$|m'| = |n| \cdot |q| + |s|.$$

With (1.10.3), this yields

$$|nq| \leq |m'| < |n|(|q| + r^{-p}),$$

SO

$$|q| \le \left|\frac{m'}{n}\right| < |q| + r^{-p}.$$

Since q is r^{-p} times an integer and $|q| < 1$, q is $\overline{m'/n}$. By the definition of m', $|m'/n| \ge r^{-1}$, so q is normalized. In general, s is not normalized. Let $c = r^g q$ and $d = r^{e+k-p}s$. Then $a = bc + d$ and $c = a \div b$. Machines such as the IBM 7090 which produce both a quotient and a remainder normally produce these values of c and d, leaving d in unnormalized form. Thus, the remainder d is characterized by

$$d = a - b(a \div b).$$

EXERCISES

1. Carry out the arithmetic in FP(10, 4, c) for each of the examples in Sections 1.6 and 1.7.

2. In FP(10, 4, c), find an equation $a * x = b$ with $a \ne 0$ which does not have a solution. Also, find an equation for which the solution is not unique.

3. Show that the associative law of addition fails to hold in FP(r, p, R).

4. It can be shown that the associative law of multiplication fails to hold in FP(r, p, R) except for the three trivial systems FP(3, 1, R), FP(2, 1, R), and FP(2, 2, R). Show that this law fails to hold in FP(r, p, R) for the following combinations of r and p:
   a. $r > 2, p \ge 2$.
   b. $r = 2, p \ge 4$.

5. It can be shown that the cancellation law fails to hold in FP(r, p, R) except for the three trivial systems FP(3, 1, R), FP(2, 1, R), and FP(2, 2, R). Show that this law fails to hold in FP(r, p, R) for the following combinations of r and p:
   a. $r > 2, p \ge 2$.
   b. $r = 2, p \ge 4$.

6. It can be shown that the distributive law fails to hold in FP(r, p, R) except for the trivial system FP(2, 1, R). Show that this law fails to hold in FP(r, p, R) for the following combinations of r and p:
   a. $r \ge 5, p \ge 3$.
   b. $r = 2, p \ge 4$.

7. It can be shown that the relation $a * (b \div a) = b$ fails to hold in FP(r, p, R) except for the trivial systems FP(2, 1, R) and FP(2, 2, R). Show that this relation fails to hold in FP(r, p, R) for the following combinations of r and p:
   a. $r > 2, p \ge 2$.
   b. $r = 2, p \ge 4$.

8. Show that the following inequalities hold in FP(r, p, R):
   a. If $a < b$, then $a \oplus c \le b \oplus c$ holds for all c.

    b. If $a < b$ and $c < d$, then $a \oplus c \le b \oplus d$.

    c. If $a < b$, then $a * c \le b * c$ holds for all $c > 0$.

9. Show that inequality a of Exercise 8 cannot be strengthened to a strict inequality.

10. For inequality b of Exercise 8, it can be shown that strict inequality holds only for the trivial system FP(2, 1, R). Show that the strict inequality fails to hold in FP(r, p, R) for the following combinations of r and p:

    a. $r > 2, p \ge 2$.

    b. $r = 2, p \ge 2$.

11. Show that inequality c of Exercise 8 cannot be strengthened to a strict inequality except for the trivial systems FP(3, 1, R), FP(2, 1, R), and FP(2, 2, R).

12. In FP(r, p, R), what is the mantissa of $a * x$ if a and x are positive and the mantissa of x is $1 - r^{-p}$?

13. If $r > 2$, show that the equation $a * x = 1$ does not have a solution in FP(r, p, R) when the mantissa of a is $1 - r^{-p}$. That is, a number whose mantissa is $1 - r^{-p}$ does not have an inverse in FP(r, p, R) if $r > 2$.

14. Show that the equation $a * x = 1$ always has a solution in FP(r, p, R) if a is positive and its mantissa m satisfies

$$r^{-1} \le m < \frac{1 + r^{-1}}{2}.$$

15. If $r > 2$ and we exclude the trivial case FP(3, 1, R), show that in FP(r, p, R) the equation $a * x = b$ has a solution for all b if and only if $|a|$ is a power of r.

16. Show that in FP(2, p, R) with $p \ge 3$ the equation $a * x = b$ has a solution for all b if and only if the absolute value of the mantissa of a is either 2^{-1} or $1 - 2^{-p}$.

17. Suppose that we use Euler's method to solve the differential equation $y' = -y$ for $0 \le x \le 1$ with $y(0) = 1$. We take N steps with step size $h = 1/N$. For this differential equation, the formula for Euler's method reduces to

$$y_{n+1} = y_n - h y_n$$

To illustrate the arithmetic involved, we solve this problem several times with different values of N, taking $N = 16, 32, 64, \ldots, 4096$. In each case we print only the final value y_N.

    We shall consider two different ways in which this formula might be coded in FORTRAN. They are

$$Y = Y - H * Y$$

and

$$Z = (1. - H) * Z.$$

Our FORTRAN program is

```
  1  FORMAT (I6,2F13.8)
     N = 16
     DO 200 K = 1,9
     Y = 1
     Z = 1
     H = 1./N
     DO 100 I = 1,N
     Y = Y−H*Y
100  Z = (1.−H)*Z
     WRITE (3,1)N,Y,Z
200  N = N*2
     STOP
     END
```

The question is whether or not the values printed for Y and Z are identical. Run this program in single-precision on whatever machine you have available and explain why the values of Y and Z are the same or different on that machine. (You may have to modify the WRITE statement to agree with the conventions at installation.)

18. The values printed for Y and Z by the program in Exercise 17 will be identical if it is run on the IBM 7090, but they will be different if it is run on the IBM System/360.
   a. Explain why the values of Y and Z are identical when the computation is performed in FP(2, 27, $c/27$) but different when the computation is performed in FP(16, 6, $c/1$).
   b. Explain in general terms how the values of Y and Z would differ if the computation were performed in FP(2, 27, $c/1$).
   c. Explain in general terms how the values of Y and Z would differ if the computation were performed in FP(16, 6, $c/6$).

19. Suppose that we have written a FORTRAN program whose input includes a temperature X measured in centigrade. The program converts X from centigrade to Farenheit by the FORTRAN statement

$$F = 1.8*X + 32.$$

Suppose that we want F to be exactly zero, so we ask what number X must be supplied as input to produce the value zero for F. Here X must be the solution of the equation

$$A * X = -32,$$

where A is the number in $S(r, p)$ to which the FORTRAN compiler converts 1.8. We assume that the integer 32 is converted exactly. Depending on the FORTRAN compiler used, A may be either $\overline{1.8}$ or $\overline{1.8}^{\circ}$. (The FORTRAN compiler for the IBM System/360 produced by the manufacturer would produce $A = \overline{1.8}$.)

Show that the following statements are true:

a. The equation $A * X = -32_D$ does not have a solution in FP(16, 6, cl1) if $A = \overline{1.8} = 1.CCCCC_H$.

b. The equation $A * X = -32_D$ has a solution in FP(16, 6, cl1) if $A = \overline{1.8}^\circ = 1.CCCCD_H$.

c. In FP(16, 14, cl1), the equation $A * X = -32_D$ has a solution if $A = \overline{1.8}^\circ$ but not if $A = \overline{1.8}$.

d. In FP(2, 27, cl27), the equation $A * X = -32_D$ has a solution if $A = \overline{1.8} = \overline{1.8}^\circ$.

e. In FP(10, 8, c), the equation $A * X = -32$ has a solution. (Here no conversion is necessary.)

20. Consider the computation of

$$c = (b \div a) * a$$

in FP(r, p, c), where r and p designate the radix and precision of the machine you are using. Find an example which shows that there are numbers a and b in $S(r, p)$ such that b and c differ by r units in the last place.

2 FLOATING-POINT OVERFLOW AND UNDERFLOW

2.1. BOUNDS FOR EXPONENTS

Up to this point we have assumed that a floating-point number was any number which could be written in the form $r^e m$, where e is any integer, $r^{-1} \leq |m| < 1$, and $|m|$ can be expressed in the base r using at most p digits. But, as we saw in Sections 1.3 and 1.4, we usually store the signed exponent as a characteristic in a few digits of the word. Thus, in a decimal machine the characteristic is often defined to be the exponent plus 50, and it is stored in two-decimal digits. This restricts the exponent to the range $-50 \leq e \leq 49$. The IBM 704, 709, 7090, and 7094 used an eight-bit characteristic which was defined to be the exponent plus 128. This restricted the exponent to $-128 \leq e \leq 127$. The IBM System/360 uses a seven-bit characteristic which is defined to be the exponent plus 64, so the exponent must lie in the range $-64 \leq e \leq 63$. The CDC 6600 uses an 11-bit field to hold the exponent, and it holds negative exponents in one's complement form. This produces a range $-1023 \leq e \leq 1023$ for the exponent. Thus, in general, the exponent is restricted to a range

$$(2.1.1) \qquad\qquad e_* \leq e \leq e^*.$$

For a machine which stores the exponent as a characteristic, we usually have

$$(2.1.2) \qquad\qquad e_* = -(e^* + 1).$$

But if the machine holds negative exponents as either one's complements or *sign and true magnitude* (as, for example, the IBM 7030 did), we may have $e_* = -e^*$. Since the CDC 6600 uses a mantissa which is a 48-bit integer, in our notation, which treats the mantissa as a fraction, we would write $e^* = 1071$ and $e_* = -975$.

Restricting the range of the exponent restricts the range of the floating-point numbers which we can represent. We shall use Ω to designate the largest positive floating-point number which can be represented subject to (2.1.1). Similarly, ω will designate the smallest normalized positive number which can be represented subject to (2.1.1). Then

$$(2.1.3) \qquad \Omega = r^{e^*}(1 - r^{-p}) < r^{e^*}$$

and

$$(2.1.4) \qquad \omega = r^{e_*}r^{-1} = r^{e_*-1}.$$

For the IBM System/360, this yields

$$(2.1.5) \qquad \begin{aligned} \Omega &= 10^{63}(1 - 16^{-6}) < 16^{63} \\ \omega &= 10^{-65}. \end{aligned}$$

This presents a slight asymmetry in our floating-point number system: There are some small numbers whose reciprocals cannot be represented because they are larger than Ω.

The fact that the bounds e^* and e_* are inherent in the machine implementation of floating-point arithmetic suggests that they should be included in the definition of floating-point numbers. Thus, instead of $S(r, p)$ we could deal with the set $S(r, p, e_*, e^*)$ which contains zero and all numbers in $S(r, p)$ which can be written in the form $r^e m$, where e is an integer satisfying (2.1.1) and $r^{-1} \le |m| < 1$. However, we shall not follow this approach. Instead, shall deal with problems related to overflow and underflow separately from problems related to rounding error and the anomalies of floating-point arithmetic. We shall perform arithmetic in the system $FP(r, p, a)$ as long as the results we obtain have exponents satisfying (2.1.1). If we try to produce a result which has an exponent outside this range, we say that we have encountered *exponent spill*. If we try to produce a number with absolute value greater than Ω, the exponent spill is called *exponent overflow* or *floating-point overflow*. Similarly, if we try to produce a nonzero number with absolute value less than ω, the exponent spill is called *exponent underflow* or *floating-point underflow*. In this book we shall not deal with fixed-point overflow, so we shall often use the simpler terms *overflow* and *underflow* to describe exponent spill.

In this chapter we shall discuss various ways of dealing with exponent spill. This often involves both the question of what the hardware does and the question of what the compiler does. Of course what the compiler can do is to some extent determined by what the hardware does.

2.2. Ω-ZERO FIXUP

One of the earliest approaches to the problem of exponent spill used what we shall call the Ω-*zero fixup*. This was used in many of the interpreters which performed floating-point arithmetic, and it is still extensively used today. With this approach, whenever we have a floating-point underflow, the result is set to zero. After a floating-point overflow, the result is set to a number which has the correct sign and whose absolute value is Ω. When Ω is used in this way, it is sometimes erroneously referred to as "infinity." But Ω is a legitimate floating-point number, and it does not act like ∞. For example, $\Omega \div 2 < \Omega$.

A more elaborate approach is used in the floating-point arithmetic of the CDC 6600. Here floating-point underflow produces zero as the answer, and floating-point overflow produces a genuine infinity. This is a special bit pattern which is treated as infinity by the hardware. Thus, for any normal floating-point number x,

$$(2.2.1) \qquad \infty \oplus x = \infty \ominus x = \infty$$

$$(2.2.2) \qquad x \div \infty = 0$$

Also, if $x \neq 0$,

$$(2.2.3) \qquad x * \infty = \infty$$

$$(2.2.4) \qquad x \div 0 = \infty$$

There is another special bit pattern which is called INDEFINITE and is produced as the result of an indeterminant form. Thus, $0 * \infty$ and $0 \div 0$ both produce INDEFINITE as the result. For any arithmetic operation, if one of the operands is INDEFINITE, the result is INDEFINITE. Thus, the CDC 6600 truly has an ∞-zero fixup. It depends on having the hardware recognize certain bit patterns as ∞ or INDEFINITE whenever they are used as operands in any floating-point operation. Thus, this fixup depends on how the hardware works. There is no reasonable way to implement it unless the hardware tests the operands in every floating-point operation.

The Ω-zero fixup and the ∞-zero fixup have the same objective: They allow the computation to proceed after exponent spill in a more or less reasonable manner. The Ω-zero fixup is easier to implement, since it does not require the testing of the operands in all floating-point operations. In fact, it is often implemented in software when the hardware produces some other result.

We shall now look at the rationale for these fixups. Producing an answer which is ∞ or INDEFINITE is usually an indication that exponent spill has occurred. With the Ω-zero fixup, if our final answer is close to Ω, it is often (perhaps usually) an indication that exponent spill occurred earlier. However, the real motivation for either the Ω-zero fixup or the ∞-zero fixup is that we may be able to produce good answers even though we have encountered exponent spill for some intermediate results.

We shall illustrate this by considering the computation of sin x. Typical coding would first reduce the problem to computing the sine or cosine of an argument x with, say, $|x| \leq \pi/4$. For sin x in this reduced range, we would set $y = x * x$ and compute $z = \sin x$ from

$$(2.2.5) \qquad z = x * (a_0 + y * (a_1 + y * (a_2 + y * (a_3 + \cdots) \cdots))).$$

Here the a_k are the coefficients for a polynomial approximation for sin x in the range $-(\pi/4) \leq x \leq \pi/4$. To produce good relative error for small x, we must have $a_0 \approx 1$. Now, suppose that $x * x$ underflows. On the IBM System/360 this means that $x^2 < 16^{-65}$, so

$$(2.2.6) \qquad\qquad\qquad |x| < .25 \cdot 16^{-32}.$$

If we set $y = 0$, we shall compute

$$z = x * a_0 \approx x.$$

Indeed, if $a_0 = 1$, then $z = x$, and for x satisfying (2.2.6) we find that the approximation sin $x \approx x$ is good to over 65 hexadecimal digits. Therefore, setting the result to zero after exponent underflow allows us to produce excellent results in this computation. If we used some other approach for underflow, say terminating the computation, we would have had to test x to see whether it satisfied (2.2.6). If it did, we would set $z = x$; otherwise we would compute z from (2.2.5). This would be faster when x satisfied (2.2.6), but it would degrade performance whenever $x^2 > 16^{-65}$, which is by far the commoner case.

The calculation described above is typical of a class of programs which will produce good answers if a quantity which underflows is replaced by zero. This does not mean that this approach is always successful. Some drawbacks to setting the underflowed result equal to zero are discussed in Sections 2.7 and 2.8, where other approaches are considered. However, there are many cases where a term which has underflowed may be ignored, so this approach enjoys wide popularity. In fact, one often hears the Ω-zero fixup referred to as the *standard fixup*.

Setting the result equal to $\pm\Omega$ after overflows is less attractive; basically, it is an attempt to approximate the ∞-zero fixup. But there are some situations in which it will produce good results. For example, suppose that we encounter overflow in computing x and that we are really interested in

$y = \arctan x$. If x is positive, arctan x and arctan Ω are both approximately $\pi/2$, so replacing x by Ω allows us to compute a good approximation for y. Similarly, we shall produce a good result for y if we replace x by $-\Omega$ whenever x is negative. Thus, the Ω-zero fixup handles this problem quite nicely. However, the example seems rather contrived.

A more typical situation is to encounter overflow in a sequence of multiplications and divisions used to compute one term in a sum. (Addition and subtraction can produce overflow only if both operands have absolute value close to Ω.) We would like to have overflow in the denominator treated as equivalent to underflow in the numerator. The ∞-zero fixup does this quite nicely. For example, consider

$$x = (a * b) \div (c * d)$$

Assume that $c * d$ overflows but that $a * b$ does not. The ∞-zero fixup will produce $x = 0$. The Ω-zero fixup will also produce $x = 0$ if $|a * b| < \Omega \cdot \omega$. [If e^* and e_* satisfy (2.1.2), $\Omega \cdot \omega = r^{-2}(1 - r^{-p}) \approx r^{-1}$.] Thus, the Ω-zero fixup produces the same result as the ∞-zero fixup as long as $|a * b| < \Omega \cdot \omega$, and it will produce a small value for $|x|$ as long as $|a * b|$ is substantially less than Ω. Suppose that we want to use the value we have computed for x to compute

$$z = x \oplus y.$$

If x has been set to zero, we shall compute $z = y$. If the Ω-zero fixup has produced a nonzero value for x and y is small, we may obtain a poor value for z. Even the ∞-zero fixup will not always yield good values for z. For example, suppose that $y \approx 1$, $a * b \approx \Omega/2$ and that $c \cdot d \approx 2\Omega$. Then x should be approximately $\frac{1}{4}$, so if we replace x by zero, we shall produce a poor value for z. This value would be even worse if y were, say, 10^{-6}. Thus, neither the Ω-zero fixup nor the ∞-zero fixup is a panacea, but there are a reasonable number of cases in which they produce good results.

In our definition of the Ω-zero fixup, the result after overflow was set to a number with the correct sign and with the absolute value Ω. Similarly, the ∞-zero fixup could have been defined to use two special bit patterns representing $+\infty$ and $-\infty$, with normal sign control and definitions such as $(+\infty) + (-\infty) = \text{INDEFINITE}$. We now consider an example which illustrates the value of retaining the sign after overflow. Suppose that we are using a gradient technique to find the maximum of a function $f(x, y)$. With such a method we vary the step size h from step to step. [See Crockett and Chernoff (1955).] If we have taken a step in gradient direction which results in a decrease in the value of $f(x, y)$, it means that we have taken too large a step. Thus, if we take a step of size h from (x_n, y_n) to (x_{n+1}, y_{n+1}) and find that

$$f(x_{n+1}, y_{n+1}) < f(x_n, y_n),$$

we reject the point (x_{n+1}, y_{n+1}) and take another step from (x_n, y_n) with smaller step size, say $h/2$. Now suppose that $f(x_{n+1}, y_{n+1})$ overflows. If $f(x_{n+1}, y_{n+1})$ $< -\Omega$, we surely want to reject the point (x_{n+1}, y_{n+1}) and reduce h. Our program will do this automatically if our computation of $f(x_{n+1}, y_{n+1})$ produces the result $-\Omega$ or any negative number close to $-\Omega$. On the other hand, if we did not preserve the sign and our computation of $f(x_{n+1}, y_{n+1})$ produced the value Ω or ∞, our program would move to the point (x_{n+1}, y_{n+1}) thinking that it had produced an increase in $f(x, y)$. Thus, there are problems in which the sign of a number which has overflowed can be vital.

If the hardware provides an interrupt on exponent spill (see Section 2.3), the actual fixup is often performed by software. Even so, the type of fixup which can be provided by the software is constrained by the way the hardware operates. We have seen that the ∞-zero fixup depends on the hardware recognizing certain bit patterns as ∞ or "indefinite" whenever they are used as operands. Even the Ω-zero fixup requires that the hardware produce the correct sign. Originally, the IBM System/360 produced a result after exponent spill which could be described as ?-zero. Here the result was set to zero after underflow, and the result after overflow was unpredictable in the sense that it varied from model to model. In some cases the result was set to zero after overflow. The Ω-zero fixup could not be performed by software because the sign of a result which had overflowed was often lost. The engineering change referred to in Section 1.8 changed the architecture of the machine so that after either overflow or underflow it now produces the *wrapped-around* characteristic described in Section 2.3. The standard software then changes this result to the Ω-zero fixup.

Other fixups are possible. For example, we might want an Ω-ω fixup, so that a result which underflowed would not appear to be zero in a test made by an IF statement. However, the Ω-zero fixup is by far the commonest now in use.

2.3. INTERRUPT

Often the programmer wants to know whether or not an exponent spill has occurred. In fact, unless the hardware provides a standard fixup, there must be some monitoring of the floating-point arithmetic, so that software can provide a fixup after exponent spill.

In many implementations of floating-point arithmetic, the result after exponent spill has a *wrapped-around characteristic*. To define this result more precisely, we shall assume that the characteristic is defined to be the exponent plus γ and that it may be any integer from 0 to $c - 1$. Then $e^* = c - 1 - \gamma$ and $e_* = -\gamma$, so

$$(2.3.1) \qquad\qquad c = e^* - e_* + 1.$$

We say that the result after exponent spill has a wrapped-around characteristic if it has the correct sign, the correct mantissa, and a characteristic in the interval $[0, c-1]$ which differs from the correct characteristic by c. Thus, instead of $c + 1$ the characteristic will be 1; instead of -3 it will be $c - 3$. This means that if we have performed an operation in which the answer is so large that it overflows, the hardware may return a very small number as the answer. Similarly, the result after underflow may be a very large number. These results are almost the worst possible numbers to use for further calculation, so some type of fixup is required if we are to proceed.

The scope of the problem may best be illustrated by looking back to the IBM 704. When exponent spill occurred in this machine, the result had a wrapped-around characteristic and an indicator was set. The only way to provide a fixup such as the Ω-zero fixup was to test an indicator after each floating-point operation—a rather obnoxious procedure. Moreover, for $\oplus$, $\ominus$, and $*$ the result appeared in the accumulator and (roughly) the low-order bits appeared in the MQ. But for $\div$, the quotient appeared in the MQ and the accumulator held the remainder. Since the accumulator and the MQ each had its own characteristic, either could overflow or underflow, and each had its own indicator which had to be tested. Unless one were programming double-precision arithmetic, underflows in the remainder, or in the MQ following $\oplus$, $\ominus$, or $*$, could usually be ignored. To determine exactly what had happened, one had to know whether the operation was $\div$, what indicators had been turned on, and, in one case, the high-order bit of the characteristic of MQ. Since it is troublesome to test the indicators after each floating-point instruction, one might be led to forego any fixup and merely test the indicators at the end of a routine to see whether spill had occurred. But such a test was ambiguous. Suppose that we found that the overflow indicator for the accumulator had been turned on. This could indicate one or more overflows or underflows of the result of $\oplus$, $\ominus$, or $*$. On the other hand, it might have been caused by the underflow of a remainder, which could be ignored. In fact, since fixed-point arithmetic was performed in the same accumulator, the overflow indicator could have been turned on by a fixed-point overflow, or even by shifting a number to extract the low-order bits. Thus, even a test at the end of the program to find out whether or not spill had occurred was often impractical. Many programs did not even bother to test for spill at all. One simply ran them and hoped for the best.

As a result of this situation, a new treatment of exponent spill was used in the IBM 709, and later, in the IBM 7090 and 7094. It was based on an approach which was referred to as *trapping* on those machines and which is now usually referred to as *interruption*. After exponent spill, the register still held the result with wrapped-around characteristic as it had before. However, the flow of the program would be interrupted, and the program would branch to a fixed location which should have a routine to handle the spill. The instruction counter would be saved, so the overflow routine could find the location of the instruction that caused the spill and return to the following instruction. This has been modified slightly on more recent machines. On the IBM System/360, there are many situations which may cause interrupts, so the interrupt handler must first determine the cause of the interrupt and then branch to the appropriate routine. Moreover, for certain of the interrupts there is a mask which can be set to specify whether the interrupt is to be taken or ignored.

The use of interruption after exponent spill provides a great deal of flexibility. If suitable programs are included in the interrupt handler, it is possible to provide a wide variety of different treatments of spill. The approaches described in Sections 2.5, 2.7, and 2.8 depend very heavily on the assumption that there is an interrupt after exponent spill and that the hardware produces the answer with wrapped-around characteristic.

It is convenient to restate the definition of the wrapped-around characteristic in a way which is not as dependent on the way in which the characteristic is stored. We use (2.3.1) as the definition of c. Then the result after exponent spill is said to have a wrapped-around characteristic if it has the correct sign, the correct mantissa, and a characteristic which is the correct characteristic plus c after underflow and the correct characteristic minus c after overflow.

2.4. MESSAGES AND TESTS

As we saw in Section 2.3, if the hardware provides an interruption after exponent spill, the fixup is often produced by the overflow routine. In addition, the overflow routine often prints error messages to indicate that exponent spill has occurred. The overflow routine is usually supplied by the compiler, and it determines what options (if any) are available to the programmer. But FORTRAN G and FORTRAN H for the IBM System/360 provide the *extended error-handling facility*, which allows the user to specify the treatment he wants for exponent spill. He can specify the number of error messages he wants printed, and he can indicate whether or not the computation should be terminated. He can even supply the name of a subroutine he wants called to provide his own fixup. This makes it easy for the user to write his own overflow routine in a higher-level language. In this section and in Section 2.10 we shall discuss some of the things we might want the overflow routine to do.

The most drastic action that can be taken after exponent spill is to terminate the program. (We would hope that if this is done the overflow routine would give us a clear indication of what happened and where in the program it occurred.) However, there are several reasons we might prefer not to terminate the execution of the program. First, as we have seen in Section 2.2, a standard fixup might allow the program to run to completion and produce the correct answer. In this case, if exponent spill terminates the program, the programmer must code tests to avoid the spill. Thus, in those cases in which a standard fixup would be successful, we clearly do not want to terminate the calculation. A second objection is that even if the exponent spill indicates a catastrophic error, terminating the program may deny the programmer the information he needs to debug the program. If the program were run to completion, it might produce output that would be helpful in tracing the error. Finally, although some of the results may be contaminated by exponent spill, others may contain meaningful information. Thus, one column in a page of output may be nonsense, while the remaining output is valid. Even more annoying, out of several cases to be run there may be one case which spills. If this happens to be the first case run, terminating the calculation denies us the results from the good cases. Thus, terminating the execution of the program is often too drastic an action to take after exponent spill, so it probably should not be adopted as the standard procedure. However, it is desirable for the overflow routine to offer the programmer the option of terminating the calculation if he wants to.

It should be pointed out that allowing the calculation to proceed with an Ω-zero fixup may produce other difficulties. For example, it may even produce an infinite loop. But many computers today use an operating system which will terminate the program when a time estimate is exceeded. This provides protection against infinite loops arising from other sources as well.

We shall assume that program execution is allowed to continue after exponent spill. Then there are four questions which must be addressed:

1. What output, if any, should be produced to indicate that spill has occurred?

2. How can the programmer insert a test in his program to find out whether he has had an overflow or an underflow?

3. What number should be produced as the result after exponent spill in order to allow the calculation to continue?

4. To what point should the overflow routine retrun after processing the interruption?

In this section we shall discuss questions 1 and 2. The other questions will be discussed in Section 2.10.

We shall first address the question of what output should be printed after exponent spill. We heartily recommend that this output use the English words *overflow* and *underflow* to describe what happened. In early versions of FORTRAN for the IBM System/360, the output printed after exponent spill was the message

IHC210I   PROGRAM   INTERRUPT ( ) OLD PSW IS

followed by 16 hexadecimal digits representing the *program status word* (PSW). By consulting the correct manual, the programmer would discover that the interrupt could have been caused by fixed-point overflow, floating-point overflow, floating-point underflow, or floating-point division by zero. To find out which it was he had to look at the eighth hexadecimal digit of the 16-digit PSW to see whether it was 9, A, C, or D. It is clearly preferable (almost mandatory) to have an English description of what happened. This is provided by the extended error-handling facility on the IBM System/360.

If a number which has overflowed is replaced by $\pm\Omega$, we are very likely to encounter further overflows. For example, if $x = \pm\Omega$, $2 * x$ overflows. Thus, if we have any overflows in a program and use the Ω-zero fixup, we are likely to have many overflows. For FORTRAN on the IBM 7090, this led to the approach of printing a line of output for each of the first five exponent spills in a program but not printing messages for spills after the first five. Since it is quite possible to encounter 2000 or 3000 overflows, something of this sort is desirable. It is better to have the overflow routine count the number of overflows and the number of underflows for which no messages were printed and print a message at the end of the program giving these counts. Also, instead of printing exactly five messages, it is desirable to allow the programmer to specify how many messages he wants.

One problem that arose on the IBM 7090 is not present on the IBM System/360. On the 7090, single-precision addition, subtraction, and multiplication produced a double-precision result in which the second word had its own characteristic. Therefore, these operations could produce underflow in either the high-order or the low-order digits of a result. In many cases, the low-order digits would never be used, so the fact that they underflowed was irrelevant. Nevertheless, if this was one of the first five spills, an underflow message would be printed even though high-order digits were valid. If some of the numbers in a calculation were getting small, we were very likely to encounter underflow first in the low-order digits. Thus, the first five messages about exponent spill might describe irrelevant underflows in the low-order digits. Then, when spills occurred which might affect the answers, no messages were printed, because five spills had already occurred. Unfortunately, messages describing these irrelevant underflows often conditioned the programmer to ignore all underflow messages. The absence of the low-order characteristic on the IBM System/360 eliminates this problem. This situation was also taken into account in the design of the extended-precision arithmetic on the models 85 and 195 and on the IBM System/370. Underflow in the characteristic of the low-order double word does not cause an interrupt on these models.

Next, we shall consider the question of how the programmer can perform a test to find out whether spill has occurred. With many of the FORTRAN compilers for the IBM System/360, this involves calling a subroutine named OVERFL. We write

$$\text{CALL OVERFL (K)}$$

and K will be set to 1, 2, or 3 by the subroutine OVERFL. K will be 2 if no spill has occurred since the last time OVERFL was called. If there have been one or more spills since the last time OVERFL was called, K will be set to 1 or 3 depending on whether the last spill was an overflow or an underflow. This may be annoying if we have a program in which we are concerned about overflows but may ignore underflows. If we call OVERFL and the value returned is 3, we do not know whether an overflow has occurred. Thus, when we call the OVERFL routine with an argument K, there should be at least four values to which K can be set. These would specify that since the last time OVERFL was called there has been no spill, overflows only, underflows only, or both overflows and underflows. With the extended error-handling facility, we can obtain even more information—we can find out the number of overflows and the number of underflows that have occurred.

2.5. ON OVERFLOW AND ON UNDERFLOW IN PL/I

A great deal of flexibility has been built into PL/I at the language level by the inclusion of the ON OVERFLOW and ON UNDERFLOW statements. We can use the construction

$$\text{ON OVERFLOW: BEGIN:}$$

and then write any PL/I statements we want to. If a floating-point overflow occurs, the interrupt handler will branch to this piece of coding, so the programmer can do whatever he wants to after overflow. In many respects this is equivalent to allowing the programmer to write his own overflow routine in a high-level language. He can print whatever output he desires, store whatever information he needs, keep track of overflows and underflows for later testing, and branch to any point in his program, including RETURN to the instruction following the one which caused the spill. The treatment of overflow may be changed by using additional ON OVERFLOW statements, and each procedure (subroutine) may have its own ON statements which modify the treatment of spill within the procedure.

The ON statements are extremely powerful, and they provide the PL/I programmer with a great deal of flexibility in the treatment of exceptional cases. However, a limitation on the power of these statements is the fact that the programmer does not have access to the contents of the register that spilled. Consequently, these statements cannot be used to provide special fixups, such as those discussed in Sections 2.7 and 2.8. In fact, they cannot even be used to produce the Ω-zero fixup after overflow.

Since the way exponent spill is handled by PL/I may depend on the implementation, we shall consider version V of the PL/I (F) compiler for the IBM System/360. First, suppose that we do not use the ON statements to provide our own treatment for spill. After an exponent spill, a message will be printed indicating whether the spill was an overflow or an underflow and where in the program it occurred. Following underflow, the result is set to zero and the calculation proceeds, but following overflow, the calculation is terminated. If we use the NO OVERFLOW prefix, the calculation will proceed after overflow, but the result will be left with a wrapped-around characteristic and no messages will be printed. Thus, there is no way to provide the Ω-zero fixup and allow the calculation to proceed after overflow.

In spite of this limitation, the ON statements provide a very powerful tool and they are a significant advance in the handling of spills at the language level. In fact, the availability of these statements might be a reason for writing a program in PL/I rather than FORTRAN.

2.6. EXAMPLE

In most implementations of floating-point arithmetic, the range of the exponents is large enough to handle the vast majority of the numbers that arise in our calculations. For example, on the IBM System/360, we have $\Omega \approx 7.23 \times 10^{75}$ and $\omega \approx 5.40 \times 10^{-78}$. Even Avogadro's number, which is 6.03×10^{23}, and Planck's constant, which is 6.55×10^{-27} erg second, are comfortably within this range. Since Ω is so large, it might appear that we would never encounter overflow unless our program contained errors in logic. We shall now consider an example which shows that this is not true.

Suppose that a manufacturer produces two models of a product, model A and model B. He has collected data which shows that 10% of his orders are for model A and 90% are for model B, and on a particular day he receives 2000 orders for this product. We assume that the orders are random and independent, so he would expect to have about 200 orders for model A. We want to compute the probability that there will be at most 200 orders for model A. Indeed, we might like to see the probability that there are exactly k orders for model A and the probability that there are at most k orders for model A, for $k = 0, 1, \ldots, 2000$. If we are going to write a program to solve this problem, we would like it to work in a more general setting in which the number of orders received is N and the probability that any order is for model A is p. Then $q = 1 - p$ is the probability that an order is for model B. Let

x_k be the probability that exactly k of the N orders are for model A, and let y_k be the probability that at most k of the N orders are for model A. Then

(2.6.1)
$$y_k = \sum_{i=0}^{k} x_i,$$

and

(2.6.2)
$$x_k = \binom{N}{k} p^k q^{N-k},$$

where $\binom{N}{k}$ is the binomial coefficient

(2.6.3)
$$\binom{N}{k} = \frac{N!}{k!(N-k)!}.$$

If we first compute the x_k's, the y_k's present no problem, so we shall consider only the problem of writing a program to compute the x_k's given by (2.6.2). The program must work for N on the order of 2000.

Clearly, the formula for x_k may be rewritten as

(2.6.4)
$$x_k = \frac{N(N-1)\cdots(N-k+1)}{k!} p^k q^{N-k}.$$

We shall consider the computation of x_{200} when $N = 2000$ and $p = .1$, and we shall assume that we are using a machine with the values of Ω and ω given above for the IBM System/360. This computation may be split into four parts, namely, computating $(.9)^{1800}$, $(.1)^{200}$, 200!, and $2000 \cdot 1999 \cdots 1801$. Then these four quantities must be combined. Now $k!$ overflows for $k \geq 57$, so 200! overflows. All the more so, $2000 \cdot 1999 \cdots 1801$ overflows. Since $(.1)^{200} = 10^{-200}$, $(.1)^{200}$ underflows. We even find that $(.9)^{1800}$ underflows, since it is approximately 4.3×10^{-83}. Thus, we cannot represent any of the four parts listed above. However, we know that each x_k satisfies $0 < x_k < 1$, and we expect y_{200} to be about one-half, so

$$\sum_{i=0}^{200} x_i \approx .5.$$

Since the sum of 201 of the x_k's is about .5, they cannot all be small. (For example, if each $x_k < .001$, we would have $y_{200} < .201$.) Therefore, many of the x_k's must be "reasonable"-sized numbers, say $.001 < x_k < 1$. Since $x_{200} > .001$ and $(.1)^{200} = 10^{-200}$, the binomial coefficient $\binom{2000}{200}$ must overflow.

If $x_k < \omega$ for some value of k, then $x_k < 10^{-77}$, so we would be quite willing to replace x_k by zero. [This will happen for $k = 0$, since $x_0 = (.9)^{2000}$.]

Thus, we shall require our program to set x_k equal to zero if it is less than ω, but otherwise we want a valid answer for x_k. There are several ways in which this might be coded, but to illustrate the handling of exponent spill we shall perform all the multiplications and divisions. For $N = 2000$ and $k = 200$, there are 2400 operations, and in the worst case, for $k = 2000$, there are 6000 operations. We shall see in Section 3.5 that this will not cost us more than 13 bits in accuracy, so if the calculation is performed in double-precision, the result will be good to better than single-precision accuracy. (These vague statements about accuracy are made more precise in Section 3.5.) For $N = 2000$, both x_0 and x_N are less than ω. To avoid the complication of branching around certain loops for these two values of k, we shall compute x_k only for $1 \leq k \leq N - 1$. Thus, our problem is

PROBLEM

Write a program which takes the values of N, p, and k with $1 \leq k \leq N - 1$ and produces x_k defined by (2.6.2). Set the answer to zero if $x_k < \omega$ (or if x_k is slightly greater than ω but the rounding error introduced by the multiplications and divisions makes our computed value for x_k less than ω.)

Since we are interested only in the computation, we shall ignore the statements necessary to type the variables as double-precision, to read in the input N, p, and k; and to write the output x_k. We shall first write crude FORTRAN and PL/I programs which ignore the problem of spill. These programs will then be modified to illustrate two different ways to cope with the spills. We shall assume that the programs are to be run on the IBM System/360.

FORTRAN Progran Ignoring Spill

```
          Q = 1−P
          X = 1
          DO  100  J = 1,K
   100    X = X*(N+1−J)
          DO 200 J = 1,K
   200    X = X/J
          DO 300 J = 1,K
   300    X = X*P
          KK = K+1
          DO 400 J = KK,N
   400    X = X*Q
```

For PL/I, we could simply translate this program. However, because of the way we shall recode it later, we shall change the DO loops to use the DO WHILE construction.

PL/I Program Ignoring Spill

```
Q = 1-P;
X = 1;
K1, K2, K3 = 1;
DO WHILE (K1 < = K);
    X = X*(N+1-K1);
    K1 = K1 + 1;
    END;
DO WHILE (K2 < = K);
    X = X/K2;
    K2 = K2+1;
    END;
DO WHILE (K3 < = K);
    X = X*P
    K3 = K3+1;
    END;
DO WHILE (K3 < = N);
    X = X*Q;
    K3 = K3+1;
    END;
```

Both these programs will produce good results if they are run in double-precision on the IBM System/360 and no spill occurs. Unfortunately, we are guaranteed to encounter spill for the data we are considering.

We first modify the FORTRAN program to prevent spill. Let BIG be a large power of r comfortably less than Ω, say 16^{50} for the IBM System/360. (We make BIG a power of r so that we shall not introduce rounding error when we use it as a scale factor.) Now if

$$(2.6.5) \qquad \frac{1.}{\text{BIG}} \leq x \leq \text{BIG},$$

multiplying or dividing X by a number less than 2000 cannot cause spill. If X gets to be larger than BIG, we shall divide it by BIG. Then the true value we are computing is represented by $X \cdot \text{BIG}$. We shall use an integer I to count the number of times we have divided by BIG, so the value of x_k will be represented by

$$(2.6.6) \qquad x_k = X \cdot (\text{BIG})^\text{I}$$

Similarly, we set $\text{SMALL} = 1./\text{BIG}$, and if X becomes less than SMALL we shall multiply X by BIG and subtract one from I. Then the value of x_k will always be given by (2.6.6). The coding is shown below.

FORTRAN Program Preventing Spill

```
      BIG  =  16.**50
      SMALL  =  1./BIG
      Q  =  1.-P
      X  =  1
      I  =  0
      DO 100 J  =  1,K
      X  =  X*(N+1-J)
      IF (X.LE.BIG) GO TO 100
      X  =  X/BIG
      I  =  I+1
  100 CONTINUE
      DO 200 J  =  1,K
      X  =  X/J
      IF (X.GE.SMALL) GO TO 200
      X  =  X*BIG
      I  =  I-1
  200 CONTINUE
      DO 300 J  =  1,K
      X  =  X*P
      IF (X.GE.SMALL) GO TO 300
      X  =  X*BIG
      I  =  I-1
  300 CONTINUE
      KK  =  K+1
      DO 400 J  =  KK,N
      X  =  X*Q
      IF (X.GE.SMALL) GO TO 400
      X  =  X*BIG
      I  =  I-1
  400 CONTINUE
```

This coding produces values of X and I such that (2.6.6) holds (except for rounding error.) If $I = 0$, we have produced the desired answer. Since (2.6.5) holds and we know that $x_k < 1$, I cannot be positive. But if I is negative, we may be able to divide X by BIG without underflow (as long as $X \geq 16^{-15}$). Thus, to complete the program, we write

```
      TEST  =  1./(16.**15)
  500 IF (I.GE.O) GO TO 700
      IF (X.LT.TEST) GO TO 600
      X  =  X/BIG
```

$$I = I+1$$
$$\text{GO TO } 500$$
$$600 \quad X = 0$$
$$700 \quad \text{CONTINUE}$$

With this added coding, we now have a solution to our problem in FORTRAN.

For the PL/I program, we shall consider a different approach which makes use of the ON statements. Of the four loops in our original PL/I program, we note that the first loop makes X larger, while the other loops make X smaller. Our approach is to start making X larger until it overflows, then go to the loops which make X smaller until it underflows, then return to the first loop, etc. Basic to this approach is an assumption about the way the PL/I compiler works. We assume that a statement such as

(2.6.7) $X = X*P;$

is compiled into a LOAD, followed by a FLOATING MULTIPLY to compute X*P, followed by a STORE to store the new value for X. Thus, if the multiplication produces an interrupt, the value of X in storage is the value X had prior to the multiplication. Similarly, we assume that the statement

$$K3 = K3+1;$$

compiles into a LOAD, ADD, STORE sequence and that this is performed after the computation in (2.6.7). Then, when an interrupt occurs, the values of X and K3 in storage are the values that these variables had prior to the computation which caused the interrupt. Our approach would not work if X and K3 were held in registers instead of being stored each time we go through the loop. (We can verify our assumption about the way PL/I compiles these statements by looking at an Assembly listing of the compiled code.) Our PL/I coding is:

PL/I Program Which Handles Spill

ON OVERFLOW GO TO SMALLER;
ON UNDERFLOW BEGIN;
        IF K1 < K THEN GO TO BIGGER;
        ELSE DO;
            X = 0
            GO TO FINIS;
            END;
        END;

```
                X = 1;
                Q = 1−P;
                K1, K2, K3 = 1;
BIGGER:         DO WHILE (K1 < = K);
                    X = X*(N+1−K1);
                    K1 = K1+1;
                    END;
SMALLER:        DO WHILE (K2 < = K);
                    X = X/K2;
                    K2 = K2+1;
                    END;
                DO WHILE (K3 < = K);
                    X = X*P;
                    K3 = K3+1;
                    END;
                DO WHILE (K3 < = N);
                    X = X*Q;
                    K3 = K3+1;
                    END;
                IF K1 < = K THEN GO TO BIGGER;
FINIS:
```

Note that in PL/I, if K2 is larger than K, then the DO loop beginning

$$DO \ WHILE \ (K2 < = K);$$

will not be executed. Thus, suppose that we overflow, go to SMALLER, and then underflow in the loop controlled on K3. If we overflow again, we go to SMALLER, but since K2 will be greater than K we shall not execute the DO loop controlled on K2. Instead, we shall skip this loop and pick up the calculation where we dropped it. Also, note that in the ON UNDERFLOW routine, we had to see whether there was any work still to be done in the loop labeled BIGGER. If we have underflowed and there is no more computing to be done in the loop which will make X larger, we set X = 0 and go to FINIS.

This PL/I program provides a solution to our problem. Here the treatment of spill required us to branch to a different point in the program rather than simply returning to the instruction after the instruction which caused the interrupt. One difficulty with this type of programming is that it makes every floating-point instruction a conditional branch. Since some of the branches in the program are not shown explicitly, the program is harder to debug.

In summary, this is an example of a problem in which the answer is of "reasonable" size, but we are exposed to spill in the calculation and a standard fixup will not help us. Yet we have been successful in solving the problem in higher-level languages.

2.7. COUNTING MODE

Probably the best treatment of exponent spill that has been described in the literature was developed by W. Kahan (1965a and 1966) for the IBM 7094 at the University of Toronto. This approach allowed the programmer to call the overflow routine and specify which one of several treatments of spill he wanted. In addition to specifying whether he wanted overflow messages printed or suppressed, he could specify one of three different modes for handling the result after spill. Naturally one of these modes called for the Ω-zero fixup. The other two modes which could be requested were the *counting mode* discussed in this section and the *gradual underflow* discussed in the next section.

Counting mode is not designed as a standard fixup to be used indiscriminantly. Rather, it is designed to be one of several options which might be requested occasionally for handling specific problems. For the problem discussed in Section 2.6, counting mode provides a solution similar to the approach we used in our FORTRAN program. However, it is a much cleaner and more elegant solution. It is based on the assumption that the result produced by the hardware after an exponent spill has a wrapped-around characteristic. We recall that this means that the result has the correct sign, the correct mantissa, and a characteristic which is the correct characteristic plus c after underflow and the correct characteristic minus c after overflow. Here c depends on the machine, and (see Section 2.3) it is defined by

$$c = e^* - e_* + 1.$$

To use the counting mode, we would call the overflow routine and tell it to begin operating in counting mode, counting in location I. It would then operate in counting mode until it was told to change. When spill occurs, the result in the register will be left unchanged, so it will still have a wrapped-around characteristic. But the number in location I is increased or decreased by 1, depending on whether the spill was an overflow or an underflow. Thus, in location I we have a count of the number of overflows minus the number of underflows. It is the responsibility of the problem programmer to store a zero in location I before he enters the counting mode and to see that I is typed as an integer. The objective of counting mode is to allow us to produce numbers X and I, such that the correct result is given by

(2.7.1) $X \cdot (r^c)^I.$

Had we had this approach available for the problem in Section 2.6, we could have used our original FORTRAN program with only slight modifications. We would have to call the overflow routine at the beginning of the program to tell it to operate in counting mode and count in location I. We would also need a statement $I = 0$ to initialize I. At the end of the program we would have found X and I such that the desired value for x_k is given by (2.7.1). We know that x_k cannot be greater than Ω, so X will hold the value of x_k unless I is negative. But if I is negative, the computed value for x_k is less than ω, so we want to set x equal to zero. Thus, at the end of the program we need only the one additional statement:

$$\text{IF (I.LT.O) } X = 0.$$

This approach is not only simpler and easier to program, but it is more efficient, because it removes the tests from the loops. We have allowed the hardware to monitor the spills instead of having to test for them ourselves.

The use of counting mode is based on several assumptions about the problem. First, at any time only one location I is used to count the spills, so the question of which variable is to be multiplied by $(r^c)^I$ must be unambiguous. For example, suppose that we wanted to compute both

(2.7.2)
$$x = \prod_{k=1}^{n} a_k$$

and

(2.7.3)
$$y = \prod_{k=1}^{n} b_k.$$

If we used the FORTRAN programming

```
        X = 1
        Y = 1
        DO 100 K = 1, N
        X = X*A(K)
100     Y = Y*B(K)
```

we would be unable to tell whether a spill affected x or y. Either we would have to modify the program to compute x and y in separate loops, reinitializing I in between, or else we would have to be able to guarantee that one of the calculations, say (2.7.2), would never cause spill.

A second requirement is that we must know whether a spill in an intermediate result affects the numerator or the denominator of a fraction. If a spill were encountered in the FORTRAN statement

$$X = (A*B)/(C*D)$$

the effect would be ambiguous. Suppose that an overflow occurred. The computed value of X should be multiplied by r^c if the overflow occurred while we were computing A*B or during the division, but it should be divided by r^c if the overflow occurred while we were computing C*D.

Finally, we require that further calculation using a spilled result must involve only multiplications and divisions, because a number which has spilled does not have the correct characteristic. If it were used in an addition or subtraction, the shifting would not be done correctly. (Of course, special routines could be written to perform the addition and subtraction correctly. The calling sequence for such a routine would specify the two terms to be added, the overflow count I, and which of the terms the overflow count applied to.) However, the most rapid change in the characteristic arises from multiplication and division, so the counting mode handles an interesting collection of cases. Indeed, it is immediately applicable to computations such as

$$x = \prod_{k=1}^{n} \frac{a_k + b_k}{c_k + d_k}$$

if we can guarantee that the addition $c_k + d_k$ never causes spill. If we cannot guarantee this, we would have to rearrange the computations as

$$x = \frac{\prod_{k=1}^{n} (a_k + b_k)}{\prod_{k=1}^{n} (c_k + d_k)}.$$

If the FORTRAN compiler provides the extended error-handling facility, it is easy for the user to produce this treatment for exponent spill. With PL/I, the ON statements enable us to produce counting mode for overflow but not for underflow. For example, we can write the statement

$$\text{ON OVERFLOW I} = \text{I}+1;$$

This would do exactly what counting mode does when an overflow occurs, because PL/I does not provide any standard fixup after overflow. (The language states that after overflow the result is undefined, and in fact the compiler for the IBM System/360 leaves the result unchanged.) However, the analogous treatment for underflow will not work, because the PL/I language requires that the result be set to zero after an underflow. Thus, we cannot use ON UNDERFLOW to do the counting, because the result is not left with wrapped-around characteristic.

2.8. GRADUAL UNDERFLOW

Gradual underflow is another treatment of underflow which was devised by Kahan (1965a and 1966). With this approach, a number which should have an exponent less than e_* will be written with the exponent e_* and an unnormalized mantissa. (For many machines this means that a number which has underflowed will be written with the characteristic zero.) Thus, if the exponent should be $e_* - k$, we shall make the exponent of the number e_* and shift the mantissa k places to the right, so the mantissa has k leading zeros. We retain only p digits to the right of the radix point in the mantissa, so the k low-order digits of the mantissa must be chopped (or rounded). If the exponent should be $e_* - p$ or less, we shall shift the mantissa p or more places to the right, leaving a zero mantissa. Since we have assumed in Section 1.4 that a normal zero is stored with the smallest allowable characteristic, this will be a normal zero.

The problem addressed by both counting mode and gradual underflow is to avoid having numbers which have spilled contaminate a final result. If the computation involves a sequence of multiplications and divisions, the counting mode provides an elegant answer. The gradual underflow is designed primarily for addition and subtraction. We want to be assured that our treatment of underflow will not contaminate the result when a number which has underflowed is added to a number which has not.

To illustrate the problem, consider the IBM System/360. Suppose that we want to compute $x \oplus y$, where x has characteristic 2 and mantissa $.123456_H$ and y has underflowed but should have had the characteristic -1 and the mantissa $.654321_H$. In FP(16, 6, cl1), where there are no bounds on the exponent, we would have shifted the mantissa of y three places to the right and performed the addition:

$$.123456$$
$$.0006543$$
$$\overline{.123AAA3_H}$$

Then the number $.123AAA3_H$ would have been chopped to six digits, so the mantissa of $x \oplus y$ would be $.123AAA$. But y underflowed, and if we use the Ω-zero fixup, we shall replace y by zero, so $x \oplus y$ will be x. Thus, the Ω-zero treatment of underflow has contaminated the fourth hexadecimal digit of our answer. In the worst case, if $x = \omega$ and y should have characteristic -1 and mantissa $.FFFFFF_H$, we have the following results:

$x + y = 16^{-64} \times .1FFFFFF_H$
$x \oplus y = 16^{-64} \times .1FFFFF_H$ in FP(16, 6, cl1) with no bounds on exponents
$x \oplus y = 16^{-64} \times .1FFFFF_H$ if the arithmetic is performed in FP(16, 6, cl1), $\omega = 16^{-65}$, and gradual underflow is used
$x \oplus y = 16^{-64} \times .1_H$ if $\omega = 16^{-65}$ and the Ω-zero fixup is used.

Thus, there are cases in which the Ω-zero fixup may produce an error of almost 1 in the first significant digit of the answer. In this situation, gradual underflow allowed us to produce the same result as we would have produced if we had been able to use -1 as a characteristic. (See Exercise 4.)

From the above examples, we see that if the exponent e of x is in the range $e_* \leq e \leq e_* + p - 2$, then $x \oplus y$ may have abnormally large error if y underflows and we use the Ω-zero fixup. For the IBM System/360, this can affect single-precision numbers with characteristic ≤ 4 and double-precision numbers with characteristic ≤ 12, and on the models 85 and 195 and the IBM System/370 this can affect extended-precision numbers with characteristic ≤ 26. Since the number 1 has characteristic 65, this means that a significant portion of the extended-precision numbers can be affected by this type of error.

Gradual underflow provides a good solution to the problem of adding a number which has underflowed to a number which has not. Even in a sequence of additions, such as the computation of

$$s = \sum_{k=1}^{n} a_k,$$

gradual underflow is attractive. Suppose that some of the a_ks underflow but that the final value of S does not. If we use gradual underflow, the bound for the error in the answer will be close to the bound that would have been obtained if no underflow had occurred. (In many cases it is the same bound.)

The penalty we pay for gradual underflow is the introduction of unnormalized numbers into our calculation. Floating-point arithmetic with unnormalized operands will be discussed in more detail in Section 12.4. As we shall see then, the use of unnormalized operands in the operations $\oplus$ and $\ominus$ may cause the wrong operand to be shifted and result in the loss of accuracy. However, this situation does not arise when the unnormalized operands have exponent e_*, so the addition or subtraction of the unnormalized numbers produced by gradual underflow does not present a problem. For many machines, the use of unnormalized operands in the operations $*$ and $\div$ may require more care. As we shall see in Section 12.4, there are machines (for example, the IBM 7090) on which the operations $*$ and $\div$ may produce unnormalized results with exponents greater than e_* and poor accuracy if the operands are unnormalized. Fortunately, this problem does not arise on the IBM System/360, because the floating-point multiply and divide operations prenormalize the operands before performing the arithmetic. Thus, if we use gradual underflow with the IBM System/360, the only unnormalized numbers we shall produce will have the exponent e_*, so we shall not encounter any difficulty with them.

2.9. IMPRECISE INTERRUPT

Some of the very fast machines, such as the IBM System/360 models 91 and 195, use the *pipeline* approach to achieve speed. With this approach, the computer will be processing different stages of several different instructions at the same time. In this case, the interrupt cannot operate as cleanly as it does on nonpipeline machines. We shall consider here the situation with regard to the models 91 and 195 of the IBM System/360. If a floating-point spill occurs and the interrupt is masked on, all instructions which have already been decoded will be completed before the interrupt is taken. When the interrupt occurs, the instruction counter will point to the next instruction to be executed. All instructions prior to this instruction have been executed, but instructions beyond it have not. However, before the interrupt is taken, we may have executed other floating-point instructions after the one which caused the interrupt. The interrupt is said to be *imprecise* in the sense that we do not know which instruction caused the interrupt. In fact, the result of an operation which spilled may be used as an operand in another floating-point operation before the interrupt occurs.

On a machine with an imprecise interrupt, we cannot count on a software fixup for exponent spill, because the interrupt may occur too late. Any fixup that is to be performed must be done by the hardware. The models 91 and 195 of the IBM System/360 provide the Ω-zero fixup in hardware, instead of leaving the result with wrapped-around characteristic. This means that some of the more elegant approaches to exponent spill, such as the counting mode or gradual underflow, cannot be used on these machines. However, we may still wish to mask the interrupt on in order to print messages or to allow a FORTRAN programmer to test whether or not spill has occurred by the use of CALL OVERFL. If we are using the ON OVERFLOW or ON UNDERFLOW statements in PL/I, care must be exercised, because the interrupt may occur slightly later in the program than we would expect it to. For this reason, the PL/I program shown in Section 2.6 would not work correctly on the models 91 and 195 of the IBM System/360.

The inability to provide a precise interrupt for exponent spill is inherent in the pipeline approach to computer design. The CDC 6600, which is also a pipeline machine, approaches the problem differently. It automatically provides the ∞-zero fixup in hardware, and it does not provide any interrupt at all for exponent spill.

2.10. CHANGING THE TREATMENT OF SPILL

As we mentioned in Section 2.4, the overflow routine is usually supplied by the compiler, and it determines what options (if any) the programmer has. While some systems are very rigid and do not allow the programmer any choice at all, other systems provide a great deal of flexibility. For example, the overflow routine developed by Kahan (1965a and 1966) allowed the programmer to specify whether he wanted the Ω-zero fixup, counting mode, or gradual underflow. In addition, he could specify the maximum number of messages he wanted printed.

The extended error-handling facility, which is available with some of the

FORTRAN compilers for the IBM System/360, provides even greater flexibility. The programmer can indicate whether or not he wants the calculation to be terminated, and he can specify the maximum number of messages he wants printed for each type of error. Also, if he does not want to use the Ω-zero fixup, he can specify the name of a subroutine he wants called to produce a special fixup. By coding the appropriate subroutine, he can produce either counting mode or gradual underflow. The ability to call his own subroutine after exponent spill also allows the programmer to supply his own error messages.

Similarly, the ON statements in PL/I enable the programmer to provide his own treatment for exponent spill. Although he cannot provide his own fixup, he can print his own overflow messages and he can perform any calculations he wants to. Moreover, the ON statements allow him to branch to any point in his program after exponent spill.

Thus, with either PL/I or the extended error-handling facility, the programmer has a great deal of freedom in specifying the action he wants taken after exponent spill. Indeed, it is almost as if he were writing his own overflow routine, subject to a few restrictions. We shall now consider how he might want to use this capability.

An important aspect of these systems is that they make it easy to change the treatment of exponent spill at any time in the program. This can be especially useful if we want to use a special fixup, such as counting mode. Although counting mode can be very effective for certain types of calculations, we have seen that we would not want to use it as the standard fixup. With either the ON statements or the extended error-handling facility, we can change the fixup to counting mode during the execution of one part of a program but use the Ω-zero fixup in the rest of the program. In particular, a subroutine can use its own fixup without altering the treatment of exponent spill in the rest of the program.

We may want to suppress error messages during part of a program but allow them to be printed during the rest of the program. For example, suppose that we are using a subroutine which has been thoroughly tested and which will produce good answers if the Ω-zero fixup is used for underflow. Then any underflows that occur in this subroutine can be ignored, so there is no need to print messages for them. In fact, it we have supplied a bound for the number of underflow messages, then printing messages for these irrelevant underflows might prevent the printing of messages for other underflows that occur later in the program but cannot be ignored. Therefore, we would like to suppress underflow messages during the execution of such a subroutine and then resume printing them when we leave the subroutine.

There are also times when we might want to supply our own error messages, so that we can make them more informative than the ones supplied by the system. For example, by printing the value of one or more variables, we might be able to indicate how far the program had progressed before the exponent spill occurred. This is easy to do with the ON statements, and it can also be accomplished with the extended error-handling facility in FORTRAN if the variables we want to print are in COMMON.

Finally, consider the point to which we want to branch after exponent spill occurs. In the PL/I solution of the problem in Section 2.6, we saw that the ability to branch to any point in the program can be extremely useful. There are other situations in which it is also convenient. Suppose that we are running several cases, only one of which produces exponent spill. It might be quite desirable to terminate the calculation of the case which spilled, write a message, and then proceed to the next case. In this way we can salvage the results of the good cases even though we cannot complete the calculation of the case that spilled.

There are some problems in which we can avoid exponent spill in the intermediate results by changing the way in which the calculation is performed. (For example, this is the case in the computation of $\sqrt{a^2 + b^2}$ discussed in Exercise 7 and in the solution of a quadratic equation discussed in Section 9.3.) For problems of this sort, it is sometimes attractive to write the program assuming that no spills occur and then use the ON statements to provide special treatment for the cases that spill. In this way, we can use the simpler procedure for the cases that do not spill, so we avoid degrading the speed of the routine. If a spill occurs, we shall use a special procedure to complete the solution of the problem, so we do not want to return to the instruction following the one that caused the spill. Thus, this approach would not be feasible without the ability to branch to some other point in the program after exponent spill.

Since we may want to change the treatment of spill many times during the execution of a program, it is important that these changes be cheap in terms of execution time. The treatment of spill can be changed in PL/I at the cost of a few loads and stores, and it can be changed with the extended error-handling facility at the cost of a subroutine call.

2.11. VIRTUAL OVERFLOW AND UNDERFLOW

We shall use the terms *virtual overflow* and *virtual underflow* to refer to situations in which a subroutine performs a test and finds that it has been asked to calculate a quantity whose absolute value is either greater than Ω or between zero and ω. In this case, no floating-point arithmetic was performed which tried to produce a number with exponent greater than e^* or less than e_*, so we do not have a genuine spill and there will be no interruption. Instead, the subroutine itself must take some action to reflect the fact that the answer cannot be represented as a legitimate floating-point number.

For example, consider the library program used to compute e^x. Now e^x

will be greater than Ω for $x > \log_e \Omega$, and $\log_e \Omega$ is a rather modest-sized number—about 174.673 for the IBM System/360. Thus, if we write a FORTRAN program for the IBM System/360 and use the expression EXP(X) with X = 200, the library program cannot compute the correct value for EXP(X), because it is greater than Ω. Similar situations arise in the programs for SINH(X), COSH(X), GAMMA(X), etc.

It is interesting to note that the subroutine for e^x would be unlikely to produce a genuine spill, even if it did not test to see whether $x > \log_e \Omega$. Typical coding to compute e^x first divides x by $\log_e r$ (or multiplies x by $\log_r e$) to produce

$$y = \frac{x}{\log_e r}.$$

Then

$$e^x = e^y \log_e r = r^y.$$

Next, y is written in the form

$$y = I - F,$$

where I is an integer and F is a fraction with $0 \leq F < 1$. This yields

$$e^x = r^I r^{-F},$$

with $r^{-1} \leq r^{-F} < 1$. Then our computed value for e^x will have an exponent equal to I and a mantissa which is an approximation for r^{-F}. The computation for r^{-F} does not produce any spill, and I is simply converted to a characteristic and inserted in the proper place in the word. Since this would produce a ridiculous answer when x is greater than $\log_e \Omega$, a test must be made. Thus, the program may produce a virtual overflow, but it will not produce a genuine overflow.

The importance of virtual overflow and underflow lies in the fact that the programmer thinks of them as being the same, or almost the same, as genuine overflow and underflow. But they do not produce an interrupt to initiate the normal treatment of spill. For example, consider the ** operation in FORTRAN or PL/I. Typically, the expression X**B is evaluated as if it were written as the FORTRAN expression EXP(B*ALOG(X)). Thus, when a real number is raised to a real power, there is an implicit use of the exponential routine. On the other hand, many compilers handle a real number raised to an integer power (X**I) by repetitive squaring. (See Section 3.6.) Suppose that $x > \sqrt{\Omega}$. Then both X*X and X**2 will produce a genuine overflow. But if the number stored in B is 2, then X**B does not. In fact, many FORTRAN compilers handle the expression X**2. as if it were written as EXP(2.*ALOG(X)), so it will not produce a genuine spill. The problem programmer does not want to have to worry about the details of this sort; he would like to have X**2 and X**2. produce the same effect.

In general, we would like to have virtual spill produce the same effects that a genuine spill would have. If we are using a standard fixup, say the Ω-zero fixup, this is easy to implement. If, say, the routine for e^x finds that the argument is greater than $\log_e \Omega$, it can simply multiply two positive numbers which are large enough to produce an overflow. This will cause an interrupt, and the overflow routine will provide a standard fixup, write the appropriate messages, and include the overflow in the counts that are tested by calling OVERFL. Control will be returned to the routine for e^x, which will deliver the standard fixup as the answer.

The problem becomes more difficult if we are using a more sophisticated treatment of spill, such as the counting mode described in Section 2.7. As a first approach, we might try to find two numbers whose product would overflow and produce the correct result with wrapped-around characteristic. But there is a bound for how large the characteristic can be after overflow. (See Exercise 3.) Even worse, in the routine for e^x we may find that e^x is greater than $r^c\Omega$. This suggests that if the counting mode is to be implemented for virtual spill, it should be possible to call the overflow routine and tell it that we have encountered virtual spill. The calling sequence would use two arguments I and X to indicate that the correct result is $X \cdot (r^c)^I$. The overflow routine would provide whatever treatment of spill was then being used and return the answer in X. It can be argued that this is unnecessary sophistication, but it would provide an elegant treatment for virtual spill.

2.12. DIVISION BY ZERO AND INDETERMINANT FORMS

Another subject that is closely related to exponent spill is division by zero. If we coded the expression $A \div B$, we did not expect B to be zero. Often A and B are approximations we have computed for numbers a and b, and we really want to form a/b. If B is zero but A is not, it is reasonable to assume that $|b|$ is less than $|a|$, so a/b has a large absolute value. The CDC 6600, which uses a special bit pattern to represent ∞, sets $A \div 0$ equal to ∞ whenever $A \neq 0$. Other machines, such as the IBM System/360, provide an interrupt when a divide operation is performed and the divisor is zero, and then they depend on the interrupt routine to write an error message and provide a standard fixup. Here the natural fixup is to set $A \div 0$ equal to Ω whenever $A \neq 0$, and we shall refer to this as the Ω fixup for division by zero. Even if we interpret $A \div 0$ to mean $A \div b$ for some b close to zero, we cannot determine the sign of $A \div b$, so the Ω fixup always sets the quotient to $+\Omega$.

Suppose that we are using the Ω fixup for division by zero and the Ω-zero fixup for exponent spill. If we wish to compute $A \div B$ and B underflows, the Ω-zero fixup will set $B = 0$. Then if $A \neq 0$, the Ω fixup for division by zero will set $A \div B$ equal to Ω. Thus, an underflow in the denominator has been treated as an overflow in the quotient. This is a reasonable approach, but it is not a panacea. For example, if $A = \omega$ and B has underflowed but should have been only slightly less than ω, this approach would set $A \div B$ equal to Ω when the correct answer is only slightly larger than 1.

Next, suppose that we try to perform a floating-point division of zero by zero. Here, there is no generally acceptable value to be used in further calculation. In fact, such a division is usually a signal of trouble in the program. The only reason for not terminating the program is that we may want to proceed to other parts of the program or other data that will not be contaminated by the result of this division. The error message printed for division by zero should distinguish the case $0 \div 0$ from the case $A \div 0$ with $A \neq 0$. Moreover, a different fixup, say 0 or 1, should be used for $0 \div 0$.

A similar situation arises with other indeterminant forms, such as $X**Y$ when X and Y are both zero. These should be handled as a special sort of virtual spill in which a special message is printed and a special fixup is used.

In FORTRAN, we may want to use different fixups for $0**0$ depending on whether it arises from $I**J$, $X**I$, or $X**Y$, where X and Y are floating-point numbers and I and J are integers. This is because integers are usually exact, but a floating-point number is often only an approximation for the number we are really interested in. Thus, if X and I are zero, we may think of $X**I$ as representing x^0 for some x close to zero, so it is natural to use one as the fixup. On the other hand, if X and Y are zero and we think of $X**Y$ as representing x^y for some x and y close to zero, there is no natural fixup.

Finally, consider the indeterminant form $0 \cdot \infty$. We mentioned in Section 2.2 that the CDC 6600 has special bit patterns to represent ∞ and INDEFI-NITE, and that $0*\infty$ is defined to be INDEFINITE. When we are using the Ω-zero fixup, we often think of Ω as infinity. But since Ω is a valid floating-point number, $0*\Omega$ is zero. In fact, $0*\Omega$ does not cause an interrupt, so we have no opportunity to provide a special fixup for it.

EXERCISES

1. The subroutine which FORTRAN uses for the $**$ operation often computes $X**N$ for $N < 0$ as if it were written as

$$1./(X**IABS(N)).$$

If $\Omega < 1/\omega$, this may produce overflow even though $X^N > \omega$. What FORTRAN coding would you use to produce the number $\omega = 16^{-65}$ on the IBM System/360?

2. For the machine you are using, find floating-point numbers a and b such that $X*X$ does not spill if $a \leq X \leq b$, but it will spill if X is any other nonzero, normalized, floating-point number.

3. Consider a machine in which the characteristic is defined to be the exponent plus γ and may be any integer from zero to $2\gamma - 1$. Then $e^* = \gamma - 1$ and $e_* = -\gamma$. Assume that arithmetic is performed in $FP(r, p, cl1)$ and that if a floating-point operation produces exponent spill, the result is left with a wrapped-around characteristic. Let a and b be nonzero, normalized, floating-point numbers. What are the upper and lower bounds for the characteristic of the result in the following situations?
   a. $a \oplus b$ after overflow.
   b. $a * b$ after overflow.
   c. $a \div b$ after overflow.
   d. $a \oplus b$ after underflow.
   e. $a * b$ after underflow.
   f. $a \div b$ after underflow.

4. Suppose that a is a normalized floating-point number, that b has underflowed, and that we use the gradual underflow fixup. Except for overflow and underflow, the calculation is performed in $FP(r, p, cl1)$.
   a. If a and b have the same sign, show that the value produced by $a \oplus b$ is the same value we would have produced if the range for the exponents had been large enough so that b did not underflow.
   b. If a and b have opposite signs, compare the value produced for $a \oplus b$ by this approach with the value we would have produced if the range for the exponent had been large enough so that b did not underflow.
   c. Let $c = a + b$ and let $\bar{c}$ be the value we would have produced for c if we had used the Ω-zero fixup instead of gradual underflow. We define the relative error by

$$\rho = \frac{\bar{c} - c}{c}.$$

   What is the largest value ρ may have?

5. For the machine that you are using, suppose that x and y are normalized floating-point numbers. What are the best bounds a and b for which you can guarantee that

$$x * x - y * y$$

will not spill if $a \leq x \leq b$ and $a \leq y \leq b$?

6. Some machines have a compare instruction which will determine whether $A > B$, $A = B$, or $A < B$ without producing exponent spill. When the hardware includes such an instruction, the FORTRAN compilers often use it to handle logical IFs, such as IF(A.LT.B). Even if there is a compare instruction, many FORTRAN compilers will implement the arithmetic IF, such as IF(A − B), by subtracting B from A and determining whether the result is positive, zero, or negative. Then the logical IF can prevent spills that

might be produced if we used the arithmetic IF. Unfortunately, some FORTRAN compilers have elected not to implement the logical IF.

Assume that A and B are normalized floating-point numbers. Write a FORTRAN program to determine whether $A > B$, $A = B$, or $A < B$ using only arithmetic IFs. This should be coded in such a way that you can guarantee that the program will never produce exponent spill.

7. Let z be the complex number $a + bi$. Suppose that we are using a version of FORTRAN which does not support complex data types or complex arithmetic. Then we shall be given the two parts of z, a and b, as normalized floating-point numbers. Write a FORTRAN program to compute

$$|z| = \sqrt{a^2 + b^2}$$

on the machine you are using. This should be coded in such a way that it will not produce exponent spill unless the answer spills. (It is clear that either $z = 0$ or else $|z| \geq \omega$, so the final answer cannot underflow. But $|\Omega + \Omega i| = \sqrt{2}\,\Omega$, so the program may have a virtual overflow. We want to get a good answer for $|z|$ whenever it does not exceed Ω. If $|z| > \Omega$, use whatever standard fixup, such as Ω or ∞, is convenient on the machine you are using.)

8. In Section 2.7, we saw that in the implementation of PL/I for the IBM System/360, the ON statements could be used to produce the counting mode for overflows but not for underflows. We may rewrite the computation of the example of Section 2.6 to change the underflows into overflows in the denominator. Let $p' = 1/p$ and $q' = 1/q$. Then

$$x_k = \frac{a}{b}$$

where

$$a = N(N - 1) \cdots (N - K + 1)$$

and

$$b = k!(p')^k(q')^{N-k}.$$

Write a PL/I program for this computation, using the ON OVERFLOW statement to produce the same effect as the counting mode.

9. A problem which can easily produce exponent spill is the computation of the determinant of a large matrix. Even if the elements of the matrix are of "reasonable" size, the determinant may be very large or very small. If A is a matrix of order n with elements of "reasonable" size, so is the matrix B obtained by multiplying every element of A by 10. But

$$\det(B) = 10^n \det(A),$$

so if n is large, at least one of these determinants must spill. A typical program to compute the determinant of A first converts A to a triangular matrix T having the same determinant (except perhaps for sign). We consider the problem of computing the determinant of a triangular matrix T. This determinant is given by

$$d = \prod_{i=1}^{n} t_{ii}.$$

a. Write a FORTRAN program for the machine you are using to compute d, given that

$$\omega \le |t_{ii}| \le \Omega$$

for all i. If d satisfies

$$\omega \le |d| \le \Omega$$

the program should produce d regardless of whether any of the intermediate results spill. If d is not in this range, a standard fixup should be used.
b. Recode this program, assuming that your system provides counting mode.
c. Code this problem in PL/I using the ON statements.

10. Suppose that we are using the counting mode to handle exponent spill. We shall consider a subroutine to perform the function alluded to in Section 2.7 of adding a number which has spilled to a number which has not. The calling sequence is

CALL ADD (A,B,I)

This is to mean that the floating-point number A is to be added to the number

$$B \cdot (r^c)^I.$$

The routine is to compute D and J, such that

$$D \cdot (r^c)^J = A + B \cdot (r^c)^I,$$

and store D and J in place of B and I.
a. Determine a scheme for computing D and J. This will involve making tests to determine the size of A, B, and I.
b. Write a FORTRAN program to implement the scheme you devised in part a.

3 ERROR ANALYSIS

3.1. SIGNIFICANT DIGITS

The idea of significant digits is familiar to any student of the physical sciences, and it is often used to motivate the idea of floating-point arithmetic. Unfortunately, the specification of which digits in a number are significant is often rather vague. Certainly leading zeros are not to be counted, but there are several different views about exactly when other digits in the number are to be considered significant. For example, a commonly used approach is to ignore leading zeros and count any other digit as significant if the error is less than one-half a unit in that radix place. On the other hand, we might be willing to consider a digit to be significant when the error is known only to be less than one unit in that radix place. We shall ignore these ambiguities and not try to give a precise definition of significant digits. Instead, we observe that in those problems in which one is led to speak of significant digits, it is usually the relative error that is the really important measure of error, in the sense that it can be handled nicely mathematically and that the way in which it is propagated can be easily understood. Thus, the number of significant digits is usually used as a rough measure of relative error.

A major disadvantage of an error analysis based on the number of significant digits is the discreteness of this measure of error. For example, in FP(r, p, a) there are usually only $p + 1$ statements that can be made about the number of significant digits, namely that $p, p - 1, \ldots, 2, 1$, or none of the digits are significant. (However, a constant such as 2 may be correct to infinitely many places.) For machines of comparable accuracy, the larger r is, the smaller p will be. Thus, the number of significant digits gives a more accurate indication of the size of the error when r is small than it does when r is large. But even on a binary machine, the discreteness of the number of significant digits is a major limitation.

This discreteness is also annoying when we try to decide whether or not a digit should be considered to be significant. For example, suppose that our definition of significant digits specifies that a digit is significant if the error is less than one-half a unit in that position. If $\tilde{x} = 12.345$ and $\hat{x} = 12.346$ are two approximations for the number $x = 12.34549981$, then we would say that $\tilde{x}$ has five significant digits while $\hat{x}$ has only four. Yet $\hat{x}$ is almost as good an approximation for x as $\tilde{x}$ is.

Another annoying aspect of the discreteness of the number of significant digits concerns its behavior in the neighborhood of a power of the radix. Suppose that $\tilde{x}$ and $\tilde{y}$ are approximations for x and y and that each of them has an error of slightly less than .0005. If $\tilde{x} = 1.002$ and $\tilde{y} = .998$, then the approximations have about the same relative error, but $\tilde{x}$ has four significant digits while $\tilde{y}$ has only three. Thus, in the neighborhood of a power of the radix there is a jump in the number of significant digits required to produce either the same relative error or the same absolute error.

Because of these difficulties, we shall not propose a precise definition for the number of significant digits. Instead, we shall view it as a crude measure of relative error. Thus, it is meaningful to discuss the distinction between, say, two significant digits and eight significant digits, but we shall not be precise about the distinction between n and $n + 1$ significant digits.

In the same vein, we shall use the expression "good to almost word length" to mean that the error is at most a "few" units in the last place. This expression will be used to distinguish between this situation and the situation in which we have only one or two significant digits.

There is a well-known rule for the number of significant digits in a product or quotient. In general terms, this states that if the factors have n' and n'' significant digits, then the product has $n = \min(n', n'')$ significant digits. In Exercise 1, we shall show that this statement of the rule is too strong and that the error in the product may be larger than is suggested by the statement that it has n significant digits. A more precise statement of this rule is given in Hildebrand (1956). However, the rule is widely remembered in the form given above, and it is often used as a justification for floating-point computation.

We shall not pursue a more precise statement of this rule; instead, our analysis of error will be based on the rules for relative error. We again observe that the number of significant digits is a crude measure of relative error, so the rule above is satisfactory when we are considering the difference between, say, two and eight significant digits rather than the distinction between seven and eight significant digits.

3.2. RELATIVE ERROR

We shall now examine the concept of relative error more closely. If $\tilde{x}$ is an approximation for x and $x \neq 0$, the relative error ρ is defined by

(3.2.1) $$\rho = \frac{\tilde{x} - x}{x}.$$

Clearly this is equivalent to

(3.2.2) $$\tilde{x} = (1 + \rho)x.$$

It turns out that it is often convenient to use (3.2.2) instead of (3.2.1), although (3.2.1) is the natural definition of relative error and it is far more familiar. However, whenever we have $\tilde{x}$, ρ, and x satisfying (3.2.2) with $x \neq 0$, we may consider $\tilde{x}$ to be an approximation for x with relative error ρ. Note that ρ is a signed number. The absolute value of $\tilde{x}$ is too large when ρ is positive and it is too small when $-1 < \rho < 0$. If $\rho < -1$, $\tilde{x}$ has the wrong sign.

When x is zero, the relative error is often left undefined. However, there are some situations in which it is convenient to extend the definition of relative error to include this case. If $\tilde{x} = x = 0$, we set $\rho = 0$, since there is no error. If x is zero but $\tilde{x}$ is not, we define ρ to be ∞.

Now suppose that x is written in the form

(3.2.3) $$x = r^e m, \qquad r^{-1} \leq |m| < 1.$$

Here we do not require that x be in $S(r, p)$, so we may use infinitely many digits in the representation of m. Let $\tilde{x}$ be an approximation for x, and write

$$\tilde{x} = r^e \tilde{m}.$$

We have used the same exponent e for x and $\tilde{x}$, so we cannot require that $\tilde{m}$ be normalized. Then

(3.2.4) $$\rho = \frac{\tilde{x} - x}{x} = \frac{\tilde{m} - m}{m}.$$

Thus, the relative error in the approximation $\tilde{x} \approx x$ is the relative error in the mantissas when the numbers are written with the same exponent.

An important special case arises if x is any real number satisfying (3.2.3) and $\tilde{x} = \check{x}$ is the approximation obtained by chopping x to p digits in the base r. Then $|\tilde{m}| \leq |m|$, so $\rho \leq 0$. Since

$$|\tilde{m} - m| < r^{-p}$$

and $r^{-1} \leq |m| \leq 1$, (3.2.4) yields

(3.2.5) $0 \geq p > -r^{-(p-1)}.$

We shall often rewrite (3.2.5) as

(3.2.6) $\bar{x} = (1 - p)x, \qquad 0 \leq p < r^{-(p-1)}.$

This is a very convenient bound for the relative error introduced by chopping a number to $S(r, p)$, and we shall use it extensively. However, we can obtain a slightly stronger bound, which we shall want to use occasionally. Suppose that x satisfies (3.2.3), and write $\bar{x} = (1 - p)x$. Now $|\bar{x}| = (1 - p)|x|$, so we may assume that $x > 0$. Let $x - \bar{x} = \epsilon$, so

$$p = \frac{\epsilon}{\bar{x} + \epsilon}.$$

Here $0 \leq \epsilon < r^{e-p}$ and $r^{e-1} \leq \bar{x} < r^e$. Let $f(t) = t/(\bar{x} + t)$. Then

$$f'(t) = \frac{\bar{x}}{(\bar{x} + t)^2} > 0,$$

so the maximum value of $f(t)$ on the interval $0 \leq t \leq r^{e-p}$ is $f(r^{e-p})$. Therefore,

(3.2.7) $0 \leq p < f(r^{e-p}) = \frac{r^{e-p}}{\bar{x} + r^{e-p}}.$

Now $\bar{x} \geq r^{e-1}$, so (3.2.7) yields

(3.2.8) $\bar{x} = (1 - p)x, \qquad 0 \leq p < \frac{r^{-(p-1)}}{1 + r^{-(p-1)}}.$

Since this bound is only slightly stronger than the simpler bound given in (3.2.6), there will be only one or two occasions when we shall find it necessary to use (3.2.8) instead of (3.2.6).

Now suppose that $\bar{x}$ in (3.2.4) is the approximation $\bar{x}^{\,o}$ obtained by rounding x to p digits in the base r. Then

$$|\tilde{m} - m| \leq \tfrac{1}{2}r^{-p},$$

so

(3.2.9) $|p| \leq \tfrac{1}{2}r^{-(p-1)}.$

In this case we do not know the sign of p.

In general, let $\bar{x} = r^e\tilde{m}$ be an approximation for x, where x is given by

(3.2.3). If $|\tilde{m} - m| = \epsilon$, then the relative error ρ satisfies

(3.2.10) $$\epsilon < |\rho| \leq r\epsilon.$$

Thus, the bound $r\epsilon$ for the relative error in x due to an error ϵ in the mantissa depends on r. If r is small, not only is the bound $r\epsilon$ small, but the range given by (3.2.10) is small. We often find that the errors introduced at each stage in the calculation have a bound which can be expressed in terms of units in the pth position. For example, each arithmetic operation introduces an error of less than one unit in the last place when the arithmetic is performed in FP(r, p, c) and an error of at most one-half a unit in the last place when the arithmetic is performed in FP(r, p, R). For other systems FP(r, p, a), we obtain bounds of the same sort. Thus, the error introduced in m by a floating-point operation is bounded by an ϵ such as r^{-p}. But the error often propagates as relative error. Therefore, we would favor a machine with small radix so that the term $r\epsilon$ in (3.2.10) is small. The choice of the radix for a machine will be discussed in more detail in Section 12.1. However, we note that the relative error introduced by chopping x to produce $\tilde{x}$ is bounded by $r^{-(p-1)}$, so for FP(16, 6, c) we get a bound $16^{-5} = 2^{-20}$, which is the same bound we would get for FP(2, 21, c).

As a special case of (3.2.10), we note that if $\tilde{x}$ is obtained from x by increasing $|x|$ by 1 in the pth place, we have $\tilde{x} = (1 + \rho)x$, with

(3.2.11) $$r^{-p} < \rho \leq r^{-(p-1)}.$$

3.3. RELATIVE ERROR IN FP(r, p, clq)

In this section we shall study both the relative error introduced by performing arithmetic in FP(r, p, clq) and the way in which relative error is propagated by arithmetic operations. We shall assume throughout that $q > 0$.

Relative Error in Multiplication and Division

In FP(r, p, clq) with $q > 0$, we have $a * b = \overline{ab}$ and $a \div b = \overline{a/b}$, so

(3.3.1)
$$a * b = (1 - \rho)ab, \qquad 0 \leq \rho < r^{-(p-1)}$$
$$a \div b = (1 - \rho)\frac{a}{b}, \qquad 0 \leq \rho < r^{-(p-1)}.$$

Now suppose that we have approximations $\tilde{x}$ and $\tilde{y}$ for x and y with

(3.3.2)
$$\tilde{x} = (1 + \sigma)x$$
$$\tilde{y} = (1 + \tau)y.$$

If $\tilde{x}$ and $\tilde{y}$ are the results of earlier calculations, we usually do not know the signs of σ and τ, and the bounds for σ and τ may be larger than $r^{-(p-1)}$. But if $\tilde{x}$ and $\tilde{y}$ are the numbers we have in the machine and we want to compute xy, the best we can do is to form $\tilde{x} * \tilde{y}$. By (3.3.1) we may write

$$\tilde{x} * \tilde{y} = (1 - p)\tilde{x}\tilde{y}, \qquad 0 \leq p < r^{-(p-1)},$$

so

$$\tilde{x} * \tilde{y} = (1 - p)(1 + \sigma)(1 + \tau)xy.$$

If we use φ to denote the relative error in $\tilde{x} * \tilde{y}$, we have

$$\tilde{x} * \tilde{y} = (1 + \varphi)xy,$$

where

(3.3.3) $$\varphi = -p + \sigma + \tau - p\sigma - p\tau + \sigma\tau - p\sigma\tau.$$

We hope that each of the relative errors p, σ, and τ is small, say less than 10^{-5}. Then the product terms in (3.3.3) are much smaller than p, σ, and τ, so

(3.3.4) $$\varphi \approx -p + \sigma + \tau.$$

Since we do not know the signs of σ and τ, we cannot get any comfort from the minus sign in (3.3.4). The signs may be such that the magnitudes of p, σ, and τ are added rather than subtracted.

Division produces similar results. If $\tilde{x}$ and $\tilde{y}$ satisfy (3.3.2), we write

$$\tilde{x} \div \tilde{y} = (1 - p)\frac{\tilde{x}}{\tilde{y}}, \qquad 0 \leq p < r^{-(p-1)},$$

so

$$\tilde{x} \div \tilde{y} = \frac{(1 - p)(1 + \sigma)}{(1 + \tau)} \frac{x}{y}.$$

If we write

$$\tilde{x} \div \tilde{y} = (1 + \delta)\frac{x}{y},$$

we have

(3.3.5) $$1 + \delta = \frac{(1 - p)(1 + \sigma)}{1 + \tau}.$$

We obtain a useful approximation for $1 + \delta$ by recalling that for $|\tau| < 1$,

$$\frac{1}{1 + \tau} = 1 - \tau + \tau^2 - \tau^3 + \cdots.$$

If $|\tau|$ is small, this yields

$$\frac{1}{1 + \tau} \approx 1 - \tau,$$

SO

(3.3.6) $\delta \approx -\rho + \sigma - \tau.$

As with (3.3.4), since we do not know the signs of σ and τ, we do not know whether these errors add or compensate.

Several comments should be made about these formulas.

In (3.3.4) we see the basis for the vague rule we referred to in Section 3.2 for the number of significant digits in a product. If $\tilde{x}$ and $\tilde{y}$ satisfy (3.3.2), then the true product $\tilde{x}\tilde{y}$ satisfies

(3.3.7) $\tilde{x}\tilde{y} = (1 + \sigma + \tau + \sigma\tau)xy,$

so the relative error in $\tilde{x}\tilde{y}$ is approximately the sum of the relative errors σ and τ in $\tilde{x}$ and $\tilde{y}$. If σ and τ are small, the relative error in $\tilde{x}\tilde{y}$ cannot be much larger than twice the maximum of $|\sigma|$, $|\tau|$. Since we use the number of significant digits as a measure of relative error, this leads to the rule that the product has as many significant digits as the least significant factor. As we have seen, this is only a rough estimate for the error in the product, and if we give a precise meaning to significant digits, it is often too strong a statement. It is too strong, not only because the relative errors are added, but also because an error of $\frac{1}{2}$ in the kth place of a factor may produce $\sigma = \frac{1}{2}r^{-(k-1)}$, while a relative error of $\frac{1}{2}r^{-(k-1)}$ in the product may correspond to an absolute error of almost $r/2$ in the kth place if its mantissa is only slightly less than 1. However, the rule does indicate correctly that the relative error does not grow rapidly when we are performing multiplications and divisions. (See Section 3.5.)

Also, in (3.3.4) we see that the relative error in the product is approximately the sum of three terms, namely the relative errors σ and τ inherited from the factors $\tilde{x}$ and $\tilde{y}$, and the new error ρ introduced because we formed the floating-point product $\tilde{x} * \tilde{y}$ instead of the true product $\tilde{x}\tilde{y}$. There is a widely remembered rule which states that it is not worth developing and retaining digits unless we can guarantee that they are significant. If we followed this rule, we would be allowed to make ρ almost as large as σ and τ. That is, we would be allowed to insert a new error whose magnitude is almost as large as the bound for the relative error due to earlier approximations. But this has very much the flavor of saying that we are willing to double the error at each stage in the calculation (or, perhaps, multiply it by 1.5 or 1.1). This may be acceptable if we only perform two or three operations, but on an automatic computer we often perform hundreds or thousands of operations, so we certainly do not want to double the error at each step.

It is quite likely that we first heard of the rule that it is not worth developing digits unless we can guarantee that they are significant when we were introduced to logarithms. At that time, we might have assumed that if we were asked to multiply four 5-digit numbers, we had to develop the full 20-digit product. Until we were persuaded that it was reasonable to retain only five digits, we could not be expected to use logarithms. But these calculations usually involved only a few operations. Even here, we would usually continue to use a five-place table of logarithms even after the error had grown to a point at which we could only guarantee that four digits of the answer were significant.

The secret of success of floating-point computation lies in the fact that we continue to do arithmetic to p digits of precision even though the accuracy of our intermediate results has degraded so that we can only guarantee that a few digits are significant. That is, we select a precision of, say, eight decimal digits, and we perform all calculations at that precision, even though we can regard only three or four of the digits as significant. Thus, the new errors introduced are small with respect to the propagated error, so our loss of accuracy is more dependent on the inherited error than on the new error introduced at the present step. This will be illustrated in Section 3.5. We shall see that the really important question is not how much earlier errors have affected the accuracy of our present result, but rather how the error we introduce now will affect the final answer.

Relative Error in Addition and Subtraction

We shall first examine the operations $\oplus$ and $\ominus$ in FP(r, p, clq), and then we shall study the way in which relative error is propagated by these operations.

In Section 1.8 we saw that in the add magnitude case we have $a \oplus b = \overline{a + b}$ and $a \ominus b = \overline{a - b}$, so in this case we have

(3.3.8)
$$a \oplus b = (1 - \rho)(a + b), \qquad 0 \leq \rho < r^{-(p-1)}$$
$$a \ominus b = (1 - \rho)(a - b), \qquad 0 \leq \rho < r^{-(p-1)}$$

We may easily reduce the subtract magnitude case to the computation of $a \oplus b$ where $a > 0 > b$ and $a \geq |b|$. The operation $a \oplus b$ will be exact if $a = -b$, so we shall assume that

$$a > 0 > b > -a.$$

Since there is no reasonable bound for the error introduced in the subtract magnitude case in FP(r, p, $cl0$), we shall discuss only FP(r, p, clq) with $q \geq 1$. By Theorem 1.8.2, the operation $a \oplus b$ produces a result which either is $\overline{a + b}$ or can be obtained by increasing $\overline{a + b}$ by 1 in the last place. Write

$$a + b = r^e m, \qquad r^{-1} \leq m < 1$$
and
$$a \oplus b = r^e \tilde{m}.$$

Then $|\tilde{m} - m| < r^{-p}$, so

(3.3.9) $a \oplus b = (1 + p)(a + b)$, $|p| < r^{-(p-1)}$.

This result produces the same bound for $|p|$ that we would have had in FP(r, p, c), but we no longer know the sign of p.

The bound in (3.3.9) may be sharpened slightly. We shall consider the case in which $a \oplus b > \overline{a + b}$. As we saw in Theorem 1.8.2, if $a \oplus b$ is not $\overline{a + b}$, then $a \oplus b$ is greater than $a + b$ is greater than $a + b$ by less than 1 in the $(p + q - 1)$st place. Then we may write

(3.3.10) $a \oplus b = (1 + p)(a + b)$, $-r^{-(p-1)} < p < r^{-(p+q-2)}$.

Thus, we have proved

THEOREM 3.3.1

In FP(r, p, clq) with $q \geq 1$,

$$a \oplus b = (1 + p)(a + b), \qquad -r^{-(p-1)} < p < r^{-(p+q-2)}$$
$$a \ominus b = (1 + p)(a - b), \qquad -r^{-(p-1)} < p < r^{-(p+q-2)}.$$

For the add magnitude case, the inequalities read $-r^{-(p-1)} < p \leq 0$.

COROLLARY

In FP(r, p, $cl1$),

$$a \oplus b = (1 + p)(a + b), \qquad |p| < r^{-(p-1)}$$
$$a \ominus b = (1 + p)(a - b), \qquad |p| < r^{-(p-1)}.$$

We shall now turn to the question of the propagation of error by addition and subtraction. We shall consider this problem in terms of relative error, although there are many cases in which it is advantageous to study it in terms of absolute error instead. (See, for example, Section 3.12.) We shall again suppose that we have approximations $\tilde{x}$ and $\tilde{y}$ for x and y satisfying (3.3.2), and we shall suppose that both σ and τ are greater than -1, so both $\tilde{x}$ and $\tilde{y}$ have the correct signs. We shall first consider the add magnitude case, so we may assume that both x and y are positive. Then

$$\tilde{x} + \tilde{y} = x + y + \sigma x + \tau y,$$

so

$$\tilde{x} + \tilde{y} = (1 + \varphi)(x + y),$$

where

$$\varphi(x + y) = \sigma x + \tau y.$$

Now

$$[\min(\sigma, \tau)](x + y) \leq \sigma x + \tau y \leq [\max(\sigma, \tau)](x + y),$$

so

(3.3.11) $\min(\sigma, \tau) \leq \varphi \leq \max(\sigma, \tau)$.

This yields

(3.3.12) $|\varphi| \leq \max(|\sigma|, |\tau|)$.

That is, in the add magnitude case the relative error in $\tilde{x} + \tilde{y}$ is at most the maximum relative error in one of the terms.

Unfortunately, in the subtract magnitude case there is no bound for the relative error in the answer due to errors in the operands. For example, suppose that we have the approximations

$$\tilde{x} = 1.2345678$$
$$\tilde{y} = 1.2345677$$

for

$$x = 1.23456776$$
$$y = 1.23456774.$$

Then $x - y = 2.10^{-8}$ while $\tilde{x} - \tilde{y} = 10^{-7}$, so even though $\tilde{x}$ and $\tilde{y}$ are good approximations for x and y, we find that $\tilde{x} - \tilde{y}$ is not a good approximation for $x - y$. Indeed, we could have good approximations $\tilde{x}$ and $\tilde{y}$ for x and y but find that $x - y$ is zero while $\tilde{x} - \tilde{y}$ is not.

In the subtract magnitude case, we are exposed to the magnification of the relative error because of the loss of leading digits. If we have lost several leading digits, then we have had to shift the answer several places to post-normalize it, and we saw in Section 1.8 that in this case the floating-point subtract operation introduces no error. That is, $\tilde{x} \ominus \tilde{y} = \tilde{x} - \tilde{y}$. However, any error in $\tilde{x}$ or $\tilde{y}$ will affect higher-order digits of $\tilde{x} - \tilde{y}$, so small errors in the operands may produce a large relative error in the answer. But the floating-point arithmetic was not at fault. To produce a better answer we must have better approximations for x and y.

3.4. APPROXIMATE LAWS OF ALGEBRA

We now return to the study of the laws of algebra discussed in Section 1.6. There, we showed that the following laws of algebra are not valid in $FP(r, p, c)$:

Associative laws:

(3.4.1) $a \oplus (b \oplus c) = (a \oplus b) \oplus c$
(3.4.2) $a * (b * c) = (a * b) * c$.

Distributive law:

(3.4.3) $a * (b \oplus c) = (a * b) \oplus (a * c).$

Cancellation law:

If $a \neq 0$ and $a * b = a * c$, then $b = c$.

We now ask whether these laws hold approximately. For example, we shall see that we can write

$$a * (b * c) \approx (a * b) * c$$

in the sense that

$$a * (b * c) = (1 + p)[(a * b) * c]$$

with a small bound for $|p|$.

For each of the laws stated above, we shall determine whether or not the law holds approximately. When it does, we shall obtain bounds for the relative difference p and for the number of units in the last place by which the two numbers may differ. As in Chapter 1, we shall study the system FP(r, p, c) in detail and relegate the study of FP(r, p, R) to the exercises.

First, we shall prove two theorems which will be helpful in the study of these questions.

THEOREM 3.4.1

Suppose that x and y are positive real numbers. In FP(r, p, c) or in FP(r, p, clq), $\bar{x} \oplus \bar{y}$ is either $\overline{x + y}$ or less than $\overline{x + y}$ by 1 in the last place.

Proof. Since we have the add magnitude case, the computation $\bar{x} \oplus \bar{y}$ produces $\overline{\bar{x} + \bar{y}}$ in either FP(r, p, c) or FP(r, p, clq). We may choose notation so that $x \geq y$ and write

$$x = r^e m, \qquad r^{-1} \leq m < 1$$
$$\bar{x} \oplus \bar{y} = r^f n, \qquad r^{-1} \leq n < 1.$$

Here f is either e or $e + 1$. Also, we have

$$x = \bar{x} + \epsilon, \qquad 0 \leq \epsilon < r^{e-p}.$$

Now $\bar{x} \oplus \bar{y} = \overline{\bar{x} + y}$, since the digits chopped from y to produce $\bar{y}$ would be chopped from $\bar{x} + y$ to produce $\overline{\bar{x} + y}$. Then

$$\bar{x} \oplus \bar{y} = \overline{\bar{x} + y} = \overline{x - \epsilon + y} \geq \overline{x + y - r^{e-p}} \geq \overline{x + y} - r^{f-p}$$

so

$$(\bar{x} \oplus \bar{y}) + r^{f-p} \geq \overline{x + y}.$$

Thus, $\bar{x} \oplus \bar{y}$ is either $\overline{x + y}$ or less than $\overline{x + y}$ by 1 in the last place.

THEOREM 3.4.2

Suppose that x and y are positive real numbers and that x is in $S(r, p)$. In FP(r, p, c) or in FP(r, p, clq) with $q \geq 1$, $x * \bar{y}$ is either $\overline{xy}$ or less than $\overline{xy}$ by at most r units in the last place.

Proof. Let

$$x = r^e m, \qquad r^{-1} \leq m < 1$$
$$y = r^f n, \qquad r^{-1} \leq n < 1$$

and

$$x * \bar{y} = r^g l, \qquad r^{-1} \leq l < 1.$$

Here

$$g = e + f - k$$

and

$$l = \overline{r^k mn},$$

where k is 1 or 0 depending on whether or not postnormalization is required. Now

$$y = \bar{y} + \epsilon_1, 0 \leq \epsilon_1 < r^{f-p}$$

and

$$x\bar{y} = x * \bar{y} + \epsilon_2, 0 \leq \epsilon_2 < r^{g-p},$$

so

$$xy = x(\bar{y} + \epsilon_1) = x * \bar{y} + \epsilon_2 + x\epsilon_1.$$

That is,

$$xy = x * \bar{y} + \epsilon,$$

where

$$\epsilon = x\epsilon_1 + \epsilon_2.$$

Then

$$\epsilon < r^{g-p} + r^e m r^{f-p} = r^{g-p}(1 + r^k m).$$

If no postnormalization is required, $\epsilon < 2r^{g-p}$, but when postnormalization is required, we have only $\epsilon < (r + 1)r^{g-p}$. Then $x * \bar{y}$ is either $\overline{xy}$ or less than $\overline{xy}$ by at most r units in the last place.

We now turn to the study of the approximate laws of algebra.

Associative Law of Addition

The example given in Section 1.6 showed that the associative law of addition (3.4.1) does not even hold approximately in FP(r, p, c).

We now ask whether (3.4.1) holds approximately in the special case in which a, b, c all have the same sign. Clearly, this may be reduced to the case in which they are all positive. We may apply Theorem 3.4.1, setting $x = a$ and

$y = b + c$. Since $\bar{y} = b \oplus c$, this shows that $a \oplus (b \oplus c)$ is either $\overline{a + b + c}$ or less than $\overline{a + b + c}$ by 1 in the last place. A similar result holds for $(a \oplus b) \oplus c$, so it can differ from $a \oplus (b \oplus c)$ by at most 1 in the last place. This yields

THEOREM 3.4.3

If a, b, and c all have the same sign, then in either FP(r, p, c) or FP(r, p, clq), the numbers $(a \oplus b) \oplus c$ and $a \oplus (b \oplus c)$ are either identical or else they differ by 1 in the last place. We can write

$$a \oplus (b \oplus c) = (1 + p)[(a \oplus b) \oplus c], \qquad |p| < r^{-(p-1)}$$

and

$$a \oplus (b \oplus c) = (1 - p)(a + b + c), \qquad 0 \le p < 2r^{-p}$$

$$(a \oplus b) \oplus c = (1 - p)(a + b + c), \qquad 0 \le p < 2r^{-p}.$$

To show that $a \oplus (b \oplus c)$ and $(a \oplus b) \oplus c$ may indeed differ by 1 in the last place even if a, b, and c all have the same sign, we may set $a = 1$ and $b = c = (r - 1)r^{-p}$. Then $(a \oplus b) \oplus c = 1$, while $a \oplus (b \oplus c) = 1 + r^{-(p-1)}$.

Associative Law of Multiplication

We shall now show that the associative law of multiplication holds approximately in FP(r, p, c) and FP(r, p, clq) for $q \ge 1$. In either of these systems, $a * b = \overline{ab}$. Let

$$(3.4.4) \qquad \beta = \frac{r^{-(p-1)}}{1 + r^{-(p-1)}},$$

so β is the bound given in (3.2.8) for the relative error introduced by chopping a number to $S(r, p)$. We may write

$$a * b = (1 - p)ab, \qquad 0 \le p < \beta,$$

and

$$(a * b) * c = (1 - \sigma)[(a * b)c], \qquad 0 \le \sigma < \beta.$$

Then

$$(a * b) * c = (1 - p - \sigma + p\sigma)abc,$$

so we may write

$$(a * b) * c = (1 - \tau)abc,$$

where $0 \le \tau < 2\beta - \beta^2$. Similarly, we may write

$$a * (b * c) = (1 - \varphi)abc,$$

where $0 \le \varphi < 2\beta - \beta^2$.

Thus, we have

$$a * (b * c) = \left(\frac{1 - \varphi}{1 - \tau}\right)[(a * b) * c],$$

so

(3.4.5) $$a * (b * c) = (1 + \delta)[(a * b) * c],$$

where

$$1 + \delta = \frac{1 - \varphi}{1 - \tau}.$$

Then

$$1 - 2\beta + \beta^2 < 1 + \delta < \frac{1}{1 - 2\beta + \beta^2},$$

so

$$-(2\beta - \beta^2) < \delta < \frac{1}{(1 - \beta)^2} - 1 = \frac{2\beta - \beta^2}{(1 - \beta)^2}.$$

This yields

(3.4.6) $$|\delta| < \frac{2\beta - \beta^2}{(1 - \beta)^2}.$$

Using the value of β in (3.4.4), (3.4.6) reduces to

(3.4.7) $$|\delta| < 2r^{-(p-1)}(1 + \tfrac{1}{2}r^{-(p-1)}).$$

We shall now try to bound the number of units in the last place by which $a * (b * c)$ and $(a * b) * c$ may differ. We may assume that a, b, and c are all positive. Since $b * c = \overline{bc}$, we may use Theorem 3.4.2 with $x = a$ and $y = bc$ to find that $a * (b * c)$ is either $\overline{abc}$ or less than $\overline{abc}$ by at most r units in the last place. A similar result holds for $(a * b) * c$, so $(a * b) * c$ and $a * (b * c)$ can differ by at most r units in the last place.

We have proved

THEOREM 3.4.4

In FP(r, p, c) and FP(r, p, clq) with $q \geq 1$, the numbers $(a * b) * c$ and $a * (b * c)$ can differ by at most r units in the last place. We may write

$$(a * b) * c = (1 + \delta)[(a * b) * c],$$

where δ satisfies (3.4.7).

We consider an example from FP(16, 6, c) to show that $a * (b * c)$ and $(a * b) * c$ can indeed differ by r units in the last place. Let $a = .FE_H$, $b = 1.000FF_H$, and $c = 1.01006_H$. Then $(a * b) * c$ is $.FF0E3F_H$, while $a * (b * c)$ is $.FF0E2F_H$.

Distributive Law

We shall now consider whether the distributive law (3.4.3) holds approximately. Since the cancellation law fails to hold in $FP(r, p, c)$ except for trivial combinations of r and p, there are positive numbers a, b, and c such that $b \neq c$ but $a * b = a * c$. Then $(a * b) \oplus [a * (-c)]$ is zero, but $a * [b \oplus (-c)]$ is not, so the distributive law (3.4.3) does not even hold approximately.

However, in $FP(r, p, clq)$ with $q \geq 1$, we know that for arbitrary a, b, and c we have

$$b \oplus c = (1 + p)(b + c), \qquad |p| < r^{-(p-1)}$$

and

$$a * (b \oplus c) = (1 - \sigma)[a(b \oplus c)], \qquad 0 \leq \sigma < r^{-(p-1)},$$

so

$$(3.4.8) \qquad a * (b \oplus c) = (1 + \tau)a(b + c),$$

where

$$1 + \tau = 1 - \sigma + p - \sigma p.$$

Then

$$(3.4.9) \qquad -2r^{-(p-1)} < \tau < r^{-(p-1)}$$

Thus $a * (b \oplus c)$ is approximately equal to $a(b + c)$. If the distributive law does not hold approximately, it means that $(a * b) \oplus (a * c)$ is not approximately equal to $ab + ac$. We note that if the arithmetic is performed in $FP(r, p, c)$ instead of $FP(r, p, clq)$, the bound for τ in (3.4.9) may be sharpened to

$$(3.4.10) \qquad -2r^{-(p-1)} < \tau \leq 0.$$

We shall now show that (3.4.3) does hold approximately in $FP(r, p, clq)$ when b and c have the same sign and $q \geq 1$. We may assume that a, b, and c are all positive. Since we have the add magnitude case, $b \oplus c = \overline{b + c}$. Then we may use Theorem 3.4.2 with $x = a$ and $y = b + c$ to find that $a * (b \oplus c)$ is either $\overline{a(b + c)}$ or less than $\overline{a(b + c)}$ by at most r units in the last place. Similarly, since $a * b = \overline{ab}$ and $a * c = \overline{ac}$, we may use Theorem 3.4.1 with $x = ab$ and $y = ac$ to find that $(a * b) \oplus (a * c)$ is either $\overline{ab + ac}$ or less than $\overline{ab + ac}$ by 1 in the last place. Then the two sides of (3.4.3) can differ by at most r units in the last place. As a consequence, we may write

$$(3.4.11) \qquad a * (b \oplus c) = (1 + p)[(a * b) \oplus (a * c)].$$
$$-r^{-(p-2)} \leq p \leq r^{-(p-1)}.$$

We have proved

THEOREM 3.4.5

In $FP(r, p, c)$ or in $FP(r, p, clq)$ with $q \geq 1$, if b and c have the same sign, then $a * (b \oplus c)$ and $(a * b) \oplus (a * c)$ can differ by at most r units in the last place and (3.4.11) holds. Regardless of the signs of b and c, we always have (3.4.8), where τ satisfies (3.4.10) when the arithmetic is performed in $FP(r, p, c)$ and τ satisfies (3.4.9) when the arithmetic is performed in $FP(r, p, clq)$ with $q \geq 1$.

To show that the two sides of (3.4.3) may indeed differ by r units in the last place when b and c have the same sign, we shall consider an example in $FP(r, p, c)$, where we assume that $r \geq 3$ and $p \geq 4$. Let $a = 1 - r^{-(p-2)}$, $b = 1$, and

$$c = r^{-(p-1)} + r^{-(p-1)}[1 - r^{-(p-2)} + r^{-(p-1)}].$$

Then the reader may verify that

$$a * (b \oplus c) = 1 - r^{-(p-2)} + (r - 1)r^{-p},$$

while

$$(a * b) \oplus (a * c) = 1 - r^{-(p-2)} + r^{-(p-1)} + (r - 1)r^{-p}.$$

A similar example may be found for a binary machine. (See Exercise 3.)

Cancellation Law

We shall now show that the cancellation law holds approximately in $FP(r, p, c)$ and $FP(r, p, clq)$ for $q \geq 1$. Suppose that $a \neq 0$ and $a * b = a * c$. Using the value of β in (3.4.4), we may write

$$a * b = (1 - \rho)ab, \qquad 0 \leq \rho < \beta$$

and

$$a * c = (1 - \sigma)ac, \qquad 0 \leq \sigma < \beta,$$

so

$$(1 - \rho)ab = (1 - \sigma)ac.$$

Then

(3.4.12) $$b = (1 + \tau)c,$$

where

$$1 + \tau = \frac{1 - \sigma}{1 - \rho},$$

so

$$1 - \beta < 1 + \tau < \frac{1}{1 - \beta}.$$

Therefore,

$$|\tau| < \frac{1}{1 - \beta} - 1 = \frac{\beta}{1 - \beta},$$

so, using (3.4.4),

(3.4.13) $$|\tau| < r^{-(p-1)}.$$

We now try to bound the number of units in the last place by which b and c may differ. Choose notation so that $|b| < |c|$ and let m be the mantissa of c. If $|b|$ is less than $|c|$ by j units in the last place, then

$$|\tau| = \frac{jr^{-p}}{|m|} > jr^{-p},$$

so (3.4.13) implies that $j < r$. Thus, b and c can differ by at most $r - 1$ units in the last place. We have proved

THEOREM 3.4.6

In FP(r, p, c) and FP(r, p, clq) with $q \geq 1$, if $a \neq 0$ and $a * b = a * c$, then b and c can differ by at most $r - 1$ units in the last place. Also, $b = (1 + \tau)c$, where $|\tau| < r^{-(p-1)}$.

We consider an example to show that b and c can indeed differ by $r - 1$ units in the last place. Assume that $p \geq 4$ and let $a = 1 + r^{-1}$, $b = 1 - r^{-2}$, and $c = b + (r - 1)r^{-p}$. Then b and c differ by $r - 1$ units in the last place, but in FP(r, p, c)

$$a * b = a * c = 1 + r^{-1} - r^{-2} - r^{-3}.$$

3.5. PROPAGATION OF ROUNDING ERROR

As a simple example of the growth of rounding error, we shall consider the problem of computing

(3.5.1) $$x = \prod_{i=0}^{n} x_i.$$

Here we shall assume that the x_i are all given exactly as floating-point numbers. Our computing procedure is to set

$$P_0 = x_0$$

and define

$$P_k = P_{k-1} x_k, \qquad k = 1, 2, \ldots, n,$$

so $x = P_n$. Since we are performing the arithmetic in FP(r, p, a), instead of computing P_k we compute $\tilde{P}_k$, $k = 1, 2, \ldots, n$. Here

$$\tilde{P}_0 = P_0$$
$$\tilde{P}_k = \tilde{P}_{k-1} * x_k, \qquad k = 1, 2, \ldots, n.$$

Then we may write

$$\tilde{P}_k = (1 + \rho_k)\tilde{P}_k x_k,$$

where the bounds for the ρ_k depend on the arithmetic used. For FP(r, p, clq) with $q \geq 1$, we know that

$$0 \leq -\rho_k < r^{-(p-1)}.$$

On the other hand, if we had used rounded arithmetic we would have had

$$|\rho_k| \leq \tfrac{1}{2} r^{-(p-1)}.$$

Now

$$\tilde{P}_1 = (1 + \rho_1)P_1,$$

and by induction, one proves that

$$\tilde{P}_k = \left[\prod_{i=1}^{k} (1 + \rho_i)\right] P_k.$$

Then, setting $\tilde{x} = \tilde{P}_n$, we have

(3.5.2) $$\tilde{x} = (1 + \sigma)x,$$

where

(3.5.3) $$1 + \sigma = \prod_{i=1}^{n} (1 + \rho_i).$$

Since an expression for relative error quite often has the form of (3.5.3), we look for a simple bound for σ. Suppose that we are given bounds ρ_* and ρ^* such that

(3.5.4) $$-\rho_* \leq \rho_i \leq \rho^*$$

holds for each i. We shall assume that ρ_* and ρ^* are nonnegative and that $\rho_* < 1$, so $1 - \rho_* > 0$. Then

(3.5.5) $$-1 + (1 - \rho_*)^n \leq \sigma \leq (1 + \rho^*)^n - 1.$$

Now

$$-1 + (1 - \rho_*)^n = -\sum_{k=1}^{n} \binom{n}{k} \rho_*^k (-1)^{k+1} > -\sum_{k=1}^{n} \binom{n}{k} \rho_*^k$$

$$= -[(1 + \rho_*)^n - 1].$$

Let $\bar{\rho}$ denote the larger of ρ_* and ρ^*. Then

(3.5.6) $$|\sigma| \leq (1 + \bar{\rho})^n - 1.$$

We shall now consider the expression $(1 + \rho)^n - 1$:

$$(1 + \rho)^n - 1 = \sum_{k=1}^{n} \binom{n}{k} \rho^k = n\rho + \frac{n(n + 1)}{2}\rho^2 + \cdots,$$

so if $n\rho$ is small, we have

(3.5.7) $$(1 + \rho)^n - 1 \approx n\rho.$$

However, we often want a bound for $|\sigma|$ instead of an approximation for a bound. Now for $\rho > 0$,

$$(1 + \rho)^n - 1 = -1 + \sum_{k=0}^{n} \binom{n}{k} \rho^k < -1 + \sum_{k=0}^{\infty} \frac{(n\rho)^k}{k!} = e^{n\rho} - 1.$$

Then (3.5.6) yields

(3.5.8) $$|\sigma| < e^{n\bar{\rho}} - 1.$$

If $n\bar{\rho}$ is small, this is approximately the bound $n\bar{\rho}$ that would be obtained using (3.5.7). In some applications, (3.5.8) is a convenient form for the bound, but in other cases, it is convenient to use the bound given by the following theorem:

THEOREM 3.5.1

If $\rho > 0$ and n is a positive integer with $n\rho < 1$, then

$$(1 + \rho)^n - 1 < n\rho + (n\rho)^2.$$

Proof. This clearly holds if n is 1 or 2, so we want to show that

$$1 + n\rho + (n\rho)^2 > (1 + \rho)^n = 1 + n\rho + \frac{n(n - 1)}{2}\rho^2 + \sum_{k=3}^{n} \binom{n}{k} \rho^k;$$

that is,

$$\left(n^2 - \frac{n^2 - n}{2}\right)\rho^2 > \sum_{k=3}^{n} \binom{n}{k} \rho^k$$

or

(3.5.9) $$\frac{n(n + 1)}{2}\rho^2 > \sum_{k=3}^{n} \binom{n}{k} \rho^k.$$

Now

$$\sum_{k=3}^{n} \binom{n}{k} \rho^k < (n\rho)^3 \sum_{k=3}^{n} \frac{(n\rho)^{k-3}}{k!} < \frac{(n\rho)^3}{6} \sum_{k=0}^{n-3} \frac{(n\rho)^k}{k!} < \frac{(n\rho)^3}{6} e^{n\rho}.$$

Since $np < 1$, this yields

$$\sum_{k=3}^{n} \binom{n}{k} \rho^k < np\frac{(np)^2}{2} < \frac{(np)^2}{2} < \frac{n(n+1)}{2}\rho^2,$$

so (3.5.9) holds.

Thus, if $n\bar{p} < 1$, we may write (3.5.6) as

$$(3.5.10) \qquad\qquad |\sigma| < n\bar{p}(1 + n\bar{p}).$$

We shall now consider

$$|-1 + (1 - \rho)^n| = 1 - (1 - \rho)^n.$$

THEOREM 3.5.2

If $0 < \rho < 1$ and n is a positive integer with $n\rho < 1$, then

$$|-1 + (1 - \rho)^n| \leq np.$$

Proof. We may assume that $n > 1$. Since

$$1 - (1 - \rho)^n = n\rho - \sum_{k=2}^{n} \binom{n}{k}(-\rho)^k,$$

it suffices to show that

$$\sum_{k=2}^{n} \binom{n}{k}(-\rho)^k > 0.$$

Let m be the largest integer $\leq [(n-1)/2]$. Then n is either $2m + 1$ or $2m + 2$, so

$$\sum_{k=2}^{n} \binom{n}{k}(-\rho)^k \geq \sum_{j=1}^{m}\left[\binom{n}{2j}\rho^{2j} - \binom{n}{2j+1}\rho^{2j+1}\right].$$

Therefore, it suffices to show that

$$(3.5.11) \qquad\qquad \binom{n}{2j}\rho^{2j} > \binom{n}{2j+1}\rho^{2j+1},$$

which reduces to

$$1 > \frac{n - 2j}{2j + 1}\rho.$$

But

$$\frac{n - 2j}{2j + 1}\rho < \frac{n\rho}{2j + 1} < \frac{1}{2j + 1} < 1,$$

so the inequality holds.

These results may be combined into the following theorem:

THEOREM 3.5.3

Let $1 + \sigma = \prod_{i=1}^{n} (1 + \rho_i)$, where $-\rho_* \leq \rho_i \leq \rho^*$ holds for $i = 1, 2, \ldots, n$. Suppose that $n\rho_*$ and $n\rho^*$ are both positive and less than 1. Then

$$-n\rho_* \leq \sigma \leq n\rho^*(1 + n\rho^*).$$

If $\bar{\rho}$ is the larger of ρ_* and ρ^*, then $|\sigma| < n\bar{\rho}(1 + n\bar{\rho})$.

These theorems are fundamental for establishing bounds for relative error. The bound for $|\sigma|$ is given by (3.5.6), and in (3.5.7) we see that $n\bar{\rho}$ is an approximation for this bound. But in Theorem 3.5.1 we see that $n\bar{\rho}$ is also a bound for the relative error in the approximation (3.5.7). Since we are bounding the relative error σ, we hope that $n\bar{\rho}$ is small, say 10^{-5} or 10^{-6}. But then the approximation (3.5.7) is very good. Indeed, if $n\bar{\rho}$ is so large that the approximation (3.5.7) is unsatisfactory, then it is almost certainly large enough to indicate that the approximation $\tilde{x} \approx x$ is unsatisfactory. Thus, (3.5.7) is a good approximation which is not likely to mislead us.

These bounds of the form $n\rho$ may be interpreted in another way. Suppose that we have performed n multiplications in FP(r, p, clq) with $q \geq 1$. Then $\rho_* = r^{-(p-1)}$ and $\rho^* = 0$, so

(3.5.12) $$-n\rho_* \leq \sigma \leq 0.$$

Now, suppose that $n = r^k$, so $n\rho_* = r^{-(p-k-1)}$. Then (3.5.12) is the bound we would have had for the relative error introduced by a single multiplication in FP($r, p - k, clq$), so we may think of the r^k multiplications as having possibly cost us k digits in accuracy. Thus, performing a million multiplications may cost us about six decimal digits, or, since $10^6 \approx 2^{20}$, about 20 bits. Then we would not want to perform a million multiplications in FP(16, 6, $cl1$), but the bound for the relative error introduced by performing a million multiplications in FP(16, 14, $cl1$) is smaller than the bound for the relative error introduced by a single multiplication in FP(16, 9, $cl1$).

This also sheds light on the precision we should use. Suppose that we want to perform 200 multiplications in FP(10, 8, c). Then we may lose between two and three decimal digits of accuracy. Instead of performing all arithmetic in FP(10, 8, c), suppose that we adhered to the "well-known rule" referred to in Section 3.3, which states that it is not worth developing digits unless we can guarantee that they are significant. This would suggest that after the first 100 multiplications we could reduce the precision of our arithmetic to six decimal digits. But then the second 100 multiplications would expose us to the loss of two more decimal digits, so the final bound for relative error would be the same as the bound for the relative error after a single multiplication in FP(10, 4, c). Here we see clearly the advantages of performing all arithmetic at the higher precision, so the new error introduced will always be small.

Reducing the precision of the arithmetic risks unnecessary damage to the final result.

Finally, suppose that instead of (3.5.1) we really wanted to compute

$$(3.5.13) \qquad\qquad y = \prod_{i=0}^{n} y_i,$$

where the x_i in (3.5.1) are floating-point numbers with $x_i \approx y_i$. If each x_i satisfies

$$x_i = (1 + \tau_i)y_i, \qquad |\tau_i| \le \tau^*,$$

then

$$x = \prod_{i=0}^{n} x_i = \left[\prod_{i=0}^{n} (1 + \tau_i) \right] y.$$

The relative error here again has the form of (3.5.3). If $(n + 1)\tau^* < 1$, we may write

$$\tilde{x} = (1 + \sigma)(1 + \tau)y,$$

where the bounds for σ are given by Theorem 3.5.3 and τ satisfies

$$-(n + 1)\tau^* \le \tau \le (n + 1)\tau^*[1 + (n + 1)\tau^*].$$

This suggests that the precision should be chosen so that the bound $\bar{\rho}$ for the ρ_i is no larger than τ^*, and we would prefer to have $\bar{\rho}$ substantially smaller than τ^*. We are sometimes interested in the special case in which $x_i = \bar{y}_i$, so the ρ_i and the τ_i have the same bounds. Then we may write $\tilde{x} = (1 + \beta)y$ with

$$-(2n + 1)\rho_* \le \beta \le 0.$$

3.6. X**N

Many higher-level languages use a special symbol, such as ** or ↑, to designate exponentiation. Since most computers do not have an instruction to perform exponentiation, X^N must be evaluated by a subroutine. Thus, in terms of the computation performed, the ** operation is similar to a function, such as SQRT or SIN.

The calculation performed for ** operation in FORTRAN often depends on the data types of the operands. For example, if X and Y are real, the calculation performed to compute X**Y is often equivalent to the evaluation of the expression EXP (Y*ALOG(X)). In this section, we shall consider the computation of X**N, where X is real and N is a positive integer. We shall assume that the computation is carried out in either FP(r, p, c) or FP(r, p, clq) with $q \ge 1$, and we shall consider three approaches for the computation of

$$Y = X**N$$

Each approach will be illustrated by a FORTRAN program, and for the sake of simplicity, we shall allow the programs to change X and N. We repeat that N is assumed to be a positive integer.

Repetitive Multiplication

Here, we shall perform the computation

```
         Y = 1
         DO 100  I = 1, N
  100    Y = Y*X
```

This approach requires N multiplications, and the growth of rounding error for this calculation was studied in Section 3.5. Using (3.5.2) and (3.5.3), we see that we compute $\tilde{Y} = (1 - \sigma)Y$, where

$$(3.6.1) \qquad 1 - \sigma = \prod_{i=1}^{N} (1 - p_i).$$

Here p_i is the relative error introduced by the ith multiplication, and it satisfies $0 \le p_i < r^{-(p-1)}$. The first multiplication forms 1*X, and since multiplication by 1 is exact, $p_1 = 0$. Then (3.6.1) may be replaced by

$$(3.6.2) \qquad 1 - \sigma = \prod_{i=2}^{N} (1 - p_i).$$

Setting $p_* = r^{-(p-1)}$, we have

$$(3.6.3) \qquad 0 \le \sigma < 1 - (1 - p_*)^{N-1},$$

or, using Theorem (3.5.2),

$$(3.6.4) \qquad 0 \le \sigma < (N - 1)p_*.$$

Repetitive Squaring

In this approach, we use the binary representation of N. Let

$$(3.6.5) \qquad N = \sum_{i=0}^{m} k_i 2^i,$$

where

$$(3.6.6) \qquad 2^m \le N < 2^{m+1}$$

and each k_i is 0 or 1. Then the binary representation of N is

$$N = (k_m k_{m-1} \cdots k_1 k_0)_B.$$

Let $P_i = X^{2^i}$, so

(3.6.7) $$P_i = P_{i-1}^2, \qquad i = 1, 2, \ldots, m,$$

and

(3.6.8) $$Y = X^{\sum_{i=0}^{m} k_i 2^i} = \prod_{i=0}^{m} X^{k_i 2^i}$$

or

(3.6.9) $$Y = \prod_{\substack{i=0 \\ k_i \neq 0}}^{m} P_i.$$

We use (3.6.9) to compute Y, and our computational procedure is

```
        P = X
        Y = 1
  100   L = N/2
        K = N−2*L
        IF(K.EQ.1) Y=Y*P
        N = L
        IF(N.EQ.0) GO TO 200
        P = P*P
        GO TO 100
  200   CONTINUE
```

Let

(3.6.10) $$k = \sum_{i=0}^{m} k_i.$$

Then this algorithm requires only $k + m$ floating-point multiplications. On a binary machine, the computation L = N/2 may be performed by shifting N one place to the right, and K = N − 2*L is the low-order bit of N prior to this shift. Thus, the fixed-point arithmetic used in this algorithm may be very easy to perform in Assembler language. In any event, we only go through the loop $m + 1$ times, so this approach produces substantial savings in computer time when N is large.

In this procedure, instead of computing the P_i and Y from (3.6.7) and (3.6.9), we compute approximations $\tilde{P}_i$ and $\tilde{Y}$ by setting $\tilde{P}_0 = X$ and using

(3.6.11) $$\tilde{P}_i = \tilde{P}_{i-1} * \tilde{P}_{i-1}, \qquad i = 1, 2, \ldots, m,$$

and

(3.6.12) $$Y = \prod_{\substack{i=0 \\ k_i \neq 0}}^{m} {}^* \tilde{P}_i,$$

where $\prod^*$ denotes the $*$ operation.

We may write

$$(3.6.13) \qquad P_i = (1 - p_i)P_{i-1}^2, \qquad 0 \le p < r^{-(p-1)}$$

and

$$(3.6.14) \qquad \tilde{Y} = (1 - \sigma_1)(1 - \sigma_2) \cdots (1 - \sigma_k) \prod_{\substack{i=0 \\ k_i \ne 0}}^{m} \tilde{P}_i,$$

with $0 \le \sigma_i < r^{-(p-1)}$. Since the first multiplication in (3.6.14) multiplies a number by 1, $\sigma_1 = 0$. Now $\tilde{P}_0 = X$, $\tilde{P}_1 = (1 - p_1)X^2$,

$$\tilde{P}_2 = (1 - p_1)^2 (1 - p_2)X^4$$

and one proves by induction that

$$\tilde{P}_i = (1 - p_i)(1 - p_{i-1})^2(1 - p_{i-2})^4 \cdots (1 - p_1)^{2^{i-1}}X^{2^i},$$

so

$$\tilde{Y} = \left[\prod_{i=1}^{k} (1 - \sigma_1)\right] \tilde{P}_0^{k_0} \prod_{i=1}^{m}[(1 - p_i) \cdots (1 - p_1)^{2^{i-1}}X^{2^i}]^{k_i}$$

$$= \left[\prod_{i=1}^{k} (1 - \sigma_i)\right] \prod_{i=0}^{m} X^{k_i 2^i} \prod_{i=1}^{m} [(1 - p_i)^{k_i} \cdots (1 - p_1)^{k_i 2^{i-1}}].$$

Thus

$$Y = (1 - \tau)\tilde{Y},$$

where

$$1 - \tau = \left[\prod_{i=1}^{k} (1 - \sigma_i)\right] \prod_{i=1}^{m} \prod_{j=1}^{i} (1 - p_j)^{k_i 2^{i-j}}$$

or

$$1 - \tau = \left[\prod_{i=1}^{k} (1 - \sigma_i)\right] \prod_{j=1}^{m} \prod_{i=j}^{m} (1 - p_j)^{k_i 2^{i-j}}.$$

Then

$$(3.6.15) \qquad 1 - \tau = \left[\prod_{i=1}^{k} (1 - \sigma_i)\right] \prod_{j=1}^{m} (1 - p_j)^{l_j},$$

where

$$(3.6.16) \qquad l_j = \sum_{i=j}^{m} k_i 2^{i-j}.$$

Thus, $1 - \tau$ is the product of factors of the form $1 - \sigma_i$ or $1 - p_i$. Let n be the number of such factors, so

$$(3.6.17) \qquad n = k + \sum_{j=1}^{m} l_j.$$

Then, using (3.6.10) and (3.6.16), we find that

$$n = \sum_{i=0}^{m} k_i + \sum_{i=1}^{m} k_i \sum_{j=1}^{i} 2^{i-j}.$$

Since

(3.6.18) $$\sum_{j=1}^{i} 2^{i-j} = 2^{i-1} + \cdots + 2 + 1 = 2^{i} - 1,$$

we have

$$n = \sum_{i=0}^{m} k_i + \sum_{i=1}^{m} k_i(2^i - 1) = \sum_{i=0}^{m} k_i 2^i = N.$$

Thus, we have N factors of the form $1 - \sigma_i$, and $1 - \rho_i$. Since $\sigma_1 = 0$, we have

(3.6.19) $$0 \leq \tau < 1 - (1 - \rho_*)^{N-1},$$

where $\rho_* = r^{-(p-1)}$. This is exactly the same bound we had in (3.6.3) for the relative error σ when we used the repetitive multiplication algorithm.

With either of these algorithms, we obtain a formula for the relative error which has the form of (3.6.1) or (3.6.15), but with the repetitive squaring algorithm there are only a few different values $1 - \rho_j$ which are repeated l_j times. Using (3.6.16), we see that

$$l_1 = \frac{1}{2} \sum_{i=1}^{m} k_i 2^i = \frac{1}{2}(N - k_0) \approx \frac{N}{2}.$$

Similarly,

$$l_2 = \frac{1}{4}(N - k_0 - 2k_1) \approx \frac{N}{4}.$$

Thus, if the first two squarings produce bad relative error, it is as if three-fourths of the multiplications in the repetitive multiplication method had produced bad relative error.

Nested Squaring

In this algorithm, we again use the binary representation of N given by (3.6.5) and (3.6.6) and the representation of Y given by (3.6.8). This time we define

$$Q_j = \prod_{i=j}^{m} X^{k_i 2^{i-j}},$$

so $Y = Q_0$. Now $Q_m = X^{k_m} = X$, and

$$Q_{j-1} = \prod_{i=j-1}^{m} X^{k_i 2^{i-j+1}} = X^{k_{j-1}} \prod_{i=j}^{m} X^{2k_i 2^{i-j}},$$

so

(3.6.20) $$Q_{j-1} = X^{k_{j-1}} Q_j^2.$$

In this approach, we first determine the value of m satisfying (3.6.6). We then set $Q_m = X$ and compute Q_j from (3.6.20) for $j = m - 1, m - 2, \ldots, 1, 0$.

Then $Y = Q_0$. The following FORTRAN program assumes only that we are given an integer L with $2^L > N$.

```
      Y  = 1
      LL = 2**L
      DO 100 I = 1, L
      Y  = Y*Y
      N  = N*2
      IF (N.LT.LL) GO TO 100
      Y  = Y*X
      N  = N-LL
100   CONTINUE
```

This approach requires $L + k$ floating-point multiplications, where k is given by (3.6.10), and even if we take $L = m + 1$, we find that this is one more floating-point multiplication than the repetitive squaring method required. But if we knew the value of m satisfying (3.6.6), we could initialize with $Y = X$ instead of $Y = 1$, and go through the loop only $m - 1$ times. In this case we would use only $k + m - 1$ floating-point multiplications, which is one fewer than we required for the repetitive squaring algorithm. Thus, this algorithm is attractive when it is easy to find m, that is, to find the location of the high-order one bit in the binary representation of N. As before, the fixed-point arithmetic is easy to perform on a binary machine.

In our computation, instead of computing Q_j given by (3.6.20), we compute an approximation $\tilde{Q}_j$ obtained by setting $\tilde{Q}_m = X$ and using

$$(3.6.21) \qquad \tilde{Q}_{j-1} = X^{k_{j-1}} * (\tilde{Q}_j * \tilde{Q}_j).$$

As before, we may write

$$(3.6.22) \qquad \tilde{Q}_{j-1} = (1 - \sigma_{j-1})^{k_{j-1}}(1 - \rho_{j-1})Q_j^2 X^{k_{j-1}},$$

where $0 \le \sigma_{j-1}, \rho_{j-1} < r^{-(p-1)}$. Proceeding by induction, one proves that

$$\tilde{Q}_{m-j} = \left[\prod_{i=0}^{j-1} (1 - \sigma_{m-j+i})^{2^i k_{m-j+i}}\right]\left[\prod_{i=0}^{j-1} (1 - \rho_{m-j+i})^{2^i}\right]\left[\prod_{i=0}^{j-1} X^{2^i k_{m-j+i}}\right] X^{2^j}$$

and

$$(3.6.23) \qquad \tilde{Y} = \tilde{Q}_0 = \left[\prod_{i=0}^{m-1} (1 - \sigma_i)^{2^i k_i}\right]\left[\prod_{i=0}^{m-1} (1 - \rho_i)^{2^i}\right]\left[\prod_{i=0}^{m-1} X^{2^i k_i}\right] X^{2^m}.$$

Using (3.6.8), (3.6.23) may be written as

$$(3.6.24) \qquad \tilde{Y} = Y\left[\prod_{i=0}^{m-1} (1 - \sigma_i)^{2^i k_i}\right]\left[\prod_{i=0}^{m-1} (1 - \rho_i)^{2^i}\right],$$

SO

(3.6.25) $$\tilde{Y} = (1 - \varphi)Y,$$

where

(3.6.26) $$1 - \varphi = \left[\prod_{i=0}^{m-1} (1 - \sigma_i)^{2^i k_i}\right]\left[\prod_{i=0}^{m-1} (1 - \rho_i)^{2^i}\right].$$

Let n' be the number of factors of the form $1 - \sigma_i$ or $1 - \rho_i$ in (3.6.26). Then

(3.6.27) $$n' = \sum_{i=0}^{m-1} k_i 2^i + \sum_{i=0}^{m-1} 2^i.$$

Using (3.6.5), (3.6.18), and the fact that $k_m = 1$, (3.6.27) reduces to $n' = N - 1$. Then, setting $\rho_* = r^{-(p-1)}$, we have

$$0 \leq \varphi < 1 - (1 - \rho_*)^{N-1},$$

which is the same bound we obtained for the relative error in the other two algorithms. As with the repetitive squaring algorithm, the relative errors from the first few multiplications have a large effect on the final answer when N is large.

Even the nested squaring algorithm does not necessarily minimize the number of floating-point multiplications needed to compute X**N. For example, Exercise 13 illustrates a faster method for computing X**15. For a detailed discussion of the problem of minimizing the number of multiplications in the computation of X**N, see Knuth (1969).

3.7. CONDITION

In studying the propagation of error, we are often confronted with the question of how an error in the input x affects the answer. For example, suppose that we want to compute x^N for some positive integer N, but that instead of x we are given an approximation $\tilde{x}$ for x with $\tilde{x} = (1 + \sigma)x$. Then we ask how close $\tilde{x}^N$ is to x^N. Now

(3.7.1) $$\tilde{x}^N = (1 + \sigma)^N x^N,$$

so Theorem 3.5.3 allows us to translate bounds for σ into bounds for the relative error in the approximation $\tilde{x}^N \approx x^N$. Indeed, comparision of (3.7.1) with Section 3.6 shows that the way in which errors in x affect the answer to this problem is quite similar to the way in which errors in the computation X**N affect the answer.

In general, we shall say that the problem of computing

(3.7.2) $y = f(x)$

is *well conditioned* if small changes in x produce only small changes in $f(x)$, while it is *ill-conditioned* or *poorly conditioned* if small changes in x can produce large changes in $f(x)$. This is still rather vague, because we have not specified the meaning of *small* and *large*. In some cases we are interested in absolute changes, while in other cases we are concerned about relative changes.

The condition of the problem (3.7.2) may depend on the data x as well as the function $f(x)$. Thus, for a given function $f(x)$, the problem of computing $y = f(x)$ may be well conditioned for certain values of x, but ill-conditioned for other values of x.

The computation in (3.7.2) may be generalized to the case in which we use

(3.7.3) $y_i = f_i(x_1, \ldots, x_n), \qquad i = 1, 2, \ldots, m$

to compute m values y_i from n input values $x_1, \ldots, x_n$. In this case, it is often convenient to measure the size of a change in the x's or y's by a norm for a vector space. Using this approach for the problem of finding a solution of a system of simultaneous equations

(3.7.4) $Ax = b,$

where the right-hand-side b is arbitrary, one can assign a *condition number* to the matrix A representing the extent of the ill-conditioning of the problem (3.7.4). [See Wilkinson (1963) or Forsythe and Moler (1967).] As Wilkinson points out, the fact that a matrix is ill-conditioned for the problem of finding the solution of (3.7.4) does not imply that it will be ill-conditioned for the problem of finding its eigenvalues or eigenvectors. Thus, we should speak of the condition of à problem, not a matrix.

We shall now consider the condition of the problem (3.7.2) in more detail. Suppose that we have a program or subroutine to perform the computation in (3.7.2), but that instead of x we are given only an approximation $\tilde{x}$ for x. In place of x, our program sees only $\tilde{x}$, so it tries to evaluate $f(\tilde{x})$ instead of $f(x)$. We now ignore the errors in computing $f(\tilde{x})$ and ask how different $f(\tilde{x})$ can be from $f(x)$. That is, we ask how much the error in the approximation $\tilde{x} \approx x$ would damage the answer if no further errors were made. This is exactly the question of how well conditioned the problem (3.7.2) is.

We let

(3.7.5) $\tilde{x} = x + \epsilon,$

and we assume that $f(x)$ is continuous and possesses a derivative at every point between x and $x + \epsilon$. Then, by the theorem of mean value,

$$\frac{f(x + \epsilon) - f(x)}{\epsilon} = f'(\xi),$$

where ξ lies between x and $x + \epsilon$. Thus,

$$(3.7.6) \qquad f(x + \epsilon) - f(x) = \epsilon f'(\xi).$$

If ϵ is small enough so that $f'(t)$ does not change very much between x and $x + \epsilon$, we have

$$(3.7.7) \qquad f(x + \epsilon) - f(x) \approx \epsilon f'(x).$$

Alternatively, suppose that we have a bound M such that

$$(3.7.8) \qquad |f'(t)| \leq M$$

holds for all t between x and $x + \epsilon$. Then

$$(3.7.9) \qquad |f(\tilde{x}) - f(x)| \leq |\epsilon M|.$$

In many cases (3.7.7) provides an adequate warning when the answer is sensitive to errors in the data, but when a bound is required we use (3.7.9).

When $f(x)$ is not zero, we may prefer to use relative error instead of absolute error. From (3.7.6) we obtain

$$(3.7.10) \qquad \frac{f(\tilde{x}) - f(x)}{f(x)} = \frac{\epsilon f'(\xi)}{f(x)}.$$

Now if $\tilde{x} = (1 + \rho)x$, we may take ρx as the value of ϵ in (3.7.5), so (3.7.10) becomes

$$(3.7.11) \qquad \frac{f(\tilde{x}) - f(x)}{f(x)} = \frac{\rho x f'(\xi)}{f(x)}.$$

As before,

$$(3.7.12) \qquad \frac{f(\tilde{x}) - f(x)}{f(x)} \approx \frac{\rho x f'(x)}{f(x)}$$

if $f'(t)$ does not change much between x and $\tilde{x}$, and if $f'(t)$ satisfies (3.7.8), we have

$$(3.7.13) \qquad \left| \frac{f(\tilde{x}) - f(x)}{f(x)} \right| \leq \left| \frac{\rho x M}{f(x)} \right|.$$

These results then give us a measure of the condition of the problem (3.7.2). We may think of $xf'(x)/f(x)$ in (3.7.12) and $xM/f(x)$ in (3.7.13) as *magnification factors* indicating how the relative error ρ in x is magnified.

As an example, we may consider the function $f(x) = e^x$. Then (3.7.11) becomes

$$\frac{f(\tilde{x}) - f(x)}{f(x)} = \rho x \frac{e^\xi}{e^x},$$

and (3.7.12) yields

$$\frac{f(\tilde{x}) - f(x)}{f(x)} \approx \rho x.$$

In place of (3.7.13), we may use the fact that $e^\xi/e^x < e^{|\rho x|}$ to obtain

$$\left| \frac{f(\tilde{x}) - f(x)}{f(x)} \right| < \left| \rho x e^{|\rho x|} \right|.$$

The results in (3.7.6) through (3.7.13) may be extended to a function

$$(3.7.14) \qquad\qquad y = f(x_1, \ldots, x_n)$$

of several variables. If $f(x_1, \ldots, x_n)$ has continuous partial derivatives with respect to $x_1, \ldots, x_n$, which we denote by $f_{x_k}(x_1, \ldots, x_n)$, then

$$(3.7.15) \quad f(x_1 + \epsilon_1, \ldots, x_n + \epsilon_n) - f(x_1, \ldots, x_n) = \sum_{k=1}^{n} \epsilon_k f_{x_k}(\xi_1, \ldots, \xi_n),$$

where each ξ_k lies between x_k and $x_k + \epsilon_k$. From (3.7.15) we find that if $\tilde{x}_k = (1 + \rho_k)x_k$, $k = 1, \ldots, n$, then

$$(3.7.16) \qquad \frac{f(\tilde{x}_1, \ldots, \tilde{x}_n) - f(x_1, \ldots, x_n)}{f(x_1, \ldots, x_n)} = \sum_{k=1}^{n} \frac{\rho_k x_k f_{x_k}(\xi_1, \ldots, \xi_n)}{f(x_1, \ldots, x_n)},$$

provided $f(x_1, \ldots, x_n)$ is not zero. As above, if the partial derivatives do not change too much, we may write (3.7.16) as

$$(3.7.17) \qquad \frac{f(\tilde{x}_1, \ldots, \tilde{x}_n) - f(x_1, \ldots, x_n)}{f(x_1, \ldots, x_n)} \approx \sum_{k=1}^{n} \frac{\rho_k x_k f_{x_k}(x_1, \ldots, x_n)}{f(x_1, \ldots, x_n)}.$$

Usually it suffices to consider the size of each of the magnification factors $x_k f_{x_k}(x_1, \ldots, x_n)/f(x_1, \ldots, x_n)$ in (3.7.17).

Finally, this approach may be extended to functions defined implicitly. If $y^*, x_1^*, \ldots, x_n^*$ satisfies

$$(3.7.18) \qquad\qquad F(y, x_1, \ldots, x_n) = 0,$$

and if F_y does not vanish at $(y^*, x_1^*, \ldots, x_n^*)$, then (3.7.18) defines y implicitly as a function $f(x_1, \ldots, x_n)$ in a neighborhood of $(x_1^*, \ldots, x_n^*)$. Here

$$F(f(x_1, \ldots, x_n), x_1, \ldots, x_n) = 0$$

and

(3.7.19) $$f_{x_k}(x_1, \ldots, x_n) = \frac{-F_{x_k}(y, x_1, \ldots, x_n)}{F_y(y, x_1, \ldots, x_n)}.$$

We may then use (3.7.15), (3.7.16), or (3.7.17) to study $f(x_1, \ldots, x_n)$.

We shall illustrate this approach by considering the problem of computing the roots of a polynomial. This problem has been studied in more detail by Wilkinson (1963). We let

$$p(x) = \sum_{k=0}^{n} a_k x^k$$

and study a simple root α of the equation $p(x) = 0$. Let

$$F(x, a_0, \ldots, a_n) = \sum_{k=0}^{n} a_k x^k,$$

so $F(\alpha, a_0, \ldots, a_n) = 0$. Now $F_{a_k}(x, a_0, \ldots, a_n) = x^k$ and $F_x(x, a_0, \ldots, a_n) = p'(x)$. If α is a simple root, then $p'(\alpha) \neq 0$, so we may write α as a function $f(a_1, \ldots, a_n)$ of the coefficients. By substituting these values for F_{a_k} and f_α in (3.7.19), we have

(3.7.20) $$\frac{\partial \alpha}{\partial a_k} = f_{a_k}(a_0, \ldots, a_n) = \frac{-\alpha^k}{p'(\alpha)}.$$

Now suppose that we change only one coefficient a_k, replacing it by $(1 + \rho)a_k$, and let $\tilde{\alpha}$ be the new value of α. If ρ is small enough so that we can use (3.7.17), we have

(3.7.21) $$\frac{\tilde{\alpha} - \alpha}{\alpha} \approx \frac{-\rho a_k \alpha^{k-1}}{p'(\alpha)}.$$

Here we see that the sensitivity of the root to changes in a coefficient depends on both which coefficient is changed and which root we are considering.

A striking example of this sensitivity was given by Wilkinson (1963). He considered the polynomial

(3.7.22) $$p(x) = (x - 1)(x - 2) \cdots (x - 20)$$

and studied the effect on the roots of changes in a_{19}. The roots near 1 are quite insensitive to such changes, but the roots near 16 are changed dramatically by small changes in a_{19}. In fact, he showed that a change of 1 in the sixty-third bit of a_{19} can produce a change of almost 1 in the thirty-second bit of the root 16. A change of 1 in the thirty-first bit of a_{19} produces so large a change in the roots that the approximation (3.7.21) cannot be used, and some of the roots become complex with imaginary parts as large as $2.8i$. For a thorough study of the condition of polynomials, the reader is referred to Wilkinson's book.

3.8. ERROR ANALYSIS OF A PROGRAM

To illustrate the error analysis of a program, we shall consider a very simple problem, namely finding the value of a linear function. Here we are given a, b, and x, and we are asked to compute

$$y = ax + b.$$

We write a subroutine LIN whose input is a, b, and x and whose output is y. The FORTRAN program for LIN is

```
SUBROUTINE LIN(A,B,X,Y)
Y = A*X+B
RETURN
END
```

We shall assume that the input A, B, and X are all normalized floating-point numbers, and we shall ignore overflow and underflow.

Suppose that we have written this program as a library subroutine and that we want to tell the user what accuracy he can expect from it. What can we say?

Before trying to answer this question, we shall consider an example. Suppose that we are using an eight-digit decimal machine and that the computation is performed in $FP(10, 8, c)$. Let

$$(3.8.1) \qquad Y = AX + B$$

and

$$(3.8.2) \qquad \tilde{Y} = (A * X) \oplus B.$$

Suppose that the input to the routine is

$$(3.8.3) \qquad \begin{aligned} A &= .56785679 \\ X &= .54325433 \\ B &= -.30849066, \end{aligned}$$

so

$$AX = .3084\ 9065\ 9987\ 4007,$$

and $A * X$ is $.30849065$. Then

$$Y = -(.00125993) \cdot 10^{-8},$$

while $\tilde{Y} = -10^{-8}$. If

$$\rho = \frac{\tilde{Y} - Y}{Y}$$

is the relative error, we find that ρ is about 792.64.

Clearly, our program is capable of producing very bad relative error. Since we are exposed to a relative error of several hundred, there is no meaningful bound on relative error that we can state to the user.

We note that for this particular set of data the absolute error is quite small. In fact, $|\tilde{Y} - Y| < 10^{-8}$. However, we might have had data with a much larger characteristic. For example, suppose that

$$A = (.56785679) \cdot 10^{40}$$
$$X = .54325433$$
$$B = -(.30849066) \cdot 10^{40}.$$

Then $|\tilde{Y} - Y| = -(.99874007) \cdot 10^{32}$, so the program may produce large absolute error. Indeed, the only bound for the absolute error is about 10^{-8} times the overflow threshold Ω.

Thus, there is no meaningful bound that we can state to the user for either the absolute error or the relative error.

We note that in the examples we have considered the error was small with respect to B. We ask whether we can promise the user that this will always be the case. But if the input to the program consists of A and X given by (3.8.3), while $B = 10^{-40}$, then we find that

$$|\tilde{Y} - Y| = (.99874007) \cdot 10^{-8} + 10^{-40},$$

which is not small with respect to B.

Next, we ask whether we can promise the user that the error will always be small with respect to AX. But if A and X have the values given in (3.8.3) and $B = 12345678$, we find that $\tilde{Y} = B$, so $|\tilde{Y} - Y| = AX$, and the error is not small with respect to AX.

In summary, these attempts to bound the error have yielded the following information:

1. Absolute error: no reasonable bound.

2. Relative error: no reasonable bound.

3. Error always small with respect to B: false.

4. Error always small with respect to AX: false.

As a final attempt, we might make the following rather vague statement:

The error is always small with respect to some quantity which appears in the calculation either as initial data, as an intermediate result, or as the final answer.

There are two objections to this statement. First, it is of no help to the user, who may be unfamiliar with the algorithm being used and who never sees the intermediate results. Second, this statement can be made about some very "sick" programs, such as the program discussed in Section 4.2 for computing e^{-x} or sin x using the power series when $x = 128$, or the calculation considered in Section 3.10 for computing sinh x from the formula

$$\sinh x = \frac{e^x - e^{-x}}{2}$$

when x is small.

Thus, all the above attempts to bound the error in our program have failed to yield a meaningful statement that can be made to a user. In the next section, we shall see that a backward error analysis will provide such a statement.

3.9. BACKWARD ERROR ANALYSIS

Suppose that we want to compute $y = f(x)$, and instead of y we have computed a value $\tilde{y}$ with $\tilde{y} \approx y$. A *forward error analysis* attempts to bound either the absolute error $\tilde{y} - y$ or the relative error $(\tilde{y} - y)/y$. A *backward error analysis* seeks a number $\tilde{x}$ with $\tilde{y} = f(\tilde{x})$ and attempts to bound either the absolute difference $\tilde{x} - x$ or the relative difference $(\tilde{x} - x)/x$.

Thus, instead of asking how well we have solved the problem, we try to find out what problem we have solved. There may be more than one value of $\tilde{x}$ with $\tilde{y} = f(\tilde{x})$, in which case we choose one close to x. Our objective is to be able to make a statement of the following form: "We have found an exact solution of the problem $\tilde{y} = f(\tilde{x})$ for some value of $\tilde{x}$ with $|\tilde{x} - x| < \delta$ or $|(\tilde{x} - x)/x| < \rho$. We do not know the specific value of $\tilde{x}$, but we have a bound (δ or ρ) for how far it can be from x."

With this approach, we view the errors in the computation as being equivalent to a perturbation of the data. That is, the computation is equivalent to first perturbing the data to produce $\tilde{x}$ and then computing $\tilde{y}$ from $\tilde{x}$ exactly. The user must then assess the effect that this perturbation of the data has on the answer. But the user already had to worry about how the answer is affected by noise in the data which arose because of the inaccuracy of measurements or even because of radix conversion in the input program. The backward error analysis merely magnifies the importance of his considering this effect.

The idea of a backward error analysis is easily extended to a function of several variables. If we wish to compute $y = f(x_1, \ldots, x_n)$, and instead we have computed $\tilde{y}$, we seek $\tilde{x}_1, \ldots, \tilde{x}_n$ with

(3.9.1) $$\tilde{y} = f(\tilde{x}_1, \ldots, \tilde{x}_n).$$

In this case there are likely to be many vectors $(\tilde{x}_1, \ldots, \tilde{x}_n)$ satisfying (3.9.1). We may try to bound $\tilde{x}_i - x_i$ for each i or $(\tilde{x}_i - x_i)/x_i$ for each i. In many applications in matrix theory, we select a norm $\|x\|$ for the vector space and try to bound $\|\tilde{x} - x\|$.

To illustrate a backward error analysis, we shall consider our FORTRAN program LIN(A,B,X,Y) discussed in Section 3.8. We wanted to compute Y given by (3.8.1), and instead we computed $\tilde{Y}$ given by (3.8.2). Now in FP(r, p, c) or in FP(r, p, clq) with $q \geq 1$, we have

$$(A * X) \oplus B = (1 + \sigma)[(A * X) + B], \qquad |\sigma| < r^{-(p-1)}$$

and

$$A * X = (1 - \rho)AX, \qquad\qquad 0 \leq \rho < r^{-(p-1)}.$$

Then

$$\tilde{Y} = (1 + \sigma)[(1 - \rho)AX + B],$$

so we may write

$$\tilde{Y} = \tilde{A}\tilde{X} + \tilde{B},$$

where

$$\tilde{A} = (1 + \sigma)A$$
$$\tilde{B} = (1 + \sigma)B$$
$$\tilde{X} = (1 - \rho)X.$$

Alternatively, we could write

$$\tilde{Y} = \tilde{A}X + \tilde{B}$$

where

$$\tilde{A} = (1 + \sigma)(1 - \rho)A$$
$$\tilde{B} = (1 + \sigma)B.$$

Thus, a backward error analysis shows that we have solved a problem close to the problem specified, so we can promise the user that the errors introduced in the calculation are equivalent to a small perturbation of the data, and we can give bounds for the perturbation. A backward error analysis has provided an answer for the question posed in Section 3.8.

We see that the subroutine LIN provides good answers for some input, but the error may be very large for other input. In fact, those cases in which LIN produces large relative error $(\tilde{Y} - Y)/Y$ are the cases in which AX and B have opposite signs and magnitudes approximately equal, so the computation

$AX + B$ involves the loss of several leading digits. But this is exactly the case in which the answer Y is extremely sensitive to changes in the input, that is, the case in which the problem (3.8.1) is ill-conditioned.

This situation is quite typical. It rather often turns out that a program produces good answers for some input and poor answers for other input. If we can perform a backward error analysis, we can view the computational errors as equivalent to perturbing the data, so the quality of the answer will depend on the condition of the problem. Since the condition of the problem may depend on the data, it is up to the user to worry about the condition of the problem he has posed.

3.10. EXAMPLES

In this section we shall consider two examples; one is ill-conditioned and the other is well conditioned.

First, consider the problem of computing $\cos x$ for x close to $\pi/2$. Now $\cos x = \sin(\pi/2 - x)$, so if $x \approx \pi/2$ we have $\cos x \approx \pi/2 - x$. Suppose that we are working in FP(10, 8, a) and x agrees with $\pi/2$ to six digits. Then a change of 1 in the eighth digit of x produces a change of about 1 in the second digit of $\cos x$. Thus, a slight change in x produces a large relative change in $\cos x$, so the problem is ill-conditioned from the point of view of relative error. This could also be verified from (3.7.12), which shows that the relative error σ in $\cos x$ due to a relative error ρ in x is approximately

(3.10.1) $\sigma \approx \rho x \tan x.$

By means of a backward error analysis, any errors introduced in computing $\cos x$ can be viewed as being equivalent to perturbing the argument x by a relative error ρ, and (3.10.1) shows that ρ will be magnified by approximately $x \tan x$.

We could compute $\cos x$ with good relative error even when x is close to $\pi/2$ if we were willing to use higher-precision arithmetic in the calculation. Thus, suppose that we are given an eight-digit number x in $S(10, 8)$ with $\pi/4 \leq x \leq \pi/2$. We could use double-precision arithmetic to subtract it from a 16-digit representation of $\pi/2$, producing a value for $\pi/2 - x$ which is good to at least eight digits. Then $\sin(\pi/2 - x)$ can be computed to about eight-digit accuracy in FP(10, 8, a), so we can produce a value for $\cos x$ with good relative error.

A typical cosine program first reduces the problem to the problem of computing the sine or cosine of an angle y with $|y| \leq \pi/4$. [See Fike (1968) or Cody (1971a).] If this reduction is performed using double-precision arithmetic, the program can produce good relative error, even for the cosine of an angle close to $\pi/2$. Since this reduction of the angle is a small part of the total work involved in computing the cosine, one might be willing to use double-precision arithmetic for the argument reduction and use single-precision arithmetic for the rest of the calculation. In this case, we can produce an answer with good relative error for the problem of finding the cosine of the argument x which was supplied to the cosine routine, but if x is close to $\pi/2$, the answer is still sensitive to any errors in x. Since most floating-point numbers that arise in computation are only approximations, one might argue that in those cases in which the problem is ill-conditioned it is not worthwhile doing extra work to produce a good answer for the cosine of the argument supplied to the routine. Which of these approaches one takes often depends on how easy it is to use double-precision arithmetic in the reduction of the argument. In the manufacturer-supplied subroutines for the FORTRAN and PL/I libraries in the IBM System/360, the single-precision SIN and COS routines use double-precision arithmetic in the reduction of the argument, but the double-precision routines do not use higher-precision arithmetic. This decision was clearly based on the fact that it was easy to include double-precision arithmetic in the single-precision program because the machine has double-precision operation codes, but on many models of the IBM System/360 there are no operation codes for extended-precision arithmetic, so it is not as easy to perform arithmetic with more than double-precision accuracy.†

As a second example, we consider the problem of computing $\sinh x$ for x close to zero. In many of the early implementations of FORTRAN, the function $\sinh x$ was not in the library, so the user had to compute it himself if he wanted to use it. He was likely to use the formula

$$(3.10.2) \qquad \sinh x = \frac{e^x - e^{-x}}{2},$$

which could be handled nicely as a *statement function* in FORTRAN. But when x is close to zero, both e^x and e^{-x} are close to 1, so many digits are lost in the subtraction in (3.10.2). For example, suppose that we are working in FP(10, 8, a) and that

$$(3.10.3) \qquad x = 10^{-5} \times .12345678.$$

Then $e^{\pm x} \approx 1 \pm x$, so

$$e^x \approx 1.0000012$$
$$e^{-x} \approx .99999877.$$

†Cody (1971a) describes a way to produce higher-precision accuracy in the argument reduction when the hardware does not provide higher-precision arithmetic. His approach uses techniques similar to those discussed in Chapter 5 for using single-precision arithmetic to program higher-precision arithmetic, but he simplifies the calculation by taking advantage of special features of the argument reduction problem.

Then (3.10.2) would yield the value

$$(3.10.4) \qquad \sinh x \approx .000001215.$$

But for x in (3.10.3), the approximation

$$(3.10.5) \qquad \sinh x \approx x$$

is good to almost 12 decimal digits, so the answer in (3.10.4) is only good to about two decimal places. However, the situation here is different from the situation with the cosine problem. We have a formula (3.10.5) which produces a good answer for small x, and this formula shows that the problem is well conditioned. Alternatively, we could use (3.7.12), which shows that the relative error σ in $\sinh x$ due to a relative error ρ in x is approximately

$$(3.10.6) \qquad \sigma \approx \rho x \frac{\cosh x}{\sinh x}.$$

For small x, $\cosh x \approx 1$ and $x/\sinh x \approx 1$, so the problem is well conditioned. Thus, we cannot blame the problem; rather, it is the algorithm (3.10.2) which is at fault. A typical program for $\sinh x$ might use a polynomial approximation for small $|x|$ and (3.10.2) when $|x|$ is large enough so that no digits are lost in the subtraction. [See Fike (1968).]

It is interesting to compare these two examples. In each case we could blame the poor relative error in the answer on a subtraction in which many leading digits were lost. But we found that the cosine problem was ill-conditioned and we either had to accept the poor relative error or use higher-precision arithmetic. On the other hand, the computation of the hyperbolic sine was well conditioned, and it was the algorithm which was at fault. It is typical that the condition of the problem can give us insight about whether or not it is worthwhile to seek another algorithm.

3.11. CHANGING THE PROBLEM

Suppose that we want to use a subroutine which requires us to supply the coefficients a and b of a linear function, written in the form

$$(3.11.1) \qquad f(x) = a(x + b),$$

which is to be used for $|x| \leq 1$. But suppose that we have been using $f(x)$ in the form

$$(3.11.2) \qquad f(x) = ax + c,$$

and we have approximations $\tilde{a}$ and $\tilde{c}$ for a and c. Let

(3.11.3) $\tilde{a} = .00000056$

                                 $\tilde{c} = .54325432,$

and suppose that $\tilde{a}$ and $\tilde{c}$ are each in error by less than $.5 \times 10^{-8}$. Then $\tilde{a}$ has two significant digits, while $\tilde{c}$ has eight. But, for $|x| \leq 1$, $f(x)$ can still be computed from (3.11.2) to an accuracy of about eight significant digits.

We have to supply a and b to the subroutine, but we do not know the value of $b = c/a$. The best that we can do is to supply $\tilde{a}$ and $\tilde{b}$, where

(3.11.4) $\tilde{b} = \tilde{c} \div \tilde{a},$

so the subroutine will evaluate the function $\tilde{f}(x)$ defined by

(3.11.5) $\tilde{f}(x) = \tilde{a} * (x \oplus \tilde{b}).$

Now $\tilde{b}$ is accurate to about two significant digits, and since $\tilde{b}$ is almost 10^6 while $|x| \leq 1$, $\tilde{b} \oplus x$ is also only accurate to two significant digits. Then $\tilde{f}(x)$ is the product of two factors, $\tilde{a}$ and $x \oplus \tilde{b}$, each of which is accurate to about two significant digits, so we might expect that $\tilde{f}(x)$ would be accurate to only two significant digits. We shall see that this badly overestimates the error. Also, it would suggest incorrectly that the division in (3.11.4) need be performed only to about two-digit accuracy.

We shall suppose that the computations in (3.11.4) and (3.11.5) are performed in FP(10, 8, c). From Theorem 3.4.5, we know that (3.11.5) may be written in the form

(3.11.6) $\tilde{f}(x) = (1 - p)[\tilde{a}(x + \tilde{b})], \qquad 0 \leq p < 2 \times 10^{-7},$

or

(3.11.7) $\tilde{f}(x) = (1 - p)\tilde{a}x + (1 + p)\tilde{a}\tilde{b}.$

Also, (3.11.4) may be written as

(3.11.8) $\tilde{b} = (1 - \sigma)\dfrac{\tilde{c}}{\tilde{a}}, \qquad 0 \leq \sigma < 10^{-7},$

so (3.11.7) becomes

(3.11.9) $\tilde{f}(x) = (1 - p)\tilde{a}x + (1 - p)(1 - \sigma)\tilde{c}.$

Thus,

                $\tilde{f}(x) = \hat{a}x + \hat{c}.$

where

(3.11.10)
$$\hat{a} = (1 - p)\tilde{a}$$
$$\hat{c} = (1 - p)(1 - \sigma)\tilde{c}.$$

The bounds for σ and p in (3.11.10) show that $\tilde{f}(x)$ is accurate to almost seven digits. In fact, for the data in (3.11.3), $\tilde{f}(x)$ is accurate to a few units in the eighth place.

Had we performed the division in (3.11.4) to only two digits of accuracy, the bound for σ in (3.11.8) would be 10^{-1} and $\tilde{f}(x)$ would be accurate only to about two digits. In fact, if we had used b instead of $\tilde{b}$ in (3.11.5), then (3.11.7) would read

$$\tilde{f}(x) = (1 - p)\tilde{a}x + (1 - p)\tilde{a}b.$$

But $\tilde{a}b$ would agree with c only to about two digits, so $\tilde{f}(x)$ would be good only to about two-digit accuracy.

Here a backward error analysis has shown that the computation in (3.11.4) and (3.11.5) produces a result $\tilde{f}(x)$ which is quite close to

(3.11.11)
$$f(x) \approx \tilde{a}x + \tilde{c},$$

and this was already found to be a satisfactory approximation for $f(x)$. Thus, we have changed the problem from (3.11.2) to (3.11.11) and then to (3.11.5) and have found that the answer $\tilde{f}(x)$ was a good approximation for $f(x)$. But a typical analysis based on significant digits would compare (3.11.5) with (3.11.1), and, finding that $\tilde{a}$ and $\tilde{b}$ agree with a and b only to two significant digits, it would conclude that $\tilde{f}(x)$ is good only to two-digit accuracy. This is a drawback of an analysis based on significant digits. A backward error analysis shows that (3.11.5) corresponds to a slight perturbation of the original data, and the fact that $\tilde{b}$ is not a good approximation for b is simply irrelevant.

As a second example of this situation, we shall consider the problem of finding the real root α of the equation

(3.11.12)
$$p(x) = ax^7 + bx + c = 0,$$

where a, b, and c are all positive, a and b are approximately 1, and $c \approx .1$. Since a and b have the same sign, it is easy to see that $p'(x)$ does not vanish for any real x, so (3.11.12) has exactly one real root. Also, if a and b are approximately 1 and $c \approx .1$, it is apparent that this root will be approximately $-.1$. We can see from (3.11.12) that $p(x)$ is not sensitive to changes in a when $|x| \approx .1$, and we could use this to show that the root α is not sensitive to changes in a. Alternatively, we could use (3.7.21) to see that α is well conditioned with respect to changes in a, b, and c and that it is much less sensitive to

changes in a than it is to changes in b or c. Thus, if we are given approximations $\tilde{b}$ and $\tilde{c}$ for b and c which are accurate to eight significant digits and an
approximation $\tilde{a}$ for a which is accurate to two significant digits, the root α
can be determined to almost eight significant digits.

Now suppose that we want to use a root finder which requires that the
lead coefficient of the polynomial be 1. We want to find the roots of the
equation

$$(3.11.13) \qquad q(x) = x^7 + dx + e = 0,$$

where $d = b/a$ and $e = c/a$. Naturally, we would compute

$$(3.11.14) \qquad \tilde{d} = \tilde{b} \div \tilde{a}, \qquad \tilde{e} = \tilde{c} \div \tilde{a},$$

and attempt to solve the equation

$$(3.11.15) \qquad \tilde{q}(x) = x^7 + \tilde{d}x + \tilde{e} = 0.$$

Now $\tilde{d}$ and $\tilde{e}$ agree with d and e only to about two places, and an analysis of
the sensitivity of the root α of (3.11.13) would lead us to expect that the root
$\tilde{\alpha}$ of (3.11.15) would agree with α only to about two places. But if the arithmetic in (3.11.14) is performed in FP(10, 8, c), we may write

$$\tilde{d} = (1 - \rho)\frac{\tilde{b}}{\tilde{a}}, \qquad \tilde{e} = (1 - \sigma)\frac{\tilde{c}}{\tilde{a}},$$

where $0 \leq \rho, \sigma < 10^{-7}$. Then (3.11.15) has the same roots as

$$(3.11.16) \qquad \hat{p}(x) = \tilde{a}x^7 + \hat{b}x + \hat{c} = 0,$$

where

$$\hat{b} = (1 - \rho)\tilde{b}, \qquad \hat{c} = (1 - \sigma)\tilde{c}.$$

Since $\hat{b}$ and $\hat{c}$ are only slight perturbations of b and c, we conclude that the
root $\tilde{\alpha}$ of (3.11.15) and (3.11.16) agrees with α to seven or eight places.

The situation described here arises rather often. There are many cases in
which the solution to a mathematical problem P involves transforming the
problem into another problem P' which has the same answer. But if we know
the data in P only approximately, we obtain a transformed problem $\tilde{P}$, which
we hope has approximately the same solution as P. Often this can be guaranteed by a backward error analysis which shows that $\tilde{P}$ has the same solution
as a problem obtained by a slight perturbation of the data in P. Our primary
concern is how close the solution to the problem $\tilde{P}$ is to the solution to P. The
fact that the data in $\tilde{P}$ differ from the data in P' is not important if we can
guarantee that we have not changed the answer much. Unfortunately, the fact

that the data in $\tilde{P}$ are not good approximations for the data in P' has sometimes been used erroneously to try to justify using less accurate arithmetic in
the solution of $\tilde{P}$, thereby needlessly contaminating the answer.

3.12. STATISTICAL ERROR ANALYSIS

So far, we have tried to produce bounds for the error. But sometimes it is
appropriate to consider the average error instead of the maximum error. As an
illustration, consider the computation of

$$(3.12.1) \qquad x = \sum_{i=1}^{n} a_i$$

in FP(10, 8, a). Suppose that the a_i are all positive and that we are given
approximations $\tilde{a}_i$ for the a_i, where the $\tilde{a}_i$ are three-digit integers. Then
$0 \leq \tilde{a}_i < 1000$. First, we suppose that the $\tilde{a}_i$ are the a_i rounded to the nearest
integer, so

$$(3.12.2) \qquad \tilde{a}_i = a_i + \epsilon_i, \qquad -\tfrac{1}{2} \leq \epsilon_i \leq \tfrac{1}{2}.$$

Then in place of (3.12.1), we attempt to compute

$$(3.12.3) \qquad \tilde{x} = \sum_{i=1}^{n} \tilde{a}_i.$$

Unless n is so large that $\tilde{x} > 10^8$, the computation of $\tilde{x}$ in (3.12.3) will be
exact, so we shall compute $\tilde{x} = x + \epsilon$, where

$$(3.12.4) \qquad \epsilon = \sum_{i=1}^{n} \epsilon_i.$$

Clearly, we have the bound $|\epsilon| \leq n/2$.

In a statistical error analysis, we assume that the ϵ_i are independent
random variables. Unless we have additional information about the a_i, it is
customary to assume that the ϵ_i are selected from a uniform distribution in
the interval $-\tfrac{1}{2} \leq \epsilon_i \leq \tfrac{1}{2}$. Then the expected value for ϵ in (3.12.4) is zero.
But, more important, the random variable ϵ has a probability density
function with a peak at zero. For example, the density function for $\epsilon_1 + \epsilon_2$ is
$p(x) = 1 - |x|$ for $|x| \leq 1$. The graph of $p(x)$ is shown in Figure 3.12.1.
Indeed, by appealing to the central limit theorem [see, for example, Mood
and Graybill (1963)] we conclude that the probability density function for ϵ
approaches that of a normal distribution as $n \to \infty$. Even for modest values
of n, the distribution for ϵ is close to a normal distribution. In fact, when the
ϵ_i are independent and uniformly distributed, the distribution of ϵ approaches

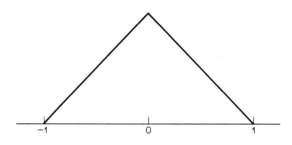

Figure 3.12.1

a normal distribution so rapidly that some of the early random number generators designed to produce numbers drawn at random from a normal distribution merely added 10 or 12 numbers selected independently from a uniform distribution. [See Muller (1959).]

Since the distribution for ϵ is approximately normal, the value $\tilde{x}$ computed in (3.12.3) is likely to be much closer to x than it is to either of the bounds $x \pm n/2$. (See Exercise 18.) In fact (see Exercise 17), it can be shown that if the ϵ_i are independent and uniformly distributed on the interval $(-\frac{1}{2}, \frac{1}{2})$, then the standard deviation of ϵ is $\sqrt{n/12}$. Therefore, for large n we would expect ϵ to be much closer to zero than to $\pm n/2$, so the error bounds are likely to give a severe overestimate of the error actually incurred. In addition to the error bounds (or perhaps instead of them) we might like to be given the expected value and standard deviation of ϵ.

Now suppose that the $\tilde{a}_i$ are produced by chopping the a_i instead of rounding them. Instead of (3.12.2), we have

$$\tilde{a}_i = a_i - \epsilon_i, \qquad 0 \leq \epsilon_i < 1,$$

and we write $\tilde{x} = x - \epsilon$, where ϵ satisfies (3.12.4). Clearly, the bounds for ϵ are given by $0 \leq \epsilon < n$. But now the expected value for ϵ is $n/2$ and the probability density function for ϵ has a peak at $n/2$. Consequently, we might wish to add $n/2$ to $\tilde{x}$ as a correction. But if we do not use such a correction, ϵ is likely to be on the order of $n/2$, that is, on the order of one-half the bound. Then using the bound instead of the average produces a difference of only one bit in our estimate of the error.

In general, when we use rounded arithmetic we may hope that the value we have computed is close to the expected value of the answer and that the error produced is considerably smaller than the error bound. Of course, we cannot guarantee that this is the case, but the discussion above suggests that this is very likely to occur. [See, however, Hartree (1949).] On the other hand, if we use chopped arithmetic instead of the rounded arithmetic, then we may expect the error to be within about one bit or so of the bound. For this reason, statistical analysis of error enjoys far more favor when the arithmetic used is rounded than when it is chopped.

We indicated that it is customary to assume that the error incurred when numbers are rounded is a random variable which is uniformly distributed between $-\frac{1}{2}$ and $\frac{1}{2}$ units in the last place and that the errors incurred at different steps in the calculation are independent. This has been supported by experiments; see, for example, Hull and Swenson (1966) and Henrici (1959). On the other hand, Hartree (1949) shows that this assumption is not always valid. His results have sometimes been cited as a reason for not using the statistical approach to error analysis, but even in the case he considers, the error incurred is still substantially smaller than the bound.

We have seen that in many cases it is relative error that is propagated. Thus, in Section 3.5 we found that for the calculation of

$$(3.12.5) \qquad x = \prod_{i=0}^{n} x_i$$

we produced a value $\tilde{x}$ which satisfied

$$\tilde{x} = x \prod_{i=1}^{n} (1 + p_i),$$

where the p_i are the relative errors introduced by the n multiplications. The bounds for the p_i are $0 \leq -p_i < r^{-(p-1)}$ when the arithmetic is performed in FP(r, p, c) and $|p_i| \leq \frac{1}{2} r^{-(p-1)}$ when it is performed in FP(r, p, R). Then $\tilde{x} = (1 + \sigma)x$, where the relative error σ satisfies

$$(3.12.6) \qquad 1 + \sigma = \prod_{i=1}^{n} (1 + p_i).$$

Since the p_i are small, we may study the approximation $\hat{\sigma}$ for σ given by

$$(3.12.7) \qquad \hat{\sigma} = \sum_{i=1}^{n} p_i.$$

First, we must consider the probability distribution for the p_i. To this end, let y be any nonzero real number and let $\tilde{y}$ be either $\bar{y}$ or $\overset{\circ}{y}$. We shall consider the relative error p given by $p = (\tilde{y} - y)/y$. Write

$$y = r^e m, \qquad r^{-1} \leq m < 1,$$

and $\tilde{y} = r^e \tilde{m}$. Here

$$\tilde{m} = m + \epsilon$$

and

$$(3.12.8) \qquad p = \frac{\epsilon}{m}.$$

As above, it is customary to assume that the absolute error ϵ is a uniformly distributed random variable. We usually want to perform an error analysis without looking at the intermediate results, so we would like to give an estimate for σ in (3.12.6) without having to look at each of the intermediate products developed in the evaluation of (3.12.5). Therefore, it is customary to assume that the mantissa m is also a random variable. Another justification of this point of view is based on thinking of the computation as being performed many times with different sets of data. Then we need to know the probability distribution for a *random mantissa*.

There is general agreement that the mantissas of floating-point numbers are not uniformly distributed. [See Hamming (1962), Pinkham (1961), or Knuth (1969).] Instead, it is customary to assume that they are distributed logarithmically, that is, that the probability density function is

$$(3.12.9) \qquad f(m) = \frac{1}{m \log_e r}, \qquad r^{-1} \leq m \leq 1.$$

This assumption is based on the following observations: First, this distribution reproduces itself under multiplication, but a uniform distribution does not. [See Hamming (1962) and Exercise 20.] A second justification is based on the fact that many of the numbers that arise in computation represent measurable quantities such as lengths, forces, etc., and it is reasonable to believe that the distribution of the mantissas of such quantities is independent of the units in which they are measured. Pinkham (1961) shows that this leads to the conclusion that the distribution of mantissas must be logarithmic. As indicated below, there is also empirical evidence to support this view. (Also, see Exercise 23.)

Let k be a positive integer which is less than r. The probability that the leading digit of m is less than k is the probability that $r^{-1} \leq m < k/r$. If we assume that (3.12.9) is the probability density function for m, we find that the probability that the leading digit of m is less than k is

$$(3.12.10) \qquad \int_{r^{-1}}^{k/r} \frac{dm}{m \log_e r} = \frac{\log_e k}{\log_e r} = \log_r k.$$

Thus, in the decimal system the probability that the leading digit is 1 is about .3, and the probability that it is 1, 2, or 3 is about .6. Hamming (1962) gives frequency counts for the leading digits of dimensioned physical constants he selected at random from the *Handbook of Chemistry and Physics*. His data agree quite well with (3.12.10).

In FP$(16, p, a)$, we are sometimes interested in the number of leading zeros in the binary representation of m. If we take k equal to 2, 4, 8 in (3.12.10), we find that the probabilities of having 0, 1, 2, or 3 leading zeros in the binary representation of m are all equal. This is validated by the experiments of Azen and Derr (1968), in which they analyzed samples of hexadecimal mantissas resulting from computation. (They also propose a simpler distribution for hexadecimal mantissas which has this property of producing a uniform distribution of the number of leading zeros in the binary representation of m.) Thus, we may expect that hexadecimal mantissas arising in computation will have three leading zeros in their binary representation about one-fourth of the time, whereas this would happen only about one-fifteenth of the time if they were uniformly distributed.

The importance of the distribution of mantissas lies in its effect on the relative error in (3.12.8). Since the probability density function in (3.12.9) has a peak at r^{-1}, random mantissas are much more likely to be close to r^{-1} than close to 1. Then for any absolute error ϵ in m, the relative error ρ is much more likely to be close to $r\epsilon$ than close to ϵ. Thus, we are more likely to experience the worst case for the relative error than we would be if the mantissas were uniformly distributed. (See Exercise 22.) This will affect our estimate of a quantity such as $\hat{\sigma}$ in (3.12.7).

EXERCISES

1. Show that even if the approximations $x \approx 10.13$ and $y \approx .9523$ are each accurate to within one-half a unit in the fourth digit, the approximation $xy \approx (10.13) \cdot (.9523)$ may be in error by more than five units in the fourth digit.

2. In FP($r, p, c l$1), show that in the subtract magnitude case we may indeed have

$$a \oplus b = (1 + \rho)(a + b),$$

where ρ is positive and close to $r^{-(p-1)}$.

3. Show that in FP($2, p, c$) for nontrivial values of p, the numbers $a * (b \oplus c)$ and $(a * b) \oplus (a * c)$ can differ by two units in the last place, even if b and c have the same sign.

4. In FP(r, p, c), find a bound for the relative error of the approximations

$$(a * b) \div a \approx b$$
$$a * (b \div a) \approx b.$$

5. Show that the following statements hold in FP(r, p, R) for interesting values of r and p.
    a. The associative law of addition does not even hold approximately.
    b. If a, b, and c all have the same sign, then

    $$a \oplus (b \oplus c) = (1 + \rho)[(a \oplus b) \oplus c],$$

    where $|\rho| \le r^{-(p-1)}$.
    c. $a * (b * c) = (1 + \rho)[(a * b) * c]$, where $|\rho| < 2r^{-(p-1)}(1 + 2r^{-(p-1)})$.

d. If $a * b = a * c$ and $a \neq 0$, then $b = (1 + p)c$, where

$$|p| < r^{-(p-1)}(1 + r^{-(p-1)}).$$

e. The distributive law does not even hold approximately.
f. For any a, b, c

$$a * (b \oplus c) = (1 + p)[a(b + c)],$$

where $|p| \leq r^{-(p-1)}(1 + r^{-(p-1)})$.

g. If b and c have the same sign, then

$$(a * b) \oplus (a * c) = (1 + p)[a * (b \oplus c)],$$

where $|p| < 2r^{-(p-1)}(1 + 2r^{-(p-1)})$.

6. Suppose that x and y are numbers in $S(r, p)$ and that we want to compute $z = x^2 - y^2$, performing the arithmetic in FP(r, p, clq) with $q \geq 1$. Show that the calculation

$$z = (x \oplus y) * (x \ominus y)$$

produces a value z with good relative error but that there is no reasonable bound for the relative error if we use the formula

$$z = (x * x) \ominus (y * y).$$

Which formula is more efficient from the point of view of computer time on the machine you are using?

7. Suppose that we are given a number x in $S(r, p)$ with $0 < x < 1$ and that we want to compute

$$y = 1 - \sqrt{1 - x},$$

performing the arithmetic in FP(r, p, clq) with $q \geq 1$. Assume that we have a library program SQRT which produces good relative error for the square root computation. Show that the formula

$$y = 1 - \text{SQRT}(1 - x)$$

may produce bad relative error when x is close to zero. By reversing the usual process of simplification of algebraic expressions, we may rationalize the numerator of $1 - \sqrt{1 - x}$ to produce

$$y = \frac{x}{1 + \sqrt{1 - x}}.$$

Show that the coding

$$y = x \div (1 + \text{SQRT}(1 - x))$$

always produces good relative error.

8. Suppose that we have N observations $x_1, x_2, \ldots, x_N$. We define the sample mean M and variance V by

$$M = \frac{1}{N} \sum_{i=1}^{N} x_i$$

$$V = \frac{1}{N} \sum_{i=1}^{N} (x_i - M)^2.$$

One readily shows that V is also given by

$$V = \frac{1}{N} \sum_{i=1}^{N} x_i^2 - M^2.$$

This formula is particularly convenient when we do not want to store all the x's. Unfortunately, it is much more sensitive to rounding error than the first formula is, particularly when V is substantially smaller than M^2.

a. Explain why the second formula for V is more sensitive to rounding error than the first formula is.

b. Perform the following computations for $N = 100$ and 1000 and $k = 0$, $1000, 2000, 3000$, and $10,000$. Let $x_i = k + i$ for $i = 1, 2, \ldots, N$. Compute M and V in both FP(16, 6, cl1) and FP(16, 14, cl1), using both formulas for V. Compare the results with the results obtained analytically from

$$\sum_{i=1}^{N} i = \frac{N(N + 1)}{2}$$

$$\sum_{i=1}^{N} i^2 = \frac{N(N + 1)(2N + 1)}{6}.$$

c. Perform the following computations for $N = 100$ and 1000 and $k = 500$, $100, 20, 10, 2$, and 1. Use a random number generator to generate N values x_i selected from a uniform distribution on the interval $500 - k$ to $500 + k$. Compute M and V in both FP(16, 6, cl1) and FP(16, 14, cl1) using each of the formulas for V. Explain the behavior of the results.

d. As above, suppose that we have N values x_i with sample mean M and sample variance V. Let

$$y_i = x_i - a, \qquad i = 1, 2, \ldots, N.$$

Then the y_i have a sample mean $M - a$ and a sample variance V. If a is close to M, using the second formula to compute V as the variance of the y's may be almost as attractive as using the first formula to compute V as the variance of the x's. (Why?) In part c, we would expect M to be close to 500. Repeat the computations in part c computing V as the variance of the y's, where

$$y_i = x_i - 500, \qquad i = 1, 2, \ldots, N,$$

and compare with the results of the previous computation.

9. Let $x = \sum_{i=1}^{N} x_i$ and $\tilde{x} = \sum_{i=1}^{N} \tilde{x}_i$, where

$$\tilde{x}_i = (1 + \rho_i)x_i, \qquad i = 1, 2, \ldots, N.$$

Suppose that all the x_i are positive. Prove that

$$\tilde{x} = (1 + \rho)x,$$

where $|\rho| \leq \max_{1 \leq i \leq N} |\rho_i|$.

10. Suppose that we want to compute $S = \sum_{i=1}^{N} x_i$, where the x_i are positive real numbers. Let $\tilde{x}_i$ be x_i chopped to $S(r, p)$, and set

$$\tilde{S}_1 = \tilde{x}_1$$
$$\tilde{S}_{i+1} = \tilde{S}_i \oplus \tilde{x}_{i+1}, \qquad i = 1, 2, \ldots, N - 1,$$

where the arithmetic is performed in FP(r, p, c) or FP(r, p, clq). Let $\tilde{S} = \tilde{S}_N$. Clearly $\tilde{S} \leq S$. Prove that S exceeds $\tilde{S}$ by less than N units in the last place of $\tilde{S}$.

11. Let A_i and B_i be positive numbers in $S(r, p)$ for $i = 1, 2, \ldots, N$, and let

$$S = \sum_{i=1}^{N} A_i B_i.$$

To compute an approximation $\tilde{S}$ for S, we set $\tilde{S}_0 = 0$ and let

$$\tilde{S}_i = \tilde{S}_{i-1} \oplus (A_i * B_i) \qquad i = 1, 2, \ldots, N,$$

where all operations are performed in FP(r, p, clq) with $q \geq 1$. Then we set $\tilde{S} = \tilde{S}_N$ and note that $\tilde{S} \leq S$. Using the result of Exercise 8, prove that S exceeds $\tilde{S}$ by less than N units in the last place of $\tilde{S}$. Show by an example that S may exceed $\tilde{S}$ by almost N units in the last place of $\tilde{S}$.

12. Let X be a number in $S(16, 6)$, so X is also in $S(16, 14)$. Let $Y = X^N$, where N is a positive integer, and suppose that we compute $\tilde{Y} = X**N$ in FP$(16, 14, cl1)$. Let $\hat{Y}$ and $\bar{Y}$ be the values of $\tilde{Y}$ and Y chopped to $S(16, 6)$. For each of the three methods for computing $X**N$ given in Section 3.6, find a bound for the relative error in the approximation $\tilde{Y} \approx Y$ and give an estimate of how large N must be before $\hat{Y}$ and $\bar{Y}$ can differ by more than one unit in the last place.

    If you have access to a machine which performs arithmetic in FP$(16, 14, cl1)$, select several values of X and N and compare the results obtained by computing $X**N$ in FP$(16, 14, cl1)$ using each of the three methods described in Section 3.6.

13. In Section 3.6 we gave a method for computing X^N requiring $m + k - 1$ floating-point multiplications. Show that X^{15} can be computed using only five floating-point multiplications, although $m + k - 1 = 6$.

14. Suppose that we want to compute $Y = X^{-N}$, where N is a positive integer and X is in $S(r, p)$. Find bounds for the relative errors in the approximations $A \approx Y$

and $B \approx Y$, where A and B are computed from the FORTRAN statements

$$A = 1./X**N$$
$$B = (1./X)**N$$

Assume that the arithmetic is performed in FP(r, p, clq) with $q \geq 1$.

15. Suppose that we want to compute the value of $p(x)$, where

$$p(x) = \sum_{k=0}^{N} a_k x^k.$$

It is customary to evaluate $p(x)$ using Horner's method:

$$p(x) = a_0 + x(a_1 + x(a_2 + \cdots + x(a_{N-1} + x \cdot a_N))) \cdots).$$

To code this in FORTRAN, we might store the coefficients in an array A, storing a_k in A($k + 1$), $k = 0, 1, \ldots$, N. Then we would use the FORTRAN coding

```
         P  =  A(N+1)
         DO 100 K  =  1,N
100      P  =  P*X+A(N+1−K)
```

Suppose that a_k and x are in S(r, p) and that we perform all the arithmetic in FP(r, p clq) with $q \geq 1$. Using a backward error analysis, show that we have computed

$$\tilde{p}(x) = \sum_{k=0}^{N} \tilde{a}_k x^k,$$

where $\tilde{a}_k = (1 + \rho_k)a_k$. Find the bounds for the ρ_k. If each $a_k \geq 0$, show that for $x > 0$

$$\tilde{p}(x) = (1 + \rho)p(x),$$

and find a bound for ρ.

16. Prove that if ϵ is a random variable which is uniformly distributed on the interval $-\frac{1}{2} \leq \epsilon \leq \frac{1}{2}$, then the variance of ϵ is $\frac{1}{12}$.

17. Suppose that $\epsilon = \sum_{i=1}^{n} \epsilon_i$ and that the ϵ_i are independent random variables which are uniformly distributed on the interval $-\frac{1}{2} \leq \epsilon_i \leq \frac{1}{2}$. Using the result of Exercise 16, show that the standard deviation of ϵ is $\sqrt{n/12}$.

18. Write a program to perform the following calculations for $N = 1000, 5000$, and 10,000 and $k = 2, 10, 20$, and 100. Use a random number generator to generate N random numbers selected from a uniform distribution on the interval $-\frac{1}{2} \leq x \leq \frac{1}{2}$ and treat these numbers as N/k observations of k random variables $\epsilon_1, \epsilon_2, \ldots, \epsilon_k$. Form $\epsilon = \sum_{i=1}^{k} \epsilon_i$, so we have N/k values for ϵ. Compute the sample mean and standard deviation for ϵ. Divide the interval $-(k/2) \leq x \leq k/2$ into 100 equal subintervals and obtain frequency counts for the number of values of ϵ which fall into each subinterval.

19. Let ϵ_1 and ϵ_2 be independent random variables which are uniformly distributed on the interval $-\frac{1}{2} \le x \le \frac{1}{2}$, and let $\epsilon = \epsilon_1 + \epsilon_2$. Prove that the probability density function for ϵ is $p(x) = 1 - |x|$ for $|x| \le 1$.

20. Let $z = xy$, where x and y are independent random variables.
  a. Show that if the mantissas of x and y have a logarithmic distribution, so does the mantissa of z.
  b. Show that if the mantissas of x and y have a logarithmic distribution, then the probability that postnormalization will be required in computing $x * y$ is .5.
  c. Find the probability density function for the mantissa of z if the mantissas of x and y are uniformly distributed on the interval $r^{-1} \le t < 1$.

21. It is often convenient to think of the mantissa of a floating-point number as a random variable. Find the mean and standard deviation of a random variable m, where
  a. m is uniformly distributed on the interval $r^{-1} \le x \le 1$.
  b. m has a logarithmic distribution on the interval $r^{-1} \le x < 1$.

22. In (3.12.8) we saw that the relative error ρ introduced by rounding or chopping a number to p digits is of the form ϵ/m. Suppose that ϵ and m are independent random variables and that $\rho = \epsilon/m$. Find the mean and variance of ρ if
  a. ϵ is uniformly distributed on the interval $-r^{-p} < x \le 0$ and m has a logarithmic distribution on the interval $r^{-1} \le x < 1$.
  b. ϵ is uniformly distributed on the interval $-\frac{1}{2}r^{-p} < x \le \frac{1}{2}r^{-p}$ and m has a logarithmic distribution on the interval $r^{-1} \le x < 1$.
  c. ϵ is uniformly distributed on the interval $-r^{-p} < x \le 0$ and m is uniformly distributed on the interval $r^{-1} \le x < 1$.
  d. ϵ is uniformly distributed on the interval $-\frac{1}{2}r^{-p} < x < \frac{1}{2}r^{-p}$ and m is uniformly distributed on the interval $r^{-1} \le x < 1$.
  e. To assess the importance of the distribution of the mantissas, compare the mean and variance computed in part a with the ones computed in part c. Similarly, compare the mean and variance computed in part b with the ones computed in part d.

23. Using the techniques described in Section 4.4, modify a matrix inversion program to extract the mantissa of every number which arises in the calculation. Divide the interval $r^{-1} \le x < 1$ into 20 subintervals and obtain frequency counts for the number of mantissas which fall in each subinterval. Compute the sample mean and standard deviation of the mantissas. Run the program for several matrices whose order is about 10.

4 EXAMPLES

4.1. QUADRATURE

In this chapter we shall consider some examples of floating-point computation. These examples will be referred to elsewhere in the book to indicate why certain approaches have been taken to implementation of features such as rounding, double-precision arithmetic, etc.

The first example we shall discuss is a quadrature problem. Suppose that we want to use Simpson's rule to compute an approximation for the value of

$$I = \int_A^B f(x)\,dx.$$

Select an even positive integer N and set $H = (B - A)/N$. Let $x_k = A + kH$, $k = 0, 1, 2, \ldots, N$. Then the approximation for I given by Simpson's rule is

$$(4.1.1) \qquad S_N = \frac{H}{3}[f(x_0) + 4f(x_1) + 2f(x_2) + 4f(x_3) + 2f(x_4) + \cdots$$
$$+ 4f(x_{N-1}) + f(x_N)].$$

If $f(x)$ has a bounded fourth derivative on $A \leq x \leq B$, it can be shown that

$$(4.1.2) \qquad I = S_N + 0(H^4).$$

[See, for example, Hildebrand (1956).]

For our example we shall try to compute

$$(4.1.3) \qquad I = \int_0^{\pi/2} \sin x\,dx.$$

123

Clearly $I = 1$, and we shall observe how well S_N approximates I as N increases. We shall perform the indicated computation using floating-point arithmetic, and we shall print S_N for $N = 2^k$, $k = 1, 2, \ldots, 16$.

For our first attempt we shall use the following rather naive FORTRAN program:

```
        B  =  1.570796
        DO   100  K  =  1,16
        N  =  2**K
        H  =  B/N
        X  =  0
        S  =  SIN(X)
        NN  =  N − 3
        IF(K.EQ.1) GO TO 200
        DO 100  I  =  1,NN,2
        X  =  X+H
        S  =  S+4.*SIN(X)
        X  =  X+H
  100   S  =  S+2.*SIN(X)
  200   X  =  X+H
        S  =  S+4.*SIN(X)
        X  =  X+H
        S  =  S+SIN(X)
        SIMP  =  S*H/3.
 1000   WRITE (  ) K,N,SIMP
```

This program was run in FP(16, 6, $cl1$), and the output is shown in Figure 4.1.1. The results are disappointing. The first three lines produce the sort of behavior we expected, but then the answer drifts below 1, with the error becoming worse and worse as the value of N increases.

To see how the precision of the arithmetic affects the result, the entire calculation was performed in FP(16, 14, $cl1$). For the IBM System/360, this meant that the variables had to be typed as double-precision, SIN had to be changed to DSIN, and B had to be set to 1.5707963267948966D0. The results of this run are shown in Figure 4.1.2. This is clearly an improvement, but we still observe the annoying drift of the answer below 1. The output suggests, correctly, that if we took still larger values of k the answer would continue to degrade. One might be tempted to dismiss this behavior with some vague comment about the growth of rounding error as the number of operations increase, but the regularity with which the answer degrades cries out for an explanation.

It turns out that the error which causes this systematic drift in the answer

K	N	SIMP	
1	2	1.00227833	
2	4	1.00013256	
3	8	1.00000668	
4	16	0.99999714	
5	32	0.99999583	
6	64	0.99999392	
7	128	0.99999267	
8	256	0.99997008	
9	512	0.99994498	
10	1024	0.99992907	
11	2048	0.99990332	
12	4096	0.99954379	
13	8192	0.99926186	
14	16384	0.99897254	
15	32768	0.99887526	
16	65536	0.99309123	Figure 4.1.1

K	N	SIMP	
1	2	1.002279877492210	
2	4	1.000134584974193	
3	8	1.000008295523968	
4	16	1.000000516684706	
5	32	1.000000032265000	
6	64	1.000000002016128	
7	128	1.000000000126000	
8	256	1.000000000007869	
9	512	1.000000000000481	
10	1024	1.000000000000017	
11	2048	0.999999999999983	
12	4096	0.999999999999901	
13	8192	0.999999999999835	
14	16384	0.999999999999767	
15	32768	0.999999999999677	
16	65536	0.999999999998398	Figure 4.1.2

arises in the accumulation of the sums S and X. In our program, X has been advanced by repetitively adding H, and S is the sum of terms of the form $2 \sin x$ and $4 \sin x$. First, consider S. For the final value of S, we have

$$S \cdot \frac{H}{3} \approx \int_0^{\pi/2} \sin x \, dx = 1,$$

so $S \approx 3/H$. Since $H = (B - A)/N$, this yields

$$S \approx \frac{6N}{\pi},$$

so the final value of S is about $2N$. Our last value for N was 2^{16}, so $2N = 20000_H$. Since S must grow to about this value, for many of the additions the six-digit hexadecimal number S will have the form $xxxxx.x_H$. But the terms added to S do not exceed 4, so the alignment of the radix points produces the situation

$$
\begin{array}{r}
xxxxx.x \\
x.xxxxx \\
\hline
xxxxx.x
\end{array}
$$

Since the numbers are positive, it is immaterial whether the low-order four digits of the number added to S are chopped before or after the addition—that is, it makes no difference whether the computation is performed in FP(16, 6, c), FP(16, 6, $cl1$), or FP(16, 6, $cl0$). Thus, we may view the computation as first chopping the terms $2 \sin x$ or $4 \sin x$, and then performing the addition exactly. This in turn may be viewed as using Simpson's rule to compute the integral of a function $\tilde{f}(x)$, where $\tilde{f}(x)$ drifts farther and farther below $\sin x$ as x increases, and it is obtained by chopping the low-order bits from $\sin x$. Since the discrepancy between $\tilde{f}(x)$ and $\sin x$ increases as N increases, this explains the systematic drift of the answer. It also shows that our program should not be sensitive to small errors introduced by the SIN routine. (See Exercise 2.)

A similar situation arises in the computation of X. X increases from 0 to B, and $H = B \div N$, so X will get to be substantially larger than H. As above, this means that the computed value for X will drift farther and farther below the correct value. Since $\sin x$ is monotone increasing in the interval, the error in X will tend to make the computed value for $\sin x$ too small, and this amplifies the effect of the errors in S. (See Exercises 2 and 8.)

The computation of X is easily corrected. Instead of advancing X by adding H, we may compute X as a multiple of H. For example, inside the loop we would use $X = I * H$ and $X = (I + 1) * H$, and similar computations can be used at the other places where X appears.

The computation of S is more troublesome. One approach is to accumulate the sum in double-precision. This could be coded so that S is the only double-precision variable in the program, and the only place where double-precision arithmetic is used is in the addition of single-precision terms such as $4 \sin x$ and $2 \sin x$ to S. Figure 4.1.3 shows the results obtained by making this change along with the change in the computation of X mentioned above. This shows that we have isolated and corrected the cause of the drift in the answer. Exercise 4 suggests explanations for the small error still remaining in these results, but we have clearly corrected the major source of error which was responsible for the continual degradation of the answer as N increases.

Correcting the computation of S in the double-precision version of our

K	N	SIMP
1	2	1.00227833
2	4	1.00013256
3	8	1.00000668
4	16	0.99999869
5	32	0.99999809
6	64	0.99999809
7	128	0.99999839
8	256	0.99999809
9	512	0.99999809
10	1024	0.99999809
11	2048	0.99999839
12	4096	0.99999809
13	8192	0.99999809
14	16384	0.99999809
15	32768	0.99999839
16	65536	0.99999809

Figure 4.1.3

program is not as simple, unless we have still higher-precision arithmetic available. However, it may be accomplished either by using the results of Section 4.3 or by exploiting our knowledge of the size of the numbers in the problem. (See Exercise 7.)

We shall now look more carefully at the computation of the sum of terms having the same sign. Suppose that we have $N + 1$ terms t_i, which we may assume are all positive, and let

$$s_n = \sum_{i=0}^{n} t_i, \qquad n = 0, 1, \ldots, N.$$

We actually compute $\bar{s}_n$ instead of s_n, where $\bar{s}_0 = t_0$ and

$$\bar{s}_{n+1} = \bar{s}_n \oplus t_{n+1}, \qquad n = 0, 1, \ldots, N - 1.$$

Write

$$\bar{s}_n = r^{e_n} m_n, \qquad r^{-1} \leq m_n < 1.$$

We shall assume that the computation is performed in FP(r, p, clq), so we have

(4.1.4) $$\bar{s}_{n+1} = \bar{s}_n + t_{n+1} - \epsilon_{n+1},$$

where $0 \leq \epsilon_{n+1} < r^{e_{n+1}} r^{-p}$. It can be proved by induction that

$$\bar{s}_n = s_n - \delta_n,$$

where

(4.1.5) $$\delta_n = \sum_{i=1}^{n} \epsilon_i.$$

Since the t_i are all positive, we have $e_{n+1} \geq e_n$ for all n. Then

$$\delta_n < nr^{e_n - p},$$

so s_N is larger than $\bar{s}_N$ by less than N units in the last place.

This bound may be sharpened. Let $l = e_N - e_0$ and suppose that for $i = 0, 1, \ldots, l$ there are k_i values of n with $e_n = e_0 + i$. Then

(4.1.6) $$\delta_N < r^{e_N - p}[k_l + k_{l-1}r^{-1} + \cdots + k_0 r^{-l}].$$

This shows that we want s_n to grow as slowly as possible. Therefore, it is desirable to arrange the order of the terms so that the smallest terms are added in first. Ideally, we would like to have

(4.1.7) $$t_0 \leq t_1 \leq t_2 \leq \cdots \leq t_N.$$

In our problem we do not satisfy (4.1.7), because we alternate between the terms $4 \sin x$ and $2 \sin x$. However, since $\sin x$ is increasing as x varies from 0 to $\pi/2$, we do have some of the effect of starting with smaller terms and proceeding to larger ones. Exercise 3 suggests some experiments which show the effect of the order in which the terms are added.

The program we have been considering so far is quite inefficient. One of the nice features of Simpson's rule is that when the number of intervals is doubled, we need not recompute $f(x)$ at the points already used. For any value of N in (4.1.1), let

$$E = f(A) + f(B)$$

$$F_N = \sum_{i=1}^{N/2} f(x_{2i-1})$$

$$T_N = \sum_{i=1}^{(N/2)-1} f(x_{2i}).$$

Then the formula for Simpson's rule is

$$S_N = \frac{H}{3}[E + 4F_N + 2T_N].$$

But when we double the number of intervals, each of the points previously used will have an even subscript, while the new points will have odd subscripts. This yields

(4.1.8) $$T_{2N} = T_N + F_N.$$

Thus if we want to compute S_{2N}, it costs only a couple of extra multiplies and adds to compute S_N as well. The following program takes advantage of this fact:

```
         B  =  1.570796
         E  =  SIN(0)+SIN(B)
         T  =  0
         DO 1000  K  =  1,16
         N  =  2**K
         F  =  0
         DO 100  I  =  1,N,2
         X  =  I*H
 100     F  =  F+SIN(X)
         S  =  (E+4.*F+2.*T)*H/3.
         T  =  T+F
1000     WRITE (  )  K,N,S
```

Figure 4.1.4 shows the output from this program, and Figure 4.1.5 shows the output from a double-precision version of this program. Not only is the program more efficient, but we see that it produces somewhat better results than our earlier program. The results of Exercise 6 will show that most of this improvement is due to the change in the way the terms are added to S. Other ways to rearrange the addition of the terms in S are also explored in Exercise 6.

Quadrature problems often require the addition of many terms having the same sign, and the behavior we have observed is quite common. The problem stems from the fact that we are using chopped arithmetic, so all the errors ϵ_i in (4.1.5) have the same sign. This suggests that it might be advantageous to use rounded arithmetic. Then the ϵ_i in (4.1.5) would tend to compensate, so we would hope that δ would be close to zero. (See Section 6.2.) To investigate the effect of rounding, the original program was run with all calculations performed in FP(16, 6, $cl1$) except for the addition of the terms $4 \sin x$ and $2 \sin x$ to S. These additions were performed in FP(r, p, R). (The way this may

K	N	SIMP
1	2	1.00227833
2	4	1.00013256
3	8	1.00000572
4	16	0.99999809
5	32	0.99999774
6	64	0.99999714
7	128	0.99999458
8	256	0.99999237
9	512	0.99999076
10	1024	0.99997044
11	2048	0.99993259
12	4096	0.99990082
13	8192	0.99987566
14	16384	0.99954891
15	32768	0.99892169
16	65536	0.99839622

Figure 4.1.4

K	N	SIMP
1	2	1.002279877492210
2	4	1.000134584974193
3	8	1.000008295523968
4	16	1.000000516684706
5	32	1.000000032265001
6	64	1.000000002016128
7	128	1.000000000126000
8	256	1.000000000007873
9	512	1.000000000000490
10	1024	1.000000000000024
11	2048	0.999999999999987
12	4096	0.999999999999978
13	8192	0.999999999999972
14	16384	0.999999999999906
15	32768	0.999999999999763
16	65536	0.999999999999642

Figure 4.1.5

K	N	SIMP
1	2	1.00227833
2	4	1.00013256
3	8	1.00000668
4	16	0.99999934
5	32	0.99999809
6	64	0.99999678
7	128	0.99999553
8	256	0.99998885
9	512	0.99998218
10	1024	0.99997073
11	2048	0.99995357
12	4096	0.99992019
13	8192	0.99984580
14	16384	0.99969101
15	32768	0.99969035
16	65536	0.99992716

Figure 4.1.6

be coded is discussed in Section 6.3.) The results are shown in Figure 4.1.6. By comparing Figure 4.1.6 with Figure 4.1.1, we see that rounded arithmetic produces significant improvement in this problem, particularly when a large number of terms are used in (4.1.1).

A final run of the original program was made using the bias removal described in Section 6.4, and the results are shown in Figure 4.1.7. As above, the only changes from the original program were that bias removal was used in the addition of the terms 2 sin x and 4 sin x to S.

4.2. POWER SERIES

Power series are familiar to anyone who has taken a course in calculus, and they seem to provide a means of computing many important functions.

K	N	SIMP
1	2	1.00227833
2	4	1.00013256
3	8	1.00000668
4	16	0.99999934
5	32	0.99999839
6	64	0.99999714
7	128	0.99999583
8	256	0.99999428
9	512	0.99998093
10	1024	0.99997294
11	2048	0.99995232
12	4096	0.99992943
13	8192	0.99985152
14	16384	0.99969864
15	32768	0.99969608
16	65536	1.00030041

Figure 4.1.7

Naturally, we would be concerned about whether the series converges, and we may have learned that some series converge so slowly that we would not want to use them directly for computational purposes, even with the aid of a digital computer. An example of such a series is

$$\log_e 2 = \sum_{k=1}^{\infty} \frac{(-1)^{k-1}}{k}.$$

But the numerical problems involved in using power series are often ignored. To illustrate the numerical difficulties, we shall consider the following problem:

PROBLEM

Use the power series to compute e^x, e^{-x}, and $\sin x$ for $x = 1, 2, 4, 8, 16, 32, 64,$ and 128.

We recall that the power series for e^x is

(4.2.1) $$e^x = \sum_{k=0}^{\infty} \frac{x^k}{k!},$$

and that the series for $\sin x$ is

(4.2.2) $$\sin x = \sum_{k=0}^{\infty} \frac{(-1)^k x^{2k+1}}{(2k+1)!}.$$

It is well known that these series converge for all values of x.

The output for this computation is shown in Figure 4.2.1. We have printed the values produced when the functions were computed by using the power series in both single-precision and double-precision. Here *single-* and *double-*

COMPUTATION OF E TO THE X BY POWER SERIES

X	SINGLE		DOUBLE		LIBRARY	
1	0.27182789E	01	0.2718281828459043D	01	0.2718281828459045D	
2	0.73890495E	01	0.7389056098930648D	01	0.7389056098930651D	
4	0.54598007E	02	0.5459815003314419D	02	0.5459815003314424D	
8	0.29809497E	04	0.2980957987041726D	04	0.2980957987041728D	
16	0.88860890E	07	0.8886110520507864D	07	0.8886110520507872D	
32	0.78962152F	14	0.7896296018268048D	14	0.7896296018268069D	
64	0.62349641E	28	0.6235149080811561D	28	0.6235149080811617D	
128	0.38875452E	56	0.3887708405994546D	56	0.3887708405994597D	
-1	0.36787921E	00	0.3678794411714422D	00	0.3678794411714423D	
-2	0.13533521E	00	0.1353335283236612 6D	00	0.1353335283236612 7D	
-4	0.18314630E-01		0.1831563888873392D-01		0.1831563888873418D-	
-8	0.41966909E-03		0.3354626279215241D-03		-0.3354626279025118D-	
-16	-0.35498899E-02		0.1125300367023338D-06		0.1125351747192591D-	
-32	-0.62804331E	06	-0.1436968486492220D-03		0.12664165549094 17D-	
-64	0.32858773E	20	0.2698891917247520D	11	-0.1603810890548638D-	
-128	0.13160663E	48	-0.3475250814162216D	38	0.2572209372642415D-	

COMPUTATION OF SIN(X) BY POWER SERIES

X	SINGLE		DOUBLE		LIBRARY	
1	0.84147096E	00	0.8414709848078965D	00	0.8414709848078965D	
2	0.90929759E	00	0.9092974268256817D	00	0.9092974268256817D	
4	-0.75680238E	00	-0.7568024953079283D	00	-0.7568024953079282D	
8	0.98923141E	00	0.9893582466233531D	00	0.9893582466233818D	
16	-0.34140092E	00	-0.2879033166747263D	00	-0.2879033166650636D	
32	0.28767938E	06	0.5516431151002459D	00	0.5514266812416899D	
64	0.20472729E	21	0.1079105327719405D	11	0.9200260381967901D	
128	0.25310850E	47	-0.2784208600181733D	38	0.7210377105017319D	

Figure 4.2.1

precision refer to FP(16, 6, *cl*1) and FP(16, 14, *cl*1), respectively. For comparison, we have also printed the values produced by the corresponding FORTRAN library routines, that is, by EXP(X) and SIN(X). In each case we terminated the calculation when the term added did not change the sum. To avoid overflows in the factorials involved, each term was computed from the preceding one. For example, in the series for e^x the kth term is given by

$$(4.2.3) \qquad\qquad t_k = \frac{x^k}{k!},$$

and we compute t_{k+1} from

(4.2.4) $$t_{k+1} = \frac{xt_k}{k+1}.$$

A similar procedure is used in the series for $\sin x$.

In Figure 4.2.1 we see that the results are quite good for small x and that the results for e^x are reasonably good for all the arguments tested. But the series computations for e^{-x} and $\sin x$ produced ridiculous answers when x was large.

To try to understand what happened, we shall consider the computation of e^x and e^{-x}. In each case, the terms added or subtracted are of the form (4.2.3). For $x = 128$, the first few terms are

$$1.$$
$$128.$$
$$8193.$$
$$349525.\,33333\cdots$$
$$11181477.\,33333\cdots.$$

We see that these terms grow quite rapidly. In fact, (4.2.4) shows that they will continue to grow until $k + 1 = x$. Thus, the terms with largest magnitude are t_{127} and t_{128}. When $k \geq x$, the terms decrease in size. Schematically, this growth of the terms when x is large may be shown by

$$1.$$
$$xx.$$
$$xxxx.$$
$$xxxxxx.xxxxxx$$
$$\vdots$$
$$\vdots$$
$$xxxxxxx \cdots xxxxxx.xxxxxx \cdots$$
$$xxxxxx \cdots xxxxxx.xxxxxx \cdots$$
$$\vdots$$
$$\vdots$$
$$.000000 \cdots 000000xxxxxx \cdots$$

But we have retained only the high-order p digits of each term. In fact, in the computation of the terms the error may grow to a point where the last two or three digits in the term are in error. Thus, instead of t_k we have an approximation $\tilde{t}_k$, where we may write

(4.2.5) $$\tilde{t}_k = t_k + \epsilon_k$$

or

(4.2.6) $$\tilde{t}_k = (1 + \rho_k)t_k.$$

Then we attempt to compute

(4.2.7) $$\tilde{S} = \sum \tilde{t}_k = \sum t_k + \sum \epsilon_k.$$

Even if we ignore the additional errors introduced because we used floating-point arithmetic to add the $\tilde{t}_k$'s in (4.2.7), our answer will be in error by the sum of the ϵ_k's. In the computation of e^x for positive x, all the t_k are positive. Then the fact that each ϵ_k is small with respect to t_k implies that $\sum \epsilon_k$ is small with respect to $\sum t_k$. In fact, by Exercise 9 of Chapter 3 we have

$$\sum \tilde{t}_k = (1 + \rho) \sum t_k,$$

where $|\rho| \leq \max_k |\rho_k|$. Since $\tilde{t}_k$ has been computed using only $2k - 2$ floating-point multiplications, the results of Section 3.5 show that if the arithmetic is performed in FP($r, p, cl1$), then

$$|\rho_k| < (2k - 2)r^{-(p-1)}.$$

Thus, our computation should produce reasonably good values for e^x when $x > 0$.

To illustrate the problems which arise in the computation of e^x when $x < 0$, we shall consider the case in which $x = -128$. The correct value for e^{-128} is about 2.57×10^{-56}, so the number we are trying to calculate is extremely small. But the t_k in (4.2.3) with largest absolute value is t_{128}, which is approximately 1.37×10^{54}. In hexadecimal, this is approximately $.E4D_H \times 16^{45}$. Since our single-precision calculation was performed in FP(16, 6, $cl1$), we retained only the high-order six hexadecimal digits of the terms t_k. In fact, as we indicated above, the errors ϵ_k may affect the last two or three digits of these terms, so we may expect the ϵ_k to be on the order of, say, 16^{39} or 16^{40}. (See Exercise 9.) Thus, even if we performed the addition in (4.2.7) exactly instead of using floating-point arithmetic, we would still have an error

(4.2.8) $$\epsilon = \sum \epsilon_k,$$

where some of the ϵ_k's have a magnitude as large as 16^{39}. We have used chopped arithmetic, so all the ρ_k's in (4.2.6) are negative. Since the t_k's alternate in sign, this implies that the ϵ_k's also alternate in sign. Thus, there may be some tendency for the ϵ_k's in (4.2.8) to compensate. However, we would still expect ϵ to be of more or less the same order of magnitude as the largest ϵ_k, say 16^{38} or 16^{39}. Exercise 9 shows that this is indeed the case.

Thus, we see that when all the t_k's are positive, the errors may accumulate, but they will affect only the low-order digits of the answer. But when the presence of terms of opposite sign produces an answer which is many orders of magnitude smaller than the largest term, we can expect our error to be larger than the answer. The only way we can produce a good result in this case is to keep enough digits in each term so that each ϵ_k is small with respect to the final answer. Since the sixth hexadecimal digit of the answer for e^{-128} is the ninety-seventh hexadecimal digit of the largest term, and since the ϵ_k may affect the low-order two or three digits of t_k, this means that if we wanted our answer for e^{-128} to be good to about six hexadecimal digits, we should perform the calculation in FP(16, 100, $cl1$) instead of FP(16, 6, $cl1$).

The situation for $\sin x$ is similar, since $|\sin x| \leq 1$ while the term with largest magnitude is the same as for e^x.

We shall next consider the way the library programs compute these functions. For e^x it is customary to begin by dividing x by $\log_e r$ (or multiplying x by $\log_r e$). Let

$$(4.2.9) \qquad y = \frac{x}{\log_e r},$$

so

$$(4.2.10) \qquad e^x = (e^{\log_e r})^y = r^y.$$

If we write

$$y = I - F,$$

where I is an integer and $0 < F \leq 1$, we have

$$e^x = r^I r^{-F}, \qquad r^{-1} \leq r^{-F} < 1.$$

Then I is the exponent of the answer and r^{-F} is the mantissa, so we have reduced the problem to the computation of r^{-x} with $0 < x \leq 1$. [Further refinements are possible. See, for example, Fike (1968) and Clark, Cody and Kuki (1971).] Naturally we shall use floating-point arithmetic in this computation, so we set

$$(4.2.11) \qquad y = x \div L, \qquad L \approx \log_e r,$$

and

$$(4.2.12) \qquad \tilde{y} = J - \tilde{F},$$

where J is an integer and $0 < \tilde{F} \leq 1$. Then we compute $r^{-\tilde{F}}$ and insert the exponent J in the answer. (For $0 < \tilde{F} \leq 1$, $r^{-\tilde{F}}$ can be computed from the power series or from a polynomial or rational approximation.)

Similarly, we may reduce the argument of the sine function by using the fact that it is periodic. A crude approach would be to write

$$y = \frac{x}{2\pi}.$$

Then if $y = I + F$, where I is an integer and $|F| < 1$, we have

$$\sin x = \sin(2\pi I + 2\pi F) = \sin(2\pi F),$$

so we can reduce the problem to the computation of the sine of an angle with absolute value less than 2π. A better approach is to divide x by $\pi/4$, writing

$$y = \frac{x}{\pi/4}$$

and

$$y = I + F,$$

where I is an integer and $0 \leq F < 1$. We then write

$$I = 8I_1 + I_2,$$

where I_1 and I_2 are integers and $0 \leq I_2 < 8$. (I_2 is particularly easy to obtain if I is written in binary.) Then, depending on the value of I_2, we may reduce the problem to the computation of either the sine or the cosine of an angle with absolute value less than $\pi/4$.

There are other functions for which it is easy to reduce the arguments. [See, for example, Fike (1968).] But if no such techniques are available, the power series may be quite treacherous.

The reduction of the argument avoids the pitfalls we have just observed in trying to use the power series for e^{-x} and $\sin x$, and it also reduces the number of terms needed. After reducing the argument, most library programs will use a polynomial or rational approximation instead of the power series, but the primary motivation for doing so is to make the routines slightly faster. Once the argument has been reduced, the use of the power series would be feasible.

Power series arise in many places in applied mathematics. For example, there are techniques for obtaining a series solution for a differential equation. But we can see that we cannot use these power series blindly for numerical calculation, and it is quite annoying that so powerful a mathematical tool can misbehave so badly.

We shall now consider the opposite situation—the case in which the power series can help us avoid numerical difficulties. Suppose that we want to find the value of

$$(4.2.13) \qquad\qquad F(x) = f(x) - g(x)$$

for small values of x and that $f(0) = g(0)$. If $f(x)$ and $g(x)$ are continuous, then as $x \to 0$ we have $F(x) \to 0$. If the common value of $f(0)$ and $g(0)$ is $c \neq 0$, then for small values of x both $f(x)$ and $g(x)$ will be approximately c, so (4.2.13) will involve cancellation of leading digits and it will produce bad relative error. In this situation, we find it very hard to compute $F(x)$ with good relative error using (4.2.13). We have already seen an example of this problem in the discussion of

$$\sinh x = \frac{e^x - e^{-x}}{2}$$

in Section 3.10.

We shall consider an alternative to (4.2.13) to be used when x is small. Suppose that $f(x)$ and $g(x)$ can be expanded in power series about the origin:

$$(4.2.14) \qquad\qquad f(x) = \sum_{k=0}^{\infty} a_k x^k$$

$$(4.2.15) \qquad\qquad g(x) = \sum_{k=0}^{\infty} b_k x^k.$$

Then

$$(4.2.16) \qquad\qquad F(x) = \sum_{k=0}^{\infty} c_k x^k,$$

where

$$c_k = a_k - b_k, \qquad k = 0, 1, 2, \ldots.$$

If the radii of convergence of the series in (4.2.14) and (4.2.15) are R_1 and R_2, then the series in (4.2.16) surely converges for $|x| < \min(R_1, R_2)$. Since $f(0) = g(0)$, we see that $c_0 = 0$. Then

$$(4.2.17) \qquad\qquad F(x) = \sum_{k=1}^{\infty} c_k x^k.$$

Where x is small enough, the series in (4.2.17) will converge very rapidly and the first few nonvanishing terms in (4.2.17) will be a good approximation for $F(x)$. In fact, if x is small enough, the first nonvanishing term in (4.2.17) is a good approximation for $F(x)$. Examples of this approach are given in Exercise 15, and Exercise 18 gives examples of the use of the Taylor series expansion about a point other than the origin.

4.3. EXACT SUMS AND DIFFERENCES IN
FP(r, p, clq)

In this section we shall consider the problem of trying to produce exact sums and differences. That is, given A and B in $S(r, p)$, we would like to produce $A + B$ and $A - B$ instead of $A \oplus B$ and $A \ominus B$. As a first step in this direction, we shall prove the following theorem:

THEOREM 4.3.1

In FP(r, p, clq) with $q \geq 1$, if

$$A \geq B \geq \frac{A}{2} \geq 0,$$

then $A \ominus B$ is exact. That is,

(4.3.1) $A \ominus B = A - B.$

Proof. Let

$$A = r^e m, \qquad r^{-1} \leq m < 1$$
$$B = r^f n, \qquad r^{-1} \leq n < 1.$$

Since $A \geq B$, we have $e \geq f$. Clearly (4.3.1) holds when $e = f$, so we may assume that $e > f$. But

$$A \leq 2B \leq rr^f n < r^{f+1},$$

so $e = f + 1$ and $B = r^e n'$, where $n' = r^{-1} n$. Since $q \geq 1$, to perform the floating-point subtraction $A \ominus B$, we first form

$$\mu' = m - n'.$$

Let k be the number of leading zeros in μ'. Then $A \ominus B = r^g \mu$, where $g = e - k$ and $\mu = \overline{r^k \mu'}$. (See Section 1.8.) Now $B \geq A/2$, so $A - B \leq B < r^f$. Then μ' has at least one leading zero, so $k \geq 1$. This yields

$$\mu = \overline{r^k \mu'} = r^k \mu',$$

and (4.3.1) follows.

Then one readily proves

COROLLARY

In FP(r, p, clq) with $q \geq 1$, if A and B have the same sign and

$$2 \cdot |A| \geq |B| \geq \frac{|A|}{2},$$

then the operations $A \ominus B$ and $B \ominus A$ are exact.

We shall now address the more complicated problem of trying to represent the exact sum of two floating-point numbers A and B. We would like to represent this sum by two floating-point numbers S and T, where S contains the high-order p digits of $A + B$ and T contains those digits of $A + B$ which do not fit in S. Thus, we would like to find S and T in $S(r, p)$ such that

1. $A + B = S + T$.
2. S and T are nonoverlapping numbers. That is, if $T \neq 0$ then

   (characteristic of T) $\leq$ (characteristic of S) $- p$.

3. If $T \neq 0$, then S and T have the same sign.

But we may not be able to achieve all three of these objectives. For example, suppose that we are using an eight-digit decimal machine and that we have $A = 1$ and $B = -10^{-50}$. Then

$$A + B = .999999 \cdots 999,$$

so it would require 50 digits to represent $A + B$. But if S and T are two positive numbers in $S(10, 8)$, they can hold only 16 of these digits. Therefore, we cannot find S and T satisfying 1, 2, and 3. But if we set $S = 1$ and $T = -10^{-50}$, then S and T will satisfy 1 and 2.

Our approach will be to try to find S and T satisfying 1 and 2, and to see in what cases we can also guarantee that 3 holds. Throughout, we shall assume that

(4.3.2) $$|A| \geq |B|.$$

In some cases this causes no difficulty because we know which of the numbers has the larger absolute value. In other cases, we would have to perform a test and interchange A and B if (4.3.2) fails to hold.

We shall study the FORTRAN coding

$$S = A + B$$
$$T = B - (S - A)$$

and we shall assume that we do not encounter exponent overflow or underflow. Thus, we shall assume that (4.3.2) holds and set

(4.3.3)
$$S = A \oplus B$$
$$C = S \ominus A$$
$$T = B \ominus C,$$

where we assume that the arithmetic is performed in FP(r, p, clq) with $q \geq 1$.

It is clear that S and T satisfy 1, 2, and 3 if B is either 0 or $-A$, so we may assume that neither B nor S vanishes. Also, changing the sign of both A and B

will simply change the signs of both S and T, so it suffices to consider the case in which A is positive. We shall write

$$A = r^e m, \qquad r^{-1} \leq m < 1$$
$$B = r^f n, \qquad r^{-1} \leq |n| < 1$$
$$S = r^g \mu, \qquad r^{-1} \leq \mu < 1.$$

Here (4.3.2) implies that $e \geq f$.

We shall first consider the case in which A and B have the same sign.

THEOREM 4.3.2

Suppose that A and B are numbers in $S(r, p)$ having the same sign and that $|A| \geq |B|$. Let $S = A \oplus B$ and $T = B \ominus (S \ominus A)$, where the arithmetic is performed in FP(r, p, clq) with $q \geq 1$. Then

1. $S + T = A + B$.
2. If $T \neq 0$, then

(characteristic of T) $\leq$ (characteristic of S) $- p$.

3. If $T \neq 0$, then S and T have the same sign.

Proof. We may assume that $A \geq B > 0$. If $e - f \geq p$, then $S = A$ and $T = B$ and the theorem follows. Therefore, we may assume that $e - f < p$. Let $B = B_1 + B_2$, where B_1 contains the high-order $p - (e - f)$ digits of B and B_2 contains the remaining digits of B. We saw in Section 1.8 that when A and B have the same sign the operation $A \oplus B$ produces the same result in FP$(r, p, cl0)$ as it does in FP(r, p, clq) for $q > 0$. Thus, B_2 has no effect on $A \oplus B$, so

(4.3.4) $$S = A \oplus B_1.$$

The characteristic g of S is either e or $e + 1$, depending on whether or not a high-order carry is produced in the addition in (4.3.4). First, suppose that $g = e$. Then no digits are lost in chopping $A + B_1$, so the addition in (4.3.4) is exact. That is, $S = A + B_1$. Since S and A have the same exponent, the computation $S - A$ is exact. Thus, $S \ominus A = B_1$, so $T = B_2$. But these values of S and T satisfy 1, 2, and 3, so the theorem holds if $g = e$.

Suppose that $g = e + 1$. Then the low-order digit of the $(p + 1)$-digit sum $A + B_1$ had to be chopped to produce $A \oplus B_1$. Write

$$A + B_1 = S + D,$$

where $D = dr^{e-p}$ and d is a single-digit number in the base r. Clearly $A \leq S \leq 2A$, so by Theorem 4.3.1

$$C = S \ominus A = B_1 - D.$$

If D is not zero, it is a single-digit number within the p digits spanned by B, so C may be represented as a p-digit floating-point number with the same characteristic as B, although it may be unnormalized when it is written in this form. Then the operation $B \ominus C$ is exact, so

$$T = B - C = B_2 + D,$$

and 1 holds. Clearly T is nonnegative, so 3 holds. Finally, $D \leq (r - 1)r^{e-p}$ and $B_2 < r^{e-p}$, so

$$T < rr^{e-p} = r^{g-p}$$

and 2 holds. This completes the proof of Theorem 4.3.2.

We shall now consider the case in which A and B have opposite signs. As we saw above, we cannot demand that S and T have the same sign. To simplify the statement of the theorem, we shall restrict our attention to FP(r, p, $cl1$).

THEOREM 4.3.3

Suppose that A and B are numbers in $S(r, p)$ having opposite signs and that $|A| \geq |B|$. Let $S = A \oplus B$ and $T = B \ominus (S \ominus A)$, where the arithmetic is performed in FP(r, p, $cl1$). Then

1. $S + T = A + B$.
2. If $T \neq 0$, then

$$(\text{characteristic of } T) \leq (\text{characteristic of } S) - p.$$

Proof. As above, we may assume that $A \geq -B > 0$. If $e = f$, we find that $S = A + B$ and $T = 0$, and the theorem follows. Also, since we have only one guard digit, if $e - f \geq p + 1$, then $S = A$ and $T = B$, so 1 and 2 hold. Therefore, we may assume that

$$(4.3.5) \qquad 1 \leq e - f < p + 1.$$

Let B_1 be the high-order $p + 1 - (e - f)$ digits of B, and let B_2 be the remaining digits of B. If $B_2 \neq 0$, then B_2 is negative and

$$(4.3.6) \qquad |B_2| < r^{e-(p+1)}.$$

Clearly $S = A \oplus B_1$. Let

$$(4.3.7) \qquad A + B_1 = S + D,$$

where D contains those digits of $A + B_1$ which must be dropped when $A + B_1$ is chopped to p digits. Here $D \geq 0$. Let k be the number of places we must left-shift $A + B_1$ to postnormalize it. Then $g = e - k$, and if $D \neq 0$, we have

$$(4.3.8) \qquad D < r^{e-(p+k)}.$$

Now if $k > 1$, we know from Theorem 1.8.2 that $A \oplus B = A + B$, so $T = 0$ and the theorem follows. This is also the case if $k = e - f = 1$. Then we may assume that we have

(4.3.9) $k < 2$

(4.3.10) $k < e - f,$

and that (4.3.5) holds.

We shall begin by showing that the computation $C = S \ominus A$ is exact; that is,

(4.3.11) $C = S \ominus A = S - A.$

If $k = 0$, this follows from the fact that S and A have the same characteristic. If $k = 1$, then by (4.3.10) we have $e - f \geq 2$, so $A \geq r^{e-1}$ and $|B| < r^{e-2}$. But then $A \geq S \geq A/2$, so (4.3.11) holds by Theorem 4.3.1.

Now (4.3.11) and (4.3.7) yield

(4.3.12) $C = B_1 - D.$

First, suppose that

(4.3.13) $e - f < p + k.$

Then $f > e - (p + k)$, so (4.3.8) shows that $D < |B|$. Since B_1 is negative and D is positive, we see from (4.3.12) that

$$|C| = |B_1| + D < 2|B_1| \leq 2|B|$$

and

$$|C| \geq |B_1| > \frac{|B|}{2}.$$

Therefore, by the corollary to Theorem 4.3.1, the floating-point subtraction $B \ominus C$ is exact, so

(4.3.14) $T = B - C = B_2 + D.$

Then 1 follows from (4.3.7) and (4.3.14). Since $D \geq 0 \geq B_2$, the characteristic of T cannot exceed the larger of the characteristics of B_2 and D. Then, since $k < 2$, 2 follows from (4.3.6) and (4.3.8).

Finally, we suppose that (4.3.5), (4.3.9), and (4.3.10) hold but that (4.3.13) does not. Then

$$p + k \leq e - f < p + 1,$$

so $k = 0$ and

(4.3.15) $$e - f = p.$$

Thus, the mantissa of B is right-shifted p places before its absolute value is subtracted from the mantissa of A, and no postnormalization is required. This implies that

$$S = A \ominus B = A - r^{e-p}$$

and that

$$C = S \ominus A = -r^{e-p}.$$

Then

$$T = B \ominus C = r^{e-p} \oplus B.$$

Here B is negative, and from (4.3.15) we see that $f = e - p$. Then the computation of T is exact, and $T < r^{e-p}$, so 1 and 2 hold. This completes the proof of Theorem 4.3.3.

We shall also need the following result:

THEOREM 4.3.4

Suppose that A and B are numbers in $S(r, p)$ and that $|A| \geq |B|$. Let $S = A \oplus B$ and $T = B \ominus (S \ominus A)$, where the arithmetic is performed in $FP(r, p, cl1)$. Then the characteristic of T does not exceed the characteristic of B.

Proof. Let the characteristics of A and B be e and f, respectively. Clearly, $e \geq f$. First, suppose that A and B have the same sign. If $f \leq e - p$, then $S = A$ and $T = B$, so the theorem holds. Suppose that $f > e - p$. The characteristic of S is at most $e + 1$, so by Theorem 4.3.2 the characteristic of T is at most $e + 1 - p \leq f$, as asserted. Next, suppose that A and B have opposite signs. If $e - f \geq p + 1$, $S = A$ and $T = B$, so the theorem holds. Suppose that $f \geq e - p$. Then the characteristic of S is at most e, so by Theorem 4.3.3 the characteristic of T is at most $e - p$, and the theorem follows.

4.4. DISMANTLING FLOATING-POINT NUMBERS

The floating-point number is comprised or three parts: the sign, the characteristic, and the absolute value of the mantissa. From time to time we want to dismantle the number into these parts so that we can work with them separately. If we are coding in Assembler language, this is usually quite easy, since we can use a logical AND or shifts to remove those parts of the number which we do not want. But if we are using a higher-level language, such as FORTRAN or PL/I, it is often harder to obtain the parts of the number. We can always test the sign of the number with an IF statement, but these languages do not provide direct access to the mantissa or the characteristic. (In Chapter 11 we shall suggest that it would be desirable for higher-level languages to allow us to extract these parts of the number.)

Since the representation of the floating-point number varies from machine to machine, any FORTRAN or PL/I coding used to extract the characteristic or mantissa will be machine-dependent. We shall describe the coding for the IBM System/360, but the modifications necessary for other machines should be quite clear. We recall that the representation of the floating-point number on the IBM System/360 consists of the sign bit, followed by the characteristic, followed by the absolute value of the mantissa. Suppose that

(4.4.1) $$x = 16^e m,$$

where $|m| < 1$. The absolute value of the mantissa m is a fraction which occupies either 24 bits or 56 bits, depending on whether the number is in single-precision or double-precision. The characteristic is defined to be the exponent e plus 64, and it is stored as a seven-bit nonnegative integer.

To begin with, we shall consider the form in which we would like to obtain the characteristic and the mantissa. Usually we would like to have the characteristic represented as an integer. It is quite easy to obtain the exponent if we have the characteristic, and vice versa, so we shall address the problem of finding the characteristic. When we are coding in FORTRAN, we might want to represent the mantissa m as a floating-point number. That is, if x is given by (4.4.1), we might want to obtain

(4.4.2) $$y = 16^0 m.$$

Alternatively, we might want to represent $2^{24} m$ as an integer I. If we are coding in PL/I, there is also the possibility of representing the fraction m as a fixed-point binary number.

We shall consider the FORTRAN coding first. The basic approach is to use the EQUIVALENCE statement to allow us to treat the floating-point number as an integer or as a logical variable. Suppose that I and A are typed by default, so I is an integer and A is a single-precision floating-point number. Write

(4.4.3)   EQUIVALENCE (A,I)

which means that A and I refer to the same word in storage. If we write A = X, then X will be stored in this word and we can refer to it as I. A difficulty arises because the IBM System/360 represents floating-point numbers as "sign and true magnitude," but it uses the 2's complement representation for negative numbers in fixed-point. If we are interested only in the characteristic, we may write

(4.4.4)   A = ABS(X)

Then I will refer to a 31-bit positive integer in which the high-order seven bits are the characteristic of X and the low-order 24 bits are 2^{24} times the absolute value of the mantissa of X. We may store the characteristic of X in J by writing

(4.4.5)   J = I/2**24

If we use this value of J and write

(4.4.6)   K = I−J*2**24

the value of K will be $2^{24}|m|$. By testing the sign of X, we can use K to obtain the appropriate representation for the mantissa of X. For example, if we want to obtain y in (4.4.2), we may write

(4.4.7)   I = 2**30+K

(4.4.8)   Y = A

(4.4.9)   IF(X.LT.0) Y = −Y

[In place of (4.4.9) we could have used the SIGN function.]

If X is a double-precision number, we can still use (4.4.3), (4.4.4), and (4.4.5) to obtain the characteristic of X, since it is only necessary to work with the high-order word of X. But on the IBM System/360, double-precision numbers have 56-bit mantissas, while integers are restricted to 32 bits. Therefore, if we want to work with the mantissa m of a double-precision number X, we shall use the floating-point representation for m shown in (4.4.2). To obtain this representation for m, we type X, Y, and A double-precision. Then (4.4.3) means that I would refer to the high-order word of A. We must also change ABS in (4.4.4) to DABS. With these modifications, (4.4.3)–(4.4.9) will produce the desired representation for the mantissa of the double-precision number X.

We do not need to employ the same subterfuge in PL/I. Instead, we may use UNSPEC to convert the floating-point number to a bit string and SUBSTR to extract the substring we want. Thus,

SUBSTR(UNSPEC(X),2,7)

is a bit string seven bits long beginning at the second bit of X, so it is a bit string comprised of the seven bits in the characteristic of X. To make it into an integer, we may concatenate it with enough zeros to make it the proper length, and then store it in the appropriate location in storage. For example, if I is typed by default as FIXED BINARY(15,0), we may write

UNSPEC(I) = '000000000'B || SUBSTR(UNSPEC(X),2,7);

As a second example from PL/I, suppose that we want to construct a floating-point number Y having the same mantissa (and sign) as X but having as its characteristic the low-order seven bits of I. To illustrate a somewhat different approach, suppose that B is declared to be a bit string of length 32. We may write

B   = UNSPEC(X);
SUBSTR(B,2,7) = SUBSTR(UNSPEC(I),10,7);
UNSPEC(Y) = B;

Of course many other approaches are possible in PL/I, all making use of the fact that we may treat the number as a bit string.

EXERCISES

1. To assess the effect of the rounding errors in the accumulation of the sum S in the quadrature problem of Section 4.1, perform the following calculations on a machine which performs arithmetic in either $FP(r, p, c)$ or $FP(r, p, clq)$:
    a. Perform the entire calculation in double-precision, except that S is typed single-precision.
    b. Perform the entire calculation in single-precision, except that S is typed double-precision. Use the original program which advances X by adding H. Compare your results with the results shown in Figure 4.1.3, which were produced by a program in which S was typed double-precision, but X was computed using $I * H$ and $(I + 1) * H$ instead of $X + H$.
    c. Perform the entire calculation in single-precision, but compute X using $I * H$ and $(I + 1) * H$ instead of $X + H$.

2. To study whether the quadrature program in Section 4.1 is sensitive to noise in the SIN routine, perform the following experiments in $FP(r, p, c)$ or $FP(r, p, clq)$. Use a version of the quadrature program in which S is double-precision but everything else is single-precision, and X is computed using $I * H$ and $(I + 1) * H$.
    a. Run the program three times with $SIN(X)$ replaced by $SIN(X) + Kr^{-p}$, $K = 1, 2, 3$.
    b. Run the program three times with $SIN(X)$ replaced by $(1 - Kr^{-p}) SIN(X)$, $K = 1, 2, 3$.

3. To study the effect of the errors in computing X in the quadrature program of Section 4.1, use the program to compute the following three integrals. Perform the calculations in FP(r, p, c) or FP(r, p clq). Use a version of the program in which S is double-precision but everything else is single-precision. Run each program twice, the first time computing X using $I * H$ and $(I + 1) * H$, and the second time computing X using $X + H$. The integrals to be computed are

a. $\int_0^{\pi/2} \sin x \, dx.$

b. $\int_0^{\pi/2} \cos x \, dx.$

c. $\int_0^1 e^{-x} \, dx.$

d. $\int_0^{1.57} \tan x \, dx.$

Explain the behavior of these programs.

4. The results in Figure 4.1.3 show that we have identified the major source of error in the quadrature program of Section 4.1. These results were produced by a program in which all calculations were performed in FP(16, 6, $cl1$) except for the accumulation of S, which was done in double-precision. In this program X was computed using $I * H$ and $(I + 1) * H$. We shall now try to identify the source of the small error still remaining in the calculation.

a. Because we used only a seven-digit representation for $\pi/2$, and because of the error introduced by conversion from decimal to hexadecimal, the value of B which is used by the program is not exactly $\pi/2$. Let $\tilde{B}$ be the value of B which is actually stored in the machine. Then our program is trying to compute $\int_0^{\tilde{B}} \sin x \, dx$ instead of $\int_0^{\pi/2} \sin x \, dx$, so the error in B introduces an error of

$$\int_{\tilde{B}}^{\pi/2} \sin x \, dx \approx \frac{\pi}{2} - \tilde{B}$$

in the final answer. First, obtain a bound for this error analytically. Then find the error in the approximation $\tilde{B} \approx \pi/2$ on the machine you are using by writing a simple FORTRAN program which obtains $\tilde{B}$ from the statement

$$B = 1.570796$$

and subtracts this value of $\tilde{B}$ from a double-precision representation of $\pi/2$.

b. In our program, H was obtained by dividing B by N. Since N is a power of 2, this division would be exact on a binary machine. But our runs were made in FP(16, 6, $cl1$), so the operation $B \div N$ need not be exact. Now B does not enter the program directly, so our program is really trying to compute $\int_0^{NH} \sin x \, dx$. Thus, the error in H produces the same effect as the error in B studied above. Find a bound for the error in the answer due to the error in the division $B \div N$. By looking at the value of $\tilde{B}$ in our problem, determine whether the division by N is exact, and if it is not exact, determine what the error in the division $B \div N$ actually is.

c. In order to use double-precision only in the accumulation of the sum S, S was declared to be double-precision and the FORTRAN statement which produces SIMP was changed to

$$\text{SIMP} = \text{SNGL(S)*H/3.}$$

If we had left the right-hand side of this statement as S*H/3., this multiplication and division would have been performed in double-precision. Obtain a bound for the error introduced by performing these two operations in single-precision. Run a version of the program which performs these operations in both single-precision and double-precision and print both answers.

d. As a final source of error, we consider the multiplications $I * H$ and $(I + 1) * H$ used to produce X. In our computation, these multiplications were performed in $FP(r, p, cl1)$. Rerun the program performing these multiplications in $FP(r, p, R)$. [The coding to produce arithmetic in $FP(r, p, R)$ is discussed in Section 6.3.]

e. Of the possible sources of error listed above, which had the largest effect on the answer?

5. Run the second version of the quadrature problem which exploits the relationship (4.1.8), but compute X by setting $X = H$ initially and advance it by $X = X + 2.*H$.

6. In the quadrature problem of Section 4.1, S is the sum of N terms t_i, $i = 1, 2, \ldots, N$. We consider the effect on the final answer of breaking up the sum S into several smaller sums. In each case, make the indicated changes in the original version of the program given in Section 4.1.

a. Let $N \geq 8$ and write

$$n_j = \frac{jN}{4}, \quad j = 0, 1, 2, 3, 4.$$

Then we may write

$$S = S_1 + S_2 + S_3 + S_4,$$

where S_j is given by

$$S_j = \sum_{i=n_{j-1}+1}^{n_j} t_i, \quad j = 1, 2, 3, 4.$$

Run the program accumulating these four sums S_j and then adding them to produce S.

b. Reprogram the problem to first compute and store the N values of t_i. By adding t_{2i-1} to t_{2i}, $i = 1, 2, \ldots, N/2$, we may reduce S to the sum of $N/2$ terms. By combining pairs of these terms, we reduce S to the sum of $N/4$ terms. Continuing in this way, we compute S. Run this program to see what effect it has on the final answer. Restrict the value of N to whatever storage size is convenient.

c. Reprogram the computation in part b to avoid having to precompute and store all the terms t_i.

d. Using Eq. (4.1.6), explain why these rearrangements of the computation of S improve the final answer.

7. We shall now try to produce good results with a double-precision version of the quadrature program of Section 4.1. For the computation of X, we can use $I * H$ and $(I + 1) * H$. But S is more difficult, since many versions of FORTRAN do not provide more than double-precision arithmetic. We shall suggest two different treatments of S below.

a. We think of S as the sum of N terms t_i, and we write each t_i as

$$t_i = t_i' + t_i'',$$

where t_i' represents the high-order digits of t_i and t_i'' is the rest of t_i. This may be accomplished by setting BIG $= r^l$ for a suitable value of l and setting

$$t_i' = (t_i \oplus \text{BIG}) \ominus \text{BIG}$$
$$t_i'' = t_i \ominus t_i'.$$

By a suitable choice of BIG,

$$S' = \sum_{i=1}^{N} t_i'$$

and

$$S'' = \sum_{i=1}^{N} t_i''$$

may be computed exactly. Determine the value of BIG to use and run the program in double-precision, computing S' and S'' and adding them to produce S.

b. First determine the exponent of 4 SIN(H). By exploiting the monotonic behavior of S as more and more terms are added, it is relatively easy to keep track of the exponent of S after each addition. Then we may accumulate the values of the k_i in Eq. (4.1.6). If we assume that the average error in the addition of a term to S is one-half a unit in the pth position of the new value of S, the expected value of the error is one-half of the value of δ_N given by (4.1.6). Run the program computing this correction and adding it to S before S is multiplied by $H/3$.

8. We shall study the effect of the computation of X on the results of the quadrature program of Section 4.1. Let $\tilde{X}$ be the value of X computed by repetitively adding H, and let $\hat{X}$ be the value of X computed using $I * H$ and $(I + 1) * H$. Run the original version of the quadrature program modified to compute X in the following ways:

a. $X = \tilde{X}$. (This is the program shown in Section 4.1.)

b. $X = \hat{X}$.

c. Use $X = \tilde{X}$ for the first half of the points, and then $X = \hat{X}$ for the rest of the points.

d. Use $X = \hat{X}$ for the first half of the points, and then $X = \tilde{X}$ for the rest of the points.

Explain the results.

9. Write a program to use the power series to compute e^{-x} for $x = 1, 2, 4, 8, \ldots,$ 128 without reducing the argument. For each value of x, print the following information in addition to that shown in Figure 4.2.1:

a. The number of terms used.

b. The value T of the term t_k with the largest absolute value.

c. We may estimate the error ϵ by comparing the result produced by the series with the result produced by the library routine. Print $\epsilon \div T$, where T is the absolute value of the largest term found in part b.

d. Let ρ be the relative error in the result produced by the power series. We may estimate ρ by dividing the ϵ computed in part c by the value for e^{-x} produced by the library routine. For each x, print this value for ρ. Also, proceeding analytically, find a bound for ρ for each x.

10. Another way to compute e^x for $x < 0$ is to represent it as $1/e^{|x|}$. For $x = 1, 2, 4, 8, \ldots,$ 128, use the power series to compute e^x without reducing the argument. Using this value for e^x, calculate $1 \div e^x$ and compare this result with the result produced by the library program for e^{-x}.

11. Suppose that we want to compute e^{-150} and e^{-170} from the power series without reducing the argument and that we want the answer to be accurate to six hexadecimal digits. By computing the term t_k in (4.2.3) with the largest absolute value and computing e^{-x} from the library program, determine how many hexadecimal digits of precision would be needed in the arithmetic to produce this result.

12. Write a program to compute e^{-x} using the power series without reducing the argument.

a. Use single-precision arithmetic to compute the terms t_k, but use double-precision arithmetic to accumulate the sum of these terms. That is, type SUM as double-precision and add the single-precision values of the terms t_k to the double-precision value of SUM. Does this improve the value of e^{-x} when x is large?

b. Use double-precision arithmetic to compute the values of the terms t_k, but accumulate the sum of the terms in single-precision. Does this improve the value of e^x when x is large?

13. We shall explore the error in the terms t_k in (4.2.3). First, proceeding analytically, obtain a bound for the relative error ρ_k in (4.2.6). Then use this bound to obtain a bound for the ϵ_k in (4.2.5) in terms of units in the last place of t_k. Write a program to compute the terms t_k in both single-precision and double-precision for $x = 128$. Let T and TT be the single-precision and double-precision values for t_k, respectively. Then $T \ominus TT$ and $(T \ominus TT) \div TT$ are

estimates for ϵ_k and ρ_k in (4.2.5) and (4.2.6) for the single-precision computation of t_k. For each of the terms t_k used in the single-precision calculation of e^{-128}, print this value for ρ_k and print this value for ϵ_k expressed in units in the last place of T. (To find the value of a unit in the last place of T, use techniques discussed in Section 4.4.)

14. Write a program to perform the argument reduction for sin x. That is, your program should take a single-precision floating-point number x and reduce the problem to the computation of the sine or cosine of an angle z with $|z| \le \pi/4$. Perform the following computations for $x = \pm 2^k$, $k = 1, 2, \ldots, 20$:

   a. Reduce the problem to the computation of the sine or cosine of an angle z with $|z| \le \pi/4$, and compute this sine or cosine using the power series.

   b. Reduce the argument as in part a, but compute the sine or cosine of z using the library program.

   c. Use double-precision arithmetic in the reduction of the argument, but compute the sine or cosine of z using the power series and single-precision arithmetic.

   d. Use double-precision arithmetic in the reduction of the argument, but compute the sine or cosine of z using the single-precision library program.

   e. Compute sin x using the library program directly.

15. Write a program to compute the following functions for small values of x, both by using the formulas directly and by using the power series. Compare the results for $x = 10^{-k}$, $k = 1, 2, \ldots, 25$. (Depending on the machine you are using, you may have to include tests to avoid underflows in the evaluation of the power series.) Also, for each function determine how small x must be before it suffices to use only the first term in the power series.

   a. $e^x - \sqrt{1 + x}.$

   b. $x \cos x - \sin x.$

   c. $\dfrac{e^x - e^{-x}}{2}$

16. For what values of x will the formula

$$\sinh x = \frac{e^x - e^{-x}}{2}$$

give a good relative error for sinh x?

17. Write a program to compute the value of the following functions for $x = 10^{-k}$, $k = 1, 2, \ldots, 25$. Each function should be computed in three ways: first, by using the formula directly; second, by using the power series; and third, by rationalizing the numerator as discussed in Exercise 7 of Chapter 3.

   a. $1 - \sqrt{1 + x}.$

   b. $\sqrt{1 + x} - \sqrt{1 - x}.$

18. Write a program to compute the value of the following functions, both by using the formula directly and by expanding the function in a power series about $x = 1$. Test the program for values of x close to 1.

   a. $x - \tan \dfrac{\pi}{4} x.$

b. $\log \dfrac{1}{x} + x \cos \dfrac{\pi}{2} x.$

19. Show that Theorem 4.3.3 need not hold when the arithmetic is performed in FP(r, p, clq) with $q > 1$ and $p + k \le e - f < p + q$.

20. Give an example to show that S and T in Theorem 4.3.3 may have opposite sign and that

$$(\text{characteristic of } T) = (\text{characteristic of } S) - p.$$

21. Suppose that A and B are in S(r, p) and that we perform all arithmetic in FP($r, p, cl1$). As in Section 4.3, set

$$S = A \oplus B$$
$$T = B \ominus (S \ominus A).$$

To avoid the situation described in Exercise 20, we perform the following cleanup:

$$S' = S \oplus T$$
$$T' = T \ominus (S' \ominus S).$$

Prove that S' and T' satisfy
a. $S' + T' = A + B$.
b. If $T' \ne 0$, then

$$(\text{characteristic of } T') \le (\text{characteristic of } S') - p.$$

c. If S' and T' have opposite signs, then

$$(\text{characteristic of } T') \le (\text{characteristic of } S') - (p + 1).$$

Also, show that if we repeat the cleanup by setting

$$S'' = S' \oplus T'$$
$$T'' = T' \ominus (S'' \ominus S'),$$

then $S'' = S'$ and $T'' = T'$.

22. Show by an example that Theorems 4.3.2 and 4.3.3 may fail to hold if $|A| < |B|$.

23. Suppose that we want to find the intersection of the circles

$$x^2 + y^2 = A^2$$
$$(x - C)^2 + y^2 = B^2,$$

where $A \ge B > 0$ and $C > 0$. If

$$C - B \le A \le B + C,$$

then the circles intersect and the coordinates of the intersections are given by

$$x = \frac{A^2 - B^2 + C^2}{2C}$$

$$y = \pm\sqrt{A^2 - x^2}.$$

When the circles are nearly tangent, the value of y may be quite sensitive to errors in A, B, and C. However, we can produce a good solution for the problem if we assume that A, B, and C are given exactly as numbers in $S(r, p)$. x may be computed with good relative error by using the approach described in Exercise 6 of Chapter 3. To compute an accurate value for y, we must compute $A^2 - x^2$ with good relative error. First, show that

$$A^2 - x^2 = \frac{(A + B + C)(A + B - C)(A + C - B)(B + C - A)}{4C^2}$$

Then, using Theorem 4.3.1, show how the right-hand side of this expression can be evaluated with good relative error whenever the circles have an intersection.

24. Write a FORTRAN program which will add a single-precision number to a double-precision number and produce a double-precision answer. Use only single-precision arithmetic to accomplish this.

25. Use the technique developed in Exercise 24 to modify a double-precision version of the quadrature program of Section 4.1 so that the sum S is accumulated in twice double-precision arithmetic.

26. Suppose that X is a single-precision floating-point number with exponent e and mantissa m. That is, $X = r^e m$, where $|m| < 1$. Write a FORTRAN program to store the value of the exponent e in J. If the value of L is an integer in the range $0 \leq L < 128$, form the floating-point number Y whose characteristic is L and whose mantissa (including sign) is the same as the mantissa of X.

27. Solve Exercise 26 when X and Y are double-precision numbers.

28. If we are using FORTRAN on the IBM System/360, we can gain access to the eight-bit bytes of a number by using EQUIVALENCE statements with variables typed LOGICAL∗1. Write a FORTRAN program using this approach to extract the characteristic of a floating-point number.

29. Let x be a positive number in $S(r, p)$, and let x' be the next larger number in $S(r, p)$. That is, x' is the smallest number in $S(r, p)$ that is greater than x. Write a program in FORTRAN or PL/I to produce x'.

5 DOUBLE-PRECISION CALCULATION

We have been considering calculations performed in a system FP(r, p, a), but from time to time we want to perform part or all of a computation in higher-precision. Most computers used for scientific computing provide two or more different precisions, either through hardware operation codes or through subroutines. Thus, we may perform arithmetic in either FP(r, p, a) or FP(r, p', a), and since the higher-precision p' is usually about $2p$, these systems are referred to as single-precision and double-precision, respectively. In fact, these terms are used even if p' is not exactly $2p$. For example, in the IBM System/360, FP(16, 6, $cl1$) is called single-precision and FP(16, 14, $cl1$) is called double-precision. The actual precision associated with the terms single-precision and double-precision varies considerably from one machine to another. For example, double-precision arithmetic on the IBM System/360 is performed in the system FP(16, 14, $cl1$), but this is roughly the same precision as the system FP(2, 48, a), which is called single-precision on the CDC 6600.

Machines designed for scientific computing usually have hardware operation codes to perform arithmetic in at least one system FP(r, p, a). Some machines, such as the IBM 7094 and System/360, also have hardware operation codes to perform double-precision arithmetic. In other cases, for example, the IBM 7090 and the CDC 6600, the hardware provides operation codes which are helpful in programming double-precision arithmetic, but the double-precision arithmetic is actually performed by calling a subroutine.

Still greater flexibility has been provided by some variable word length machines. For example, the IBM 1620 had hardware operation codes to perform floating-point arithmetic in FP(10, p, a), where p could be any integer from 2 to 100.

5.1. PROGRAMS USING DOUBLE-PRECISION ARITHMETIC

As we saw in the quadrature program discussed in Section 4.1, it is often desirable to perform one or two arithmetic operations in a program with greater precision than is used in the rest of the program. In other problems, one may decide at the outset to perform all calculations in double-precision, so the problem will be solved in FP(r, $2p$, a) instead of FP(r, p, a).

There have been a few implementations of FORTRAN, especially those for variable word length machines, in which one can specify at the outset the precision to be used in the whole problem. But the situation in which we want to insert a few double-precision operations in a program which is otherwise single-precision arises often enough so that languages such as FORTRAN and PL/I usually implement double-precision in a manner designed to support this use. The commonest approach [see American Standards Association (1964)] is for each variable to have its precision defined independently, either by default or by an explicit declaration. The precision of a constant is determined by its appearance. In FORTRAN, any constant which contains a decimal point is considered to be a floating-point number. Its precision can be specified explicitly by writing it with an E or D exponent, where E designates single-precision and D designates double-precision. Thus, 2. and 2.E0 are single-precision numbers, while 2.D0 is a double-precision number. With many FORTRAN compilers, the precision of a constant which is written without an E or D exponent is determined by the number of digits it has. There is a number N, which depends on the machine, such that the constant will be considered to be a single-precision number if it has N or fewer digits, but it will be considered to be a double-precision number if it has more than N digits.† (Compilers may differ as to whether or not leading zeros should be counted in applying this rule.)

The precision of the operands determines the precision of the floating-point arithmetic which will be used. If both operands have the same precision, this precision will be used in the arithmetic and the result will have the same precision as the operands. But if one of the operands is single-precision and the other operand is double-precision, the single-precision operand will be extended to double-precision by appending zeros to it, the arithmetic will be performed in double-precision, and the result will be typed as double-precision.

There are slight differences in the details of the implementation of

---

†The original versions of the manufacturer-supplied FORTRAN compilers for the IBM System/360 used $N = 7$. However, later versions of these compilers consider any floating-point constant written without an exponent to be a single-precision number. The only way to make a constant double-precision is to use a D exponent.

compilers for different machines. We shall discuss the problems associated with writing double-precision programs for the manufacturer-supplied FORTRAN for the IBM System/360. Other compilers may treat these problems in a slightly different way.

To illustrate the problem of converting a program from single-precision to double-precision, we refer to the program for the quadrature problem discussed in Section 4.1. First, we must type each of the floating-point variables as double-precision. This means that each floating-point variable must appear in either a DOUBLE PRECISION statement or a REAL*8 statement. Here we are immediately faced with the clerical problem of assuring that we have not accidentally omitted a variable. It is very easy to slip up and produce a program in which almost everything is done in double-precision but which produces results that are good to only single-precision accuracy because we forgot to type one variable as double-precision. For example, suppose that we want to interchange A and B. This might be coded as

$$\text{TEMP} = \text{B}$$
$$\text{B} = \text{A}$$
$$\text{A} = \text{TEMP}$$

If TEMP is not used elsewhere in the program, we might overlook it and forget to type it as double-precision. Then this coding would store the high-order digits of B in A and store zeros in the low-order digits of A.

Many compilers produce a list of the variables used in the program. This list should be consulted to see if there are any variables which we forgot to type as double-precision. It would be convenient if the compiler provided a list of the variables sorted by type, so that we could easily spot any single-precision variables.

Next, we have to worry about converting the constants from single-precision to double-precision. This may require us to look up some constants again to find their values more accurately. For example, in the quadrature problem in Section 4.1, the value for $\pi/2$ had to be changed from 1.570796 to 1.5707963267948966. A more annoying situation concerns constants which can be specified with fewer than N digits. For example, if a formula calls for 23/17, instead of computing this fraction by hand we might write the statement

(5.1.1) $\text{C} = 23./17.$

and let the computer do the division. This takes a little more computer time, but if the computation is not inside a loop, it is not very expensive. But according to the rules of FORTRAN, the numbers 23. and 17. in (5.1.1) are single-precision numbers, so the division of 23 by 17 will be performed in single-precision. If C is typed double-precision, this quotient will be extended to double-precision by appending zeros to it before it is stored in C. Thus, even though C is typed double-precision, its value will be accurate only to single-precision. To produce a value for C which is good to double-precision accuracy, one of the constants in (5.1.1) must be changed to a double-precision number. The easiest way to accomplish this is to use a D exponent;† for example, we may write

$$C = 23.D0/17.$$

As the expressions become more complicated, the pitfalls become more subtle. For example, suppose that we want to set $Y = 23X/17$, where X and Y are typed double-precision. We might code this as

(5.1.2) $Y = 23./17.*X$

Since the implied order of the operations is left to right, 23 will be divided by 17 in single-precision, and then the quotient will be extended with zeros and multiplied by X in double-precision. Thus, Y will be accurate to only single-precision accuracy. But if we had written

(5.1.3) $Y = X*23./17.$

the entire calculation would have been performed in double-precision. That is, 23 would have been extended to double-precision and multiplied by X in double-precision, and then the result would have been divided by 17 in double-precision. Thus, (5.1.3) produces double-precision accuracy, while (5.1.2) produces only single-precision accuracy. In either case the rules of FORTRAN allow us to determine how the calculation will be performed. But, because we can use single-precision constants in expressions such as (5.1.3) without ill effects, we often neglect converting constants to double-precision when it is necessary. It might be better to discipline oneself to convert all floating-point constants to double-precision.

A final problem with constants concerns statements such as

(5.1.4) $X = .1$

FORTRAN considers .1 to be a single-precision number, so the conversion from decimal to the radix r of the machine will be performed to only single-precision accuracy. If X is typed double-precision, this single-precision value for .1 will be extended with zeros and stored in X. Although $.1_D$ may be expressed with one decimal digit, it requires infinitely many digits in the base

---

†With many FORTRAN compilers, we can force a constant to be double-precision by appending enough zeros to the right of the decimal point so that it has more than N digits.

r if r is a power of 2. For example,

$$.1_D = .1999999999 \cdots_H$$

Thus, the FORTRAN statement in (5.1.4) produces only single-precision accuracy on a machine such as the IBM System/360, even if X is typed double-precision. The obvious correction is to write .1 as .1D0.

Next, we have to worry about the functions used in the program. If we have used any FORTRAN functions, the names must be changed to refer to the double-precision versions of these functions. For example, in the quadrature problem in Section 4.1, we must change SIN to DSIN. (Some compilers, such as WATFOR, even require that the name DSIN appear in a double-precision statement.) Since a function such as SIN may appear at many points in the program, it is often simpler to do this conversion by using an arithmetic statement function such as

(5.1.5)   SIN(X) = DSIN(X)

at the beginning of the program.

We also have to worry about user-supplied functions. Suppose that we have coded a subroutine to compute a function $F(x)$, using the statement

(5.1.6)   FUNCTION F(X)

Clearly, the function subprogram must be changed to compute a double-precision value for F(X). In addition, we shall change the FUNCTION statement in (5.1.6) to

DOUBLE PRECISION FUNCTION F(X)

This tells the subroutine that it is to return a double-precision value as the answer. However, it does not tell the calling program to look for a double-precision value for F(X). To do this, F must be typed as double-precision in the calling program by means of a DOUBLE PRECISION or REAL*8 statement. Otherwise the calling program would take the high-order part of F and extend it with zeros. Similarly, if we have used (5.1.5) to change SIN to the double-precision sine routine, then SIN must appear in either a DOUBLE PRECISION or REAL*8 statement.

All subroutines called must be replaced by double-precision versions. Naturally, the appropriate variables in the subroutines must be typed double-precision, but this is especially important for arrays. Suppose that an array A which is typed as double-precision in the calling program is passed to a subroutine as an argument. If we neglect to type this array as double-precision in the subroutine, the subroutine will treat the low-order part of A(1) as A(2), etc., which is likely to produce a ludicrous answer.

Similarly, a floating-point variable or array which appears in COMMON must be given the same precision in all programs which refer to COMMON. Failure to do so would change the layout of COMMON. For example, suppose that we have used

COMMON   A(4),B(10)

where A and B are typed double-precision in the calling program. If a subroutine uses the same COMMON statement but types only B as double-precision, then A will be allowed only four words instead of eight words (four double words). Then the subroutine will think that B starts four words (two double words) earlier, so when it tries to refer to, say, B(6) it will actually be referring to B(4). This is likely to produce disastrous results. Thus, by affecting the layout of COMMON, the precision of A may affect a subroutine which does not use A explicitly.

Changing the layout of COMMON is one of several ways in which converting a program from single-precision to double-precision may affect the management of storage. A similar situation concerns the use of the EQUIVALENCE statement. For example, suppose that we have written

DIMENSION   A(10),I(10)
EQUIVALENCE   (A,I)

Then, for example, A(6) and I(6) occupy the same location. But if A is declared to be double-precision, then I(5) and I(6) occupy the same location as A(3). This change may cause the program to fail to execute correctly.

But by far the most serious problem with the management of storage arises if the conversion of the program from single-precision to double-precision causes us to use so much storage that the data no longer fit in main memory. In some cases, in a multiprogramming environment, this may be remedied by requesting more storage. But in other cases it may mean that the problem must be partitioned, using auxiliary storage such as tape, disk, or drum. This may change the flow of the problem, and it certainly is not a trivial change. In this case the management of storage might be a major reason for trying to avoid the use of double-precision. This is particularly true with some of the very fast machines which provide double-precision arithmetic that is almost as fast as single-precision. Indeed, on the IBM System/360 models 91 and 195, most of the double-precision operations take about the same amount of time as the single-precision operations do, and on the model 85, double-precision addition is faster than single-precision addition. With these machines, we might very well plan at the outset to use double-precision unless it would require us to use too much storage.

Thus, even though the language supports double-precision arithmetic, it may be a nontrivial task to convert a program from single-precision to double-precision. For the most part, the individual changes are easy enough to make, the only difficulty being that we may overlook a change that is necessary. But the problem becomes more formidable if converting to double-precision changes the management of storage, or if we have used some single-precision library subroutines for which no double-precision versions are available.

We shall now turn to the problem of inserting a few double-precision statements in an otherwise single-precision program. In the quadrature problem in Section 4.1, we saw the advantage of using double-precision to accumulate the sums. This is easily accomplished. If we type S as double-precision, then a statement such as

$$S \ = \ S + SIN(X)$$

compiles a single-precision evaluation of sin x, then extends this number to double-precision, and adds it to S using double-precision addition.

A more annoying problem concerns the accumulation of an inner product (x, y) defined by

(5.1.7) $$(x, y) = \sum_{i=1}^{n} x_i y_i.$$

This is an extremely important operation, which constitutes the inner loop of many matrix operations. There are sound reasons for evaluating (5.1.7) by forming the double-precision product of the single-precision numbers x_i and y_i and performing the addition in double-precision. In fact, there are some algorithms for which it is absolutely essential that this approach be used. [See the discussion of iterative refinement in Wilkinson (1963) or Forsythe and Moler (1967).]

First, consider the coding

```
        S = 0
        D0 100 I = 1,N
100     S = S+X(I)*Y(I) .
```

where S is typed double-precision. The result produced will depend on the particular implementation of FORTRAN we are using. In many cases, X(I) would be multiplied by Y(I) using single-precision multiplication, and then the result would be extended to double-precision and added to S using double-precision addition. It is then a question of how the product X(I)*Y(I) is extended to double-precision. On many machines—for example, the IBM 7094 and System/360—the single-precision multiply command automatically produces the double-precision product of the single-precision operands and stores it in two registers. When we are working in single-precision, we take the high-order part of this product and ignore the low-order part. Then in statement 100 above, we would like the compiler to add the double-precision value of X(I)·Y(I) to S using double-precision addition. (This is what the manufacturer-supplied FORTRAN for the 7094 did.)

The manufacturer-supplied FORTRAN for the IBM System/360 uses a different approach. Evoking the rule that the product of two single-precision operands is single-precision, it interprets X(I)*Y(I) to mean the high-order part of the product X(I)·Y(I), and it extends this with zeros before adding it to S with a double-precision add operation. Ironically, it takes two or three extra instructions to replace the low-order digits with zeros.

To produce the answer we want on the IBM System/360, we have to force one of the factors to be double-precision. This may be done by replacing statement 100 with the statement

(5.1.8)   100   S = S+DBLE(X(I))*Y(I)

Then X(I) and Y(I) will be extended to double-precision by appending zeros, the product X(I)*Y(I) will be computed using a double-precision multiply command, and it will be added to S in double-precision. This produces the answer we want, but it requires the execution of unnecessary instructions. This loss of efficiency is annoying because the calculation of an inner product (5.1.7) appears in the inner loop of many matrix programs.

A common mistake is to use the coding

                   100   S = S+DBLE(X(I)*Y(I))

instead of (5.1.8). This produces exactly the same result as we produced with our original coding. That is, the single-precision number X(I)*Y(I) is extended to double-precision by appending zeros to it before it is added to S.

Thus, we find that the rule that the result is single-precision whenever both operands are single-precision may be troublesome. It would be much more convenient if the rule stated the following: When the result of a single-precision operation is to be extended to double-precision, if the single-precision operation automatically produces a double-precision result, then the low-order digits are to be retained instead of being replaced by zeros.

Finally, we note that PL/I solves some, but not all, of the problems discussed in this section. The variables must still be converted to double-precision, but the concept of generic functions eliminates the problem of changing the names of functions. We do not have to change SIN to DSIN as we did in FORTRAN, because PL/I does not use the name DSIN. Instead, SIN(X) will call either the single-precision or double-precision sine routine depending on the precision of X, and the result will have the same precision as X has.[†]

[†]Some FORTRAN compilers also provide generic functions. The FORTRAN H Extended compiler for the IBM System/360 has a GENERIC statement which can be used to specify that the common function names are to be treated as generic names.

Converting the constants is still a problem in PL/I, and the problem is amplified by the fact that PL/I has no D exponent. The only way to force a constant to be double-precision is to use more than N digits. But in some respects, the way PL/I treats constants is helpful. For example, if X is double-precision, then .1*X produces the result we want. The constant .1 is treated as a fixed-point decimal number. When it is to be used in an operation where the other operand is a double-precision floating-point number, it will be converted to a double-precision floating-point number, and the radix conversion will be good to double-precision accuracy. FORTRAN would have converted .1 to only single-precision accuracy, so .1*X would be good to only single-precision.

Finally, we note that PL/I presents the same problem FORTRAN does in the computation of inner products.

5.2. IMPLICIT TYPING OF NAMES

We have seen that compilers provide us with the facility to produce double-precision calculation wherever we want it, thereby making it easy to insert a few double-precision operations in an otherwise single-precision program. But when we wanted to perform the entire calculation in double-precision, we found that we were faced with the clerical task of assuring that all floating-point variables had been declared to be double-precision. The IMPLICIT statement in FORTRAN is designed to alleviate this difficulty. It allows us to establish different conventions for the data type of names beginning with certain letters. Thus, if we wanted all floating-point variables to be double-precision on the IBM System/360, we would write

(5.2.1)   IMPLICIT   REAL*8   (A—H,O—Z)

This means that variables or functions whose names begin with one of the letters A through H or O through Z will be typed REAL*8 (double-precision), unless this implicit typing is overridden by an explicit declaration. If we wanted, say, TEMP to be single-precision but all other floating-point variables to be double-precision, we would use (5.2.1) along with

REAL*4   TEMP

Similarly, if only a few variables were to be typed double-precision, we might decide that we wanted all variables whose names begin with D to be typed double-precision. Then, in place of (5.2.1) we would use

IMPLICIT   REAL*8   (D)

It is clearly easier to use (5.2.1) than to type each double-precision variable explicitly by including it in a DOUBLE PRECISION statement. But by far the most important advantage of the IMPLICIT statement is that it eliminates the careless clerical errors of forgetting to type a variable as double-precision. Those FORTRAN compilers which have implemented the IMPLICIT statement make it much easier to perform the entire calculation in double-precision, although the IMPLICIT statement does not handle all the problems discussed in Section 5.1. We still have to change the names of the library functions, replacing SIN by DSIN, etc., and we still have to worry about the constants and the layout of storage. However, the IMPLICIT statement removes one common source of errors.

The DEFAULT statement, which has been implemented by some PL/I compilers, is similar to the IMPLICIT statement in FORTRAN. It allows us to specify the default attributes of variable names, so we may specify that all names beginning with A through H or O through Z are to be double-precision.

The FORTRAN H Extended compiler for the IBM System/360 provides even greater assistance when we want to write the entire program in double-precision. It has a feature called *automatic precision increase* which allows us to specify that everything that appears to be single-precision is to be "promoted" to double-precision. When this feature is used, the floating-point variables will be typed double-precision, the floating-point constants will be treated as if they were written with D exponents, and double-precision versions of the FORTRAN library programs will be called. While this approach is not foolproof, it eliminates most of the clerical problems that arise in the conversion of a single-precision program to double-precision.

5.3. ROUTINES TO PERFORM DOUBLE-PRECISION ARITHMETIC

We shall now consider how we can write subroutines to perform double-precision arithmetic on a machine which does not have double-precision operation codes in hardware. For example, in Section 5.4 we shall discuss a subroutine to multiply two double-precision numbers to produce a double-precision result. One approach would be to use only fixed-point arithmetic in the subroutine, just as we would if we wanted to program floating-point arithmetic in a machine which did not have any floating-point operation codes. This is a distinct possibility, and on some machines it might be the best way to proceed. But in this book we shall discuss only the programming of double-precision arithmetic using the single-precision operations. Thus, we shall assume that we have operation codes to perform arithmetic in $FP(r, p, a)$ and that we want to write subroutines to perform arithmetic in $FP(r, 2p, a)$.

At this point we shall introduce abbreviations for certain operations. For each of the arithmetic operations $\oplus$, $\ominus$, $*$, and $\div$, there are two operands and one result. We shall use the letters S and D to indicate whether these numbers are to be single-precision or double-precision. Thus single-precision operations in FP(r, p, a) will be called SSS operations, and full double-precision arithmetic in FP($r, 2p, a$) will be referred to as DDD arithmetic. However, we may want to refer to an operation which, say, multiplies two single-precision numbers to produce a double-precision product. This will be called an SSD multiply. Similarly, an SDD add is an operation which adds a single-precision number to a double-precision number to produce a double-precision result.

When we use floating-point arithmetic to code the double-precision operations, we represent a double-precision number as the sum of two single-precision numbers.† Thus, the double-precision number A is given by

$$(5.3.1) \qquad\qquad A = A_1 + A_2$$

where A_1 and A_2 are numbers in S(r, p) and are usually stored in consecutive words in memory.‡

In the representation of A in the form (5.3.1), several decisions must be made. First, since we are thinking of A_1 and A_2 as the high-order and low-order digits of A, it is common to require that they not overlap. That is, we require

$$(5.3.2) \qquad\qquad \text{characteristic } (A_2) \leq \text{characteristic } (A_1) - p.$$

Some implementations have required that equality hold in (5.3.2) unless $A_2 = 0$. If we make this requirement, A_2 may be unnormalized. Other implementations have required that A_2 be normalized, so the inequality may hold in (5.3.2). Also, we might wish to require that A_1 and A_2 have the same sign unless $A_2 = 0$, although some implementations have allowed A_1 and A_2 to have opposite signs.

After these decisions about the representation of A have been made, we may assume that the input to our routines has these specifications and we must guarantee that the output does also. Thus, the specifications for the representation of A in (5.3.1) represent a trade-off between the advantages of requiring, say, sign agreement in the input and the extra work required to produce this feature in the output. In discussing the programming of double-precision arithmetic, we shall often be rather vague about the exact specifi-

---

†Even when the subroutines for double-precision arithmetic are coded using only fixed-point operations, we still usually use two words to store double-precision numbers. But in this case, we might elect not to store a characteristic with the low-order digits, so the mantissa might be more than twice as long as the mantissa of a single-precision number.

‡In Section 5.7, we shall discuss how the compiler can be coerced into handling these numbers if it does not support the double-precision data type.

cations of A_1 and A_2, because the trade-offs depend on the details of the single-precision arithmetic available to us.

An annoying problem related to the representation in (5.3.1) arises when A_2 underflows but A_1 does not. If the double-precision arithmetic were performed by hardware, it might be implemented in such a way that it would ignore the characteristic of A_2. Then underflow in A_2 would not present a problem. This is also possible when the subroutines for double-precision arithmetic use only fixed-point operations. But when floating-point operations are used in the routines for double-precision arithmetic, A_2 must be a valid floating-point number. One approach would be to give A_2 a wrapped-around characteristic when it underflows and modify the double-precision arithmetic routines to handle operands of this sort. But it is far more common to set A_2 to zero when it underflows, thereby giving up double-precision accuracy.

5.4. DOUBLE-PRECISION MULTIPLICATION

Suppose that we are given double-precision representations

$$(5.4.1) \qquad \begin{aligned} A &= A_1 + A_2 \\ B &= B_1 + B_2 \end{aligned}$$

for A and B and that we want to produce a double-precision representation

$$C = C_1 + C_2$$

for the product of A and B. For simplicity, we shall assume that if the low-order part of A or B does not vanish, then it has the same sign as the high-order part and that the characteristics differ by at least p. We shall also assume that we have available an SSD multiply.

Since double-precision multiplication presents no difficulty if either A or B vanishes, we shall assume that $AB \neq 0$. Clearly

$$(5.4.2) \qquad AB = A_1 B_1 + A_1 B_2 + A_2 B_1 + A_2 B_2,$$

and each of the four terms on the right-hand side can be computed exactly using an SSD multiply. Moreover, our assumption about sign agreement implies that the nonzero terms in (5.4.2) all have the same sign. Write

$$(5.4.3) \qquad \begin{aligned} A_1 &= r^e m_1, \qquad r^{-1} \leq |m_1| < 1 \\ B_1 &= r^f n_1, \qquad r^{-1} \leq |n_1| < 1. \end{aligned}$$

Then

$$(5.4.4) \qquad r^{e+f-2} \leq |A_1 B_1| < r^{e+f},$$

and A_1B_1 may be represented as a $2p$-digit number with exponent $e + f$. (Although it may be unnormalized when it is written with this exponent, its mantissa has at most one leading zero.)

We may write

$$A_2 = r^{e-p}m_2, \qquad |m_2| < 1$$
$$B_2 = r^{f-p}n_2, \qquad |n_2| < 1.$$

Then we can represent A_1B_2 and A_2B_1 as $2p$-digit numbers with exponent at most $e + f - p$, and A_2B_2 as a $2p$-digit number with exponent at most $e + f - 2p$.

To illustrate the alignment of the terms on the right of (5.4.2), we take $p = 6$ and show A_iB_j/r^{e+f}. In each case we show the minimum number of leading zeros:

(5.4.5)

$$
\begin{aligned}
A_1B_1 &= .xxxxxx \quad xxxxxx \\
A_1B_2 &= .000000 \quad xxxxxx \quad xxxxxx \\
A_2B_1 &= .000000 \quad xxxxxx \quad xxxxxx \\
A_2B_2 &= .000000 \quad 000000 \quad xxxxxx \quad xxxxxx
\end{aligned}
$$

We must combine these four terms and preserve the high-order $2p$ digits of the product. Since $m_1n_1 \geq r^{-2}$, we want either the first $2p$ digits to the right of the radix point in this sum, or else digits 2 through $2p + 1$. In any event, the low-order p digits of the $2p$-digit product A_2B_2 will not affect the result, so we may use an SSS multiply to form A_2B_2.

A commonly used approach is to develop only the first $2p$ digits to the right of the radix point in the products shown in (5.4.5). We shall call this "coarse double-precision" multiplication. With this approach, we ignore A_2B_2 and compute A_2B_1 and A_1B_2 using SSS multiplies. Our answer is then produced by adding these two products to the SSD product A_1B_1, using SDD adds. The digits dropped from each of the products A_1B_2, A_2B_1, and A_2B_2 have an absolute value less than r^{e+f-2p}, so the absolute value of the error is less than $3r^{e+f-2p}$. Thus, if the product does not require postnormalization, the error is less than three units in the last place. But if postnormalization is required, the bound for the error is $3r$ units in the last place. Similarly, we could define coarse triple-precision multiplication or coarse n-fold-precision multiplication. In general we shall call the arithmetic "coarse" if the error can be as large as a few units in the last digit or two.

By contrast, "clean double-precision multiplication" would develop an answer in which the error is less than one unit in the last place. To produce this result, we would not only have to use SSD multiplies to compute A_1B_2 and A_2B_1, but we would also have to perform the addition of the terms in

(5.4.2) in such a way that we achieve this accuracy. This can be accomplished by using the procedures for double-precision addition; they will be described in the next section, so we shall leave the details of the program to Exercise 3. An annoying problem that may arise in this program is that A_2B_2 and the low-order parts of A_1B_2 and A_2B_1 may underflow even though C_1 and C_2 do not. One way to handle this problem is to multiply A or B by a suitable scale factor before computing the product and then divide the answer by the scale factor. If we choose the scale factor be to a power of r, the scaling will not introduce any error and it can be performed by adjusting the characteristics of the numbers.

We often find that the work required to produce clean double-precision multiplication lies more or less midway between the work required for coarse double-precision multiplication and the work required for coarse triple-precision multiplication. That is, it may require a significant amount of extra work to gain a few extra bits of accuracy. This has often led the developers of routines for double-precision arithmetic to provide coarse double-precision arithmetic rather than clean double-precision arithmetic.

When double-precision arithmetic is provided by the hardware, the trade-offs may be quite different. In many cases, clean arithmetic may be produced at little extra cost. But on some machines, the hardware performs double-precision arithmetic using an algorithm similar to the one described above for programming it. In this case, the hardware designer may elect to provide coarse double-precision arithmetic.

We note that the crucial point is the availability of an SSD multiply and either an SSD or an SDD add. In some cases these operations are available in hardware, but in other cases they may have to be programmed. The operations which are available in hardware vary substantially from one machine to another. The IBM System/360 provides a full set of DDD operations for double-precision arithmetic. [Here double-precision means arithmetic in the system FP(16, 14, $cl1$).] On the other hand, most models of the IBM System/360 do not provide extended-precision arithmetic—that is, arithmetic in FP(16, 28, $cl1$). If we wanted to program extended-precision arithmetic, we would use the approaches discussed here, thinking of FP(16, 14, $cl1$) as single-precision and FP(16, 28, $cl1$) as double-precision. In this sense, we would have only SSS operations available. But the models 85 and 195 of the IBM System/360 have DDD operations for addition, subtraction, and multiplication in FP(16, 28, $cl1$). On the IBM 7090, the single-precision operations $\oplus$, $\ominus$, and $*$ were really SSD operations. One simply ignored the low-order word of the answer when arithmetic was to be performed in FP(2, 27, $cl27$). The CDC 6600 uses a somewhat different approach. To produce an SSD add, subtract, or multiply, one must execute two instructions—one to produce the high-order part of the answer and the other to produce the low-order part of the answer. Thus, an SSD multiplication requires two multiplications. But this is a pipeline machine with two multipliers, so we may be able to perform the two multiplications simultaneously.

Only rarely have machines provided an SDD add, although the IBM 7030 did. But it is quite common to have an SSD add.

5.5. DOUBLE-PRECISION ADDITION AND SUBTRACTION

We shall discuss only the programming of double-precision addition. Double-precision subtraction may be performed either by changing the sign of one of the operands or by making the obvious modifications in the procedures described below. However, we shall distinguish between the add magnitude and subtract magnitude cases.

We shall describe the use of SSD addition to program double-precision addition. If an SSD add operation is not provided by the hardware, it must be programmed. When the arithmetic is performed in $FP(r, p, cl1)$, this may be accomplished by the coding described in Section 4.3.

The details of a routine for double-precision addition depend on the representation of double-precision numbers and on the manner in which SSD addition is performed. We shall assume that the double-precision number A is represented as

$$(5.5.1) \qquad\qquad A = A_1 + A_2,$$

where A_1 and A_2 are two nonoverlapping, normalized, single-precision numbers. We shall also assume that arithmetic is performed in $FP(r, p, cl1)$ and that SSD addition is performed by using the coding described in Section 4.3. That is, to compute the SSD sum of A and B, we first interchange A and B, if necessary, to make $|A| \geq |B|$. Then we use the formulas (4.3.3). Thus SSD addition produces the results described in Theorems 4.3.2, 4.3.3, and 4.3.4.

We shall begin by discussing the programming of SDD addition. Let A have the representation (5.5.1) and let B be a single-precision number. We would like to produce a double-precision number

$$(5.5.2) \qquad\qquad S = S_1 + S_2,$$

which contains the high-order $2p$ digits of $A + B$. But since we do not require sign agreement in (5.5.1), this is hard to attain. Instead, we shall require that the error in the approximation $S_1 + S_2 \approx A + B$ be less than one unit in the last place of S_2, but we shall not require that $|S_1 + S_2| \leq |A + B|$.

We shall first consider the special case in which we assume that the characteristic of B is at least as large as the characteristic of A_1 and $B \neq -A$. Let $C_1 + C_2$ be the SSD sum of B and A_1. Since A_1 and A_2 do not overlap, there will be no overlap of A_2 with nonzero digits of C_2. Let D be the SSS sum $A_2 \oplus C_2$, and set $S_1 = C_1$ and $S_2 = D$. Then the SDD sum of A and B is given by S in (5.5.2).

Next, we consider the add magnitude case, which we define to be the case in which A_1 and B have the same sign. For this case, we can use a special algorithm which is slightly simpler than the general procedure for SDD addition. If the characteristic of B is less than the characteristic of A_1, the coding described above might produce an overlap of A_2 with nonzero digits of C_2. Then there could be a high-order carry in the addition $A_2 \oplus C_2$, and this carry could make S_1 and S_2 overlap. The overlap of S_1 and S_2 could be cleared up by combining S_1 and S_2 with an SSD add, but then we would have only $2p - 1$ digits of the answer. Although this answer might be acceptable for coarse SDD addition, we shall try to produce a cleaner result.

PROCEDURE 1

To produce the SDD sum of A and B, when A_1 and B have the same sign,

1. Let $C_1 + C_2$ be the SSD sum of B and A_2.
2. Let $D_1 + D_2$ be the SSD sum of C_1 and A_1.
3. Let E be the SSS sum $D_2 \oplus C_2$.
4. Let $S_1 = D_1$ and $S_2 = E$. Then S in (5.5.2) is the SDD sum of A and B.

To show that this procedure produces the desired result, we shall consider two cases. First, suppose that the characteristic of A_1 is at least as large as the characteristic of B. The first step reduces the problem to computing the SDD sum of $C_1 + C_2$ and A_1, where the characteristic of A_1 is at least as large as the characteristic of C_1. This is the special case considered above, and steps 2 and 3 are exactly the approach we used there. On the other hand, suppose that the characteristic of B is greater than the characteristic of A_1. Then the characteristics of B and A_2 differ by at least $p + 1$, so $C_1 = B$ and $C_2 = A_2$. Again, this is the special case considered above, and the arithmetic in steps 2 and 3 is exactly the computation used there to add B to A. Thus, the procedure handles the add magnitude case.

But Procedure 1 is not adequate for the subtract magnitude case. Suppose that

$$A_1 = r^e(1 - r^{-p})$$
$$A_2 = r^{e-p}(1 - r^{-p})$$
$$B = -r^{e+1}(r^{-1} + r^{-p}).$$

With our assumption about the way SSD addition is performed, step 1 in

Procedure 1 would produce $C_1 = B$ and $C_2 = A_2$. In step 2 we would obtain $D_1 = A_1 + B$ and $D_2 = 0$, so step 3 would produce $E = A_2$. But then

$$S_1 = A_1 + B = -r^{e+2-p}(r^{-1} + r^{-2})$$

and $S_2 = A_2$, so the characteristics of S_1 and S_2 differ by 2. That is, S_1 and S_2 overlap unless $p \leq 2$. (Further complications would arise with other definitions of SSD addition.)

For the general case, we use the following procedure:

PROCEDURE 2

To produce the SSD sum of A and B,

1. Let $C_1 + C_2$ be the SSD sum of B and A_1.
2. Let $D_1 + D_2$ be the SSD sum of C_2 and A_2.
3. Let $E_1 + E_2$ be the SSD sum of C_1 and D_1.
4. Let F be the SSS sum of E_2 and D_2.
5. Let $S_1 = E_1$ and $S_2 = F$. Then S in (5.5.2) is the SSD sum of A and B.

Clearly, the first two steps of Procedure 2 reduce the problem to computing the SSD sum of C_1 and $D_1 + D_2$. Unless $B = -A_1$, the characteristic of C_1 is at least as large as the characteristic of D_1, so steps 3 and 4 produce the desired result in either the add magnitude or subtract magnitude case. It is easy to see that this procedure also handles the case in which $B = -A_1$, so it is a general procedure for SSD addition. It is quite expensive, since it requires four additions, three of which are SSD additions. Therefore, we might be willing to accept a coarse version of SSD addition.

We shall now consider the programming of DDD addition. If the hardware provides SSD addition, it would be natural to use two SSD adds. But if SSD addition must be programmed, it is preferable to program DDD addition directly.

Suppose that $B = B_1 + B_2$ is a double-precision number.

PROCEDURE 3

To produce the clean DDD sum of A and B,

1. Let $C_1 + C_2$ be the SSD sum of A_1 and B_1.
2. Let $D_1 + D_2$ be the SSD sum of A_2 and B_2.
3. Let $E_1 + E_2$ be the SSD sum of C_2 and D_1.
4. Let $F_1 + F_2$ be the SSD sum of C_1 and E_1.
5. Let G be the SSS sum $F_2 \oplus E_2$.
6. Let H be the SSS sum $G \oplus D_2$.
7. Let $S_1 + S_2$ be the SSD sum of F_1 and H. Then S in (5.5.2) is the DDD sum of A and B.

We shall now show that this procedure produces the desired result. Clearly steps 1 and 2 reduce the problem to computing the DDD sum of the double-precision numbers $C_1 + C_2$ and $D_1 + D_2$. Also, it follows from Theorem 4.3.4 that D_2 does not overlap with either D_1 or any nonzero digits of C_2. Similarly, E_2 does not overlap with E_1 or C_1. Then the first four steps reduce the problem to finding the sum of four nonoverlapping numbers F_1, F_2, E_2, and D_2. This is accomplished by steps 5, 6, and 7. In the add magnitude case, we would not need to perform the addition in step 7. It would suffice to set $S_1 = F_1$ and $S_2 = H$. But in the subtract magnitude case, F_1 and H might overlap, so the SSD addition in step 7 is needed to clean up the answer.

A subject closely related to addition and subtraction is the comparison of two double-precision numbers A and B. One approach would be to use a DDD operation to compute $A - B$ and then perform a test to see whether $A - B$ is positive, negative, or zero. But other approaches may be used if we make certain assumptions about the representation of double-precision numbers. We have required that the high-order and low-order parts of the double-precision number be normalized and nonoverlapping. Suppose that in addition we require that the two parts of the number have the same sign unless the low-order part vanishes. Then the comparison may be simplified. We first compare the high-order parts of A and B. If they are unequal, the number with the larger high-order part is larger. If the high-order parts are equal, we compare the low-order parts. The result of this comparison determines whether the numbers are equal and which number is larger when they are unequal.

Of course, one reason for using a comparison is that it is likely to be faster than DDD subtraction. But another advantage is that on many machines it may be performed without underflow. The subtraction exposes us to underflow when A and B are both small and nearly equal.

5.6. DOUBLE-PRECISION DIVISION

If the hardware provides some, but not all, of the double-precision operations, division is likely to be omitted. For example, on the models 85 and 195 of the IBM System/360, there are hardware operation codes for extended-precision addition, subtraction, and multiplication but not for division. Thus, it is quite common to have to program double-precision division. We shall assume that the arithmetic is performed in FP(r, p, $cl1$) and that SSD addition produces the results described for the coding in Section 4.3.

We shall begin by discussing SSD division. That is, we want to develop the double-precision quotient of two single-precision numbers A and B. We use SSS division to compute $Q_1 = A \div B$. The remainder, which was discussed in Section 1.10, is given by

$$(5.6.1) \qquad\qquad R = A - BQ_1.$$

In Section 1.10 we saw that R can be represented as a single-precision number. On some machines, for example the IBM 7090, the single-precision divide command produces R as well as Q_1. But if the hardware does not provide R, we must compute it. To do so, we first let $C_1 + C_2$ be the SSD product BQ_1. Theorem 1.9.3 shows that C_1 can differ from A by at most $r - 1$ units in the last place, so we may use an SSS subtraction to compute the one-digit number $D = A \ominus C_1$. Since we know that R can be represented exactly as a single-precision number, we can compute R by using an SSS subtraction to form $R = D \ominus C_2$. Let $Q_2 = R \div B$. Then Q_1 and Q_2 are nonoverlapping numbers, so $Q_1 + Q_2$ is the desired representation for the double-precision quotient.

We shall now extend this approach to produce DSD division. Here DSD division means the division of a double-precision number $A_1 + A_2$ by a single-precision number B to produce a double-precision answer. We shall consider clean DSD division first, and then we shall see how this procedure can be shortened to produce a coarse version.

Let C_1 be the SSS quotient $A_1 \div B$, and compute the remainder $R' = A_1 - BC_1$ as above. Let $R = A - BC_1$, so

$$R = R' + A_2.$$

It is possible that we may require more than p digits to express the sum of R' and A_2, so we let $R_1 + R_2$ be the SSD sum of R' and A_2. Let C_2 be the SSS quotient $R_1 \div B$. Unlike the situation in SSD division, C_1 and C_2 may overlap, so we may have fewer than $2p$ digits of the quotient. (The extent of this overlap will be studied below.) We continue the process by computing the remainder $R'' = R_1 - BC_2$. Let S be the remainder $A - B(C_1 + C_2)$. Clearly

$$S = R - BC_2,$$

so

$$S = R'' + R_2.$$

If we wished to compute a triple- or quadruple-precision answer, we would compute S exactly and continue this process. But for DSD division, we merely compute $S_1 = R'' \oplus R_2$ using SSS addition, and then we form the quotient $C_3 = S_1 \div B$. To produce the answer, we form the double-precision sum of C_1, C_2, and C_3.

We shall now study the extent of the overlap of the C_i's. Without loss of generality we may assume that A_1 and B are positive. Let

$$A_1 = r^e m, \qquad r^{-1} \le m < 1$$
$$B = r^f n, \qquad r^{-1} \le n < 1.$$

We saw in Section 1.10 that the exponent of the quotient C_1 is $e - f + k$, where k is 0 if $m < n$ and 1 if $m \ge n$. The remainder R' may be written in the form

$$R' = r^{e+k-p}l'.$$

Here l' may be unnormalized, but it is a p-digit fraction and $0 \leq l' < n$. It follows that Q_1 and Q_2 in our procedure for SSD division do not overlap. But in our procedure for DSD division, the addition of A_2 to R' may make the mantissa of R' greater than n. In fact, the exponent of R_1 might be $e + k - p + 1$. But the largest R can be is

$$r^{e+k-p}l' + r^{e-p}(1 - r^{-p}) < r^{e+k-p+1}(r^{-1} + r^{-1}l') < r^{e+k-p+2}n.$$

It follows that the exponent of C_2 is at most $e + k - p + 2 - f$, so there is at most a two-digit overlap of C_2 with C_1. Similarly, there is at most a two-digit overlap of C_3 with C_2.

We shall now prove that there cannot be both a two-digit overlap of C_1 with C_2 and a two-digit overlap of C_2 with C_3. To this end, suppose that there is a two-digit overlap of C_1 with C_2. Clearly, this implies that A_2 is positive. Let $A_2 = r^g m_2$, where $r^{-1} \leq m_2 < 1$. Since $g \leq e - p$, we may write $A_2 = r^{e+k-p}m_2'$. Here m_2' is positive and less than 1, but it may have more than p digits to the right of the radix point. Then

$$R = r^{e+k-p}(l' + m_2'),$$

and since there is a two-digit overlap of C_2 with C_1, we must have

(5.6.2) $$l' + m_2' \geq rn.$$

Then $n + m_2' > rn$, so $m_2' > (r - 1)n$. Thus, the exponent of A_2 must be $e + k - p$, so we must have $k = 0$ and $g = e - p$. But then R can be represented as a $(p + 1)$-digit number, so R_2 is a one-digit number. We may write $R_2 = r^{e-2p+1}d$, where d is either zero or a one-digit number in the range $r^{-1} \leq d < 1$. From (5.6.2) we see that the mantissa of R_1 is at least n, so we may write $R'' = r^{e+2-2p}l''$, where $l'' < n$. Then

(5.6.3) $$S = r^{e+2-2p}(l'' + r^{-1}d),$$

so either $S < r^{e+2-2p}$ or else $S \leq r^{e+3-2p}l''$. But in either case there can be at most a one-digit overlap of C_3 with C_2. Therefore, the sum $C_1 + C_2 + C_3$ spans at least $3p - 3$ digits.

To see how large the error in the approximation $C_1 + C_2 + C_3 \approx A/B$ may be, we shall consider what would happen if the division process were continued to produce C_4. There can be at most a two-digit overlap of C_4 with C_3, so the error in $C_1 + C_2 + C_3$ is less than 1 in the $(3p - 5)$th digit. To obtain a better bound for this error, suppose that $C_1 + C_2 + C_3$ spans only $3p - 3$ digits. If there is a two-digit overlap of C_2 with C_3, there can be at most a one-digit overlap of C_3 with C_4. Suppose that there is a two-digit overlap of C_1 with C_2 and a one-digit overlap of C_2 with C_3. Then S is given by (5.6.3), and it may be expressed as a $(p + 1)$-digit number. Proceeding as above, one may show that there is at most a one-digit overlap of C_3 with C_4. Thus, in all cases the error in the approximation $C_1 + C_2 + C_3 \approx A/B$ is less than one unit in the $(3p - 4)$th digit.

The final operation in clean DSD division is combining C_1, C_2, and C_3. We use SSD addition to combine C_2 and C_3, and then we add C_1 to this sum using SDD addition. This is the special case of SDD addition in which the characteristic of the single-precision number is at least as large as the characteristic of the double-precision number.

For coarse DSD division, it is typical not to develop C_3. We first compute C_1 and R' as above; then C_2 is computed as $(R' \oplus A_2) \div B$ using SSS operations. Our answer $Q_1 + Q_2$ is the SSD sum of C_1 and C_2. There may be a two-digit overlap of C_1 and C_2, so we may have developed only $2p - 2$ digits. Moreover, if we were to develop C_3, it might overlap with C_2. Proceeding as above, it is easy to show that the error in the approximation $Q_1 + Q_2 \approx A/B$ is less than one unit in the $(2p - 3)$rd digit.

We now turn to DDD division. There are several ways in which this can be programmed, and we begin by discussing two approaches often used for coarse DDD division.

The first of these methods is based on the power series

$$(5.6.4) \qquad \frac{1}{1 + x} = 1 - x + x^2 - x^3 + \cdots.$$

This series converges when $|x| < 1$, and it converges very rapidly when x is small. Write

$$\frac{A_1 + A_2}{B_1 + B_2} = \frac{A_1 + A_2}{B_1[1 + (B_2/B_1)]}.$$

Using $x = B_2/B_1$ in (5.6.4), we have

$$(5.6.5) \qquad \frac{A_1 + A_2}{B_1 + B_2} = \frac{A_1 + A_2}{B_1}\left[1 - \frac{B_2}{B_1} + \left(\frac{B_2}{B_1}\right)^2 - \cdots\right].$$

If B_1 and B_2 do not overlap and B_1 is normalized, we have

$$\left|\frac{B_2}{B_1}\right| < r^{-(p-1)}.$$

Then we may surely ignore the terms in (5.6.5) after $(B_2/B_1)^2$, and we usually ignore this term as well. Thus, we use DSD division to divide $A_1 + A_2$ by B_1, and then we multiply this quotient by $1 - B_2/B_1$.

Let $C_1 + C_2$ be the result obtained when $A_1 + A_2$ is divided by B_1 using DSD division, and let D be the SSS quotient $B_2 \div B_1$. We form an approximation for

$$(C_1 + C_2)(1 - D) = C_1 + C_2 - C_1 D - C_2 D$$

by ignoring the term $C_2 D$ and using SSS multiplication to form $E = C_1 * D$. Then our answer will be obtained by subtracting E from $C_1 + C_2$ using an SDD operation.

A second method sometimes used for coarse DDD division first computes the reciprocal of B and then multiplies this reciprocal by A using DDD multiplication. We compute the reciprocal of B by using Newton's method to solve the equation

(5.6.6)
$$\frac{1}{x} - B = 0.$$

Newton's method for solving an equation

(5.6.7)
$$f(x) = 0$$

requires us to select a first approximation x_0 and form $x_1, x_2, \ldots$ using

(5.6.8)
$$x_{n+1} = x_n - \frac{f(x_n)}{f'(x_n)}.$$

Suppose that x_* is a solution of (5.6.7), and let $\epsilon_n = x_n - x_*$. It is well known that if $f(x)$ is twice differentiable, then

$$\epsilon_{n+1} = \frac{\epsilon_n^2 f''(\xi)}{2 f'(x_n)}$$

for some ξ lying between x_n and x_*. [See, for example, Hildebrand (1956).] If p_n is the relative error ϵ_n / x_*, we have

(5.6.9)
$$p_{n+1} = \frac{p_n^2 x_* f''(\xi)}{2 f'(x_n)}.$$

Applying this method to Eq. (5.6.6), we obtain from (5.6.8)

$$x_{n+1} = x_n(2 - Bx_n).$$

We shall use $1 \div B_1$ for our first approximation x_0, so x_0 is a good approximation for $1/B$. From (5.6.9) we find (see Exercise 12)

(5.6.10)
$$p_{n+1} \approx p_n^2.$$

Since x_0 is accurate to almost single-precision accuracy, x_1 will be accurate to almost double-precision accuracy. We shall take x_1 to be our approximation for $1/B$.

We now consider the details of this computation. Let C be the SSS quotient $1 \div B_1$, so

$$(5.6.11) \qquad\qquad x_1 = C(2 - BC).$$

Here $BC \approx 1$, and this approximation is good to almost p digits. Then it is reasonable to approximate the expression $2 - BC$ by $1 + D$, where D is a single-precision number. We want

$$1 + D \approx 1 + (1 - BC),$$

so we take D to be the high-order p digits of $1 - BC$. Let $E_1 + E_2$ be the SDD product of B and C. Then we need an SDS operation to compute D. This is accomplished by the SSS operations

$$D = (1 \ominus E_1) \ominus E_2.$$

We now replace (5.6.11) by

$$x_1 \approx C(1 + D) = C + CD.$$

Our approximation for $1/B$ will be the SSD sum of C and F, where F is the SSS product $C * D$. The analysis of the accuracy of this operation forms Exercise 12.

We now address the problem of producing clean DDD division. Our approach will be based on the procedure used above to produce the clean DSD quotient. In general terms, our procedure for the clean division of the double-precision number $A_1 + A_2$ by the double-precision number $B_1 + B_2$ will be

1. Let C_1 be the SSS quotient $A_1 \div B_1$.
2. Compute the remainder

$$R = (A_1 + A_2) - (B_1 + B_2)C_1.$$

3. Let R_1 be the high-order word of R.
4. Let C_2 be the SSS quotient $R_1 \div B_1$.
5. Compute the remainder

$$S = R - (B_1 + B_2)C_2.$$

6. Let S_1 be the high-order word of S.

7. Let C_3 be the SSS quotient $S_1 \div B_1$.
8. Let $Q_1 + Q_2$ be the double-precision approximation for $C_1 + C_2 + C_3$.

If R and S were computed exactly in steps 2 and 5, this process could be continued to produce a triple- or quadruple-precision answer.

In considering the details of this procedure, we may assume that A_1 and B_1 are positive. Let

$$A_1 = r^e m_1, \qquad r^{-1} \leq m_1 < 1,$$
$$B_1 = r^f n_1, \qquad r^{-1} \leq n_1 < 1.$$

As we saw in the case of DSD division, there may be an overlap of C_1 with C_2 and of C_2 with C_3. But DDD division differs from DSD division in that B_1 may be less than B, so we may have $C_1 > A/B$. If this happens, R will be negative.

We shall now consider the computation of R in step 2. First, let $D_1 + D_2$ be the SSD product of B_1 and C_1, and let E be the remainder $A_1 - B_1 C_1$. This is exactly the computation used in DSD division, so we know that E may be represented as a single-precision number and that it may be computed as

$$E = (A_1 \ominus D_1) - D_2$$

using SSS operations. Now C_1 has the exponent $e + k - f$, where k is 0 if $m < n$ and k is 1 if $m \geq n$. Then E may be written as $r^{e+k-p}m'$, where m' is a fraction with $0 \leq m' < n$. Now

$$R = E + A_2 - B_2 C_1.$$

Here A_2 has an exponent $g \leq e - p$, and the exponent of $B_2 C_1$ is at most $e + k - f + (f - p) = e + k - p$. Since we want only $2p$ digits of the quotient, we develop only a double-precision representation for R. Thus, we let $F_1 + F_2$ be the SSD product $B_2 C_1$ and let $R_1 + R_2$ be the double-precision representation for $E + A_2 - F_1$. Then

$$R = R_1 + R_2 - F_2.$$

We use this value of R_1 in step 4 to compute C_2.

To compute S in step 5, we form the remainder

$$G = R_1 - C_2 B_1$$

as above. Then

$$S = R - C_2 B = G + R_2 - F_2 - C_2 B_2.$$

Let H be the SSS product $C_2 * B_2$ and let S_1 be the sum $G \oplus R_2 \ominus F_2 \ominus H$ computed with SSS operations. We use this value of S_1 to compute C_3 in step 7.

With this procedure we cannot guarantee that we shall produce the properly chopped answer, since the remainder at the final step may be negative. The analysis of the overlap of C_1 with C_2 and of C_2 with C_3 forms Exercise 14, and the shortening of this procedure to produce coarse DDD division forms Exercise 16.

One problem that arises in this computation is underflow in the remainder. The characteristics of R_1 and S_1 are about p and $2p$ less than the characteristic of A_1. In fact, they may be even smaller unless the remainders are unnormalized. Then if A is small, these remainders may underflow. But if B is also small, the quotient may be on the order of 1, so it is annoying not to be able to compute it. We note that the quotient will underflow when $|A|$ is small and $|B|$ is extremely large. This suggests that we can perform a test, and when the quotient does not underflow we can scale A and B before division.

A minor irritant is that we may have $|A_1 \div B_1| > \Omega > |A/B|$. Then the quotient A/B does not overflow, but the computation of $C_1 = A_1 \div B_1$ does. For a coarse division routine, we might allow the answer to overflow when this happens. But for clean DDD division, we would have to test for this case and introduce scale factors.

Finally, we repeat our earlier warning. The details of a program to perform double-precision arithmetic will depend on the representation of the double-precision numbers and on the results produced by the SSD operations. The procedures discussed here show the general approaches, but modifications may be necessary for a specific implementation of them.

5.7. WRITING DOUBLE-PRECISION PROGRAMS WITHOUT LANGUAGE SUPPORT

We shall now treat the problem of writing a program using higher-precision arithmetic than the compiler supports. To be specific, suppose that we are using FORTRAN and that the compiler does not support the double-precision data type. (More generally, we may interpret double-precision to mean twice the highest precision supported by the compiler.)

We shall have to have subroutines which we can call to perform the double-precision arithmetic. For each arithmetic operation we shall write a call, such as

(5.7.1)   CALL ADD (A1,A2,B1,B2,C1,C2)

To avoid having to write six arguments in each of these calls, the subroutines might require that the double-precision number be thought of as a subscripted variable with dimension 2. Then instead of (5.7.1), the call would have the form

(5.7.2)   CALL ADD (A,B,C)

With this approach, the subroutine would require that the high- and low-order parts of the double-precision number be stored in adjacent memory locations. If we wanted A, B, and C to be 10-by-10 matrices, we would have to make them three-dimensional arrays dimensioned (2,10,10). The subscript which determines which part of the double-precision number we are referring to must be the first subscript, so that the two parts of the number are stored in adjacent locations. To add the (i, j) elements of A and B and store the result in the (i, j) position of C, we would have to write

CALL ADD (A(1,I,J),B(1,I,J),C(1,I,J))

It is clear that our program will be more tedious to write and more difficult to read than it would be if we could use the double-precision data type. In fact, the "higher-level language" no longer seems to be as high-level. But we can still use its indexing capabilities, and we can use DO loops and IF statements to control the flow of the program.

The introduction of an additional subscript to handle double-precision numbers may make the program rather cumbersome. There is a trick which can sometimes be used to simplify the representation of double-precision numbers. Suppose that we are using a version of FORTRAN which suppose the COMPLEX data type and that our program uses only real numbers. If we type our double-precision numbers as COMPLEX, the proper storage will allocated and the right arguments will be passed to subroutines. Since we are not using complex arithmetic in the program, we may be able to exploit the COMPLEX data type still further. Complex multiplication and division are often performed by subroutines, so the operations $*$ and $\div$ are compiled as subroutine calls. We could replace the routines which perform complex arithmetic by new routines which have the same names but which perform double-precision arithmetic. The double-precision multiplication could be coded as C = A$*$B instead of using a CALL statement. Unfortunately, complex addition and subtraction are so simple that they are usually compiled directly, so we would still have to code double-precision addition and subtraction in the form (5.7.2).

A more serious problem arises when we want to enter constants to double-precision accuracy. For example, suppose that we want to enter π to 32 decimal places but that the version of FORTRAN we are using supports only arithmetic in FP(16, 14, a). Suppose that we wrote

(5.7.3)
$$A1 = 3.141592653589793$$
$$A2 = .2384626433832795E\text{-}15$$

and then set

$$(5.7.4) \qquad\qquad PI = A1 + A2$$

Here PI would be accurate to only 14 hexadecimal digits, because the conversion of A1 is performed to that accuracy. To see that this is so, we shall consider the simpler case of a statement such as B = .2. The hexadecimal equivalent of the decimal number .2 is .333 $\cdots_H$, so the statement B = .2 will produce a value of B which is accurate to only 14 hexadecimal digits. Thus, the addition of A2 to A1 in (5.7.4) cannot compensate for the fact that the value of A1 produced by (5.7.3) is accurate to only 14 hexadecimal digits.

Instead of using (5.7.3) and (5.7.4), our approach will be based on the fact that integers less than 16^{14} may be entered exactly. Now $16^{14} \approx 7.206 \times 10^{16}$, so we may enter any 16-digit integer exactly. In place of (5.7.3) we write

$$A1 = 3141592653589793.$$
$$A2 = 2384626433832795.$$

Then the number we want to store in PI is

$$X = (A1 + A2/10^{16})/10^{15}$$

We compute X using double-precision arithmetic and store the value in PI.

The same approach could be used for input. Fortunately, the treatment of input is usually simplified by the fact that the numbers are shorter. Counted numbers seldom involve more than 10 or 12 digits, and measured quantities are usually known to only a few digits of accuracy. Mathematical constants, such as π, are usually entered as constants at compile time rather than entered as input.

For output, the problem is more complicated. We may want to print our answers to double-precision accuracy, either for use in some other program or for testing the program. One approach is to print them without converting them. (We could use the Z format with the FORTRAN compilers for the IBM System/360.) This is often the best approach for numbers to be used as constants in another program. But if we want to print the double-precision representation for a number A, we first find an integer k such that

$$10^k > |A| \geq 10^{k-1},$$

and then we reverse the process described above for input.

5.8. USES OF DOUBLE-PRECISION

As we have mentioned, some machines have hardware operation codes to perform double-precision arithmetic, while other machines use subroutines to perform these operations. Regardless of how the arithmetic is performed, it may or may not be supported by the higher-level languages. There are many machines on which FORTRAN supports the double-precision data type even though the hardware has no operation codes for double-precision arithmetic. On the other hand, there are machines on which the hardware can perform higher precision arithmetic than that supported by the compilers. We would prefer to have the double-precision arithmetic performed by the hardware, because it is significantly faster. Also, we would find it much more convenient to have the compilers support the double-precision data type. In some situations, the manner in which double-precision is supported is crucial; in other cases it is more a matter of convenience. In this section we shall discuss several different situations in which we would want to use some double-precision arithmetic. In each case, we shall consider how important it is to have the arithmetic performed by the hardware and to have the double-precision data type supported by the compilers.

1. Development and Testing of Single Precision Subroutines

Here we consider single-precision subroutines which are used as library programs. They may be part of the compiler or they may be programs written at a given installation, but because of their extensive use, we would like them to be both fast and accurate. They are expected to be carefully written and carefully tested.

a. Development of an Approximation

Consider the problem of developing a library program to compute a function $f(x)$. For example, we might want to write a single-precision exponential routine. It is typical to begin by reducing the range of the argument to some interval $a \leq x \leq b$ and then use a polynomial or rational approximation $\varphi(x)$ for $f(x)$ in this interval. Thus, we seek a function $\varphi(x)$ with

$$(5.8.1) \qquad \varphi(x) \approx f(x) \qquad \text{for } a \leq x \leq b.$$

Our program for $f(x)$ will evaluate $\varphi(x)$ using single-precision arithmetic, so the value it will produce for $f(x)$ will be $\tilde{\varphi}(x)$ instead of $\varphi(x)$, where

$$(5.8.2) \qquad \tilde{\varphi}(x) \approx \varphi(x).$$

Typically, we would like the approximation (5.8.1) to be good to, say, two or three bits beyond the word length, so that the error in (5.8.1) is small with respect to the error in (5.8.2).

We usually have to write a program to compute the coefficients of $\varphi(x)$. Since our objective is to produce an approximation $\varphi(x)$ which is good to at least single-precision accuracy, we use higher-precision arithmetic to compute these coefficients. They will then be rounded to single-precision and used as constants in our program for $f(x)$.

Thus, as part of the development of a single-precision program to compute $f(x)$, we have to use a double-precision program to compute the coefficients of our approximation. We may want our final program which computes $f(x)$ to be very efficient, but the speed of the program which computes the coefficients of $\varphi(x)$ is not important. This program is run only once as a development tool, so it is acceptable to have the double-precision arithmetic performed by subroutines. In fact, there have been cases in which the coefficients of $\varphi(x)$ were computed on a different machine from the one for which the program for $f(x)$ was being written.

In developing the approximation, it would be extremely convenient to have a compiler which supports the double-precision data type. But these approximations have often been computed without such support. Indeed, the approximations in the library routines for the highest precision supported by the compiler have to be produced without such support.

b. Testing a Single-Precision Subroutine

To test a subroutine which computes the value of a function $f(x)$, we generate some test values for x and use the subroutine to compute $f(x)$. One of the best ways to test these results is to extend each test value of x to double-precision by appending zeros and then compute $f(x)$ using a double-precision routine. (See Chapter 10.) Then we can estimate the error by comparing these values for $f(x)$. Since we are interested in only the first two or three decimal digits of the error, the double-precision program for $f(x)$ need be accurate to only two or three digits beyond single-precision. This accuracy is usually quite easy to achieve.

Since our double-precision program will be run for many different test cases, the speed of the double-precision arithmetic is a little more important than it was under heading a. The time required for each test case may determine how extensive our testing will be.

As under heading a, it is convenient, but not mandatory, to have the compiler support the double-precision data type.

2. Inserting a Few Double-Precision Operations in an Otherwise Single-Precision Program

There are many programs in which it is desirable to perform a few operations in higher precision than that used in the rest of the program. One example of this situation is the quadrature problem discussed in Section 4.1. There we saw that it was quite attractive to use higher-precision arithmetic in the accumulation of the sum. Another well-known example is the use of higher-precision arithmetic in the accumulation of inner products in matrix programs.

These examples are quite typical. We can often produce a more accurate answer by the judicious use of a few double-precision operations. Then we have a trade-off between speed and accuracy, and the speed of the double-precision arithmetic may determine whether or not we shall perform these operations in double-precision. It may be crucial to have the double-precision arithmetic performed by the hardware.

For example, in writing a library subroutine for a function such as sin x or e^x, we might want to use double-precision arithmetic in the reduction of the argument. [See Cody (1971b) or Kuki (1971).] If the double-precision arithmetic is performed by the hardware, this may produce better accuracy with negligible loss of speed. But if the hardware does not provide double-precision arithmetic, we may be unwilling to accept the degradation in speed. In other situations, we might be willing to have the program run significantly slower if it will produce better accuracy, so it would be acceptable to have the arithmetic performed by subroutines.

When the double-precision arithmetic is performed by the hardware, it can be used in programs written in Assembler language, even if the compilers do not support the double-precision data type. This can be of benefit to the user of higher-level languages by providing him with more accurate single-precision library programs.

However, a great many of today's programs and subroutines are written in higher-level languages, and we would like to have these programs use double-precision arithmetic where it is appropriate. Some of them may do so by calling subroutines, as described in Section 5.7, but the use of double-precision will be much more extensive if the compiler supports the double-precision data type.

3. Increasing the Precision of a Program To Determine Its Accuracy

It is quite common for a subroutine intended to be used as a library program to be subjected to the testing described under heading 1.*b*. But such testing is less often applied to an application program; the user is more likely to wait until he has reason to question the accuracy of the results produced. Then he would like to run the program in higher-precision to assess the accuracy of his single-precision program. Usually the objective is to determine whether single-precision arithmetic will suffice, so only modest changes will be made in the mathematics.

The double-precision program is being used for test runs rather than production runs, so the speed of the double-precision arithmetic is usually not vital. It is acceptable to have the arithmetic performed by subroutines.

But the conversion of the program from single-precision to double-precision may be a formidable task, so compiler support for the double-precision data type is often crucial.

4. Programs Requiring Double-Precision Calculation

There are some programs in which the calculation must be performed in double-precision in order to produce a good answer. Since this situation is commonly misunderstood, we shall discuss the way it can arise. One often hears the comment that high-precision arithmetic is unnecessary because the data are known to only a few digits of accuracy and we want only a few digits of accuracy in the answer. To be specific, suppose that the data are accurate to 1 part in 10,000 and that we want to know the answer to within 1 part in 1000. That is, we hope that neither the errors in the measurement of the data nor the errors in the calculation will produce an error in the answer greater than 1 part in 1000. To have any hope of achieving this result, the problem must be well conditioned. A relative error of .0001 in the data must produce a relative error less than .001 in the answer. But even though the problem is well-conditioned, we may be using an algorithm which is not. Consider the use of the power series in Section 4.2 to compute e^{-x} and $\sin x$. These problems were reasonably well conditioned, but as the value of x increased, the precision needed to produce a good answer grew rapidly. Another example of this situation is the use of the formula

$$\sinh x = \frac{e^x - e^{-x}}{2}$$

when x is small. (See Section 3.10.) Thus, even though the problem is well conditioned, the algorithm we are using may require that we use high-precision arithmetic.

In the examples mentioned above, it is easy to see how to revise the algorithm so that the calculation can be performed in single-precision. This might suggest that when high-precision arithmetic is needed for a well-conditioned problem, one should look for a better algorithm instead of increasing the precision. This approach should certainly be considered. But an appropriate algorithm may be far from obvious, and the user is interested in getting an answer to his problem. Therefore, he would often prefer to increase the precision and use the algorithm at hand rather than undertaking the research necessary to discover a better algorithm.

When one realizes that more than normal precision is required for his calculation, he will do whatever is necessary to produce the answer. Except for extremely large problems, he will usually not demand that the arithmetic be performed by hardware. While he would like to have the compiler support the double-precision data type, he will usually do whatever coding is necessary to solve his problem.

5. Programs Requiring Calculations in Higher-than-Double-Precision

In Section 4.2 we saw that the use of the power series to compute sin x when x is large is an example of an algorithm which is so sensitive to rounding error that more than double-precision accuracy is required. Higher-precision arithmetic will be discussed in the next section.

5.9. HIGHER-PRECISION ARITHMETIC

Of the uses of double-precision considered in the previous section, those discussed in Sections 5.8.1, 2, and 3 involved the use of double-precision arithmetic to support programs written primarily in single-precision. Now the meaning of single-precision varies considerably from one machine to another. But if a significant number of problems are run in a given precision, the discussion in the previous section shows the desirability of hardware and software support for arithmetic at twice that precision. Experience shows that if single-precision is on the order of 20 or 25 bits, there will be a significant number of programs which will be run in double-precision. But if single-precision is on the order of 50 bits, the vast majority of today's programs can be run primarily in single-precision. Thus, precision on the order of 50 bits seems to be adequate as the normal precision for most of todays programs, and precision on the order of 100 bits is desirable as support for programs written primarily in 50-bit precision.

As we mentioned at the end of the previous section, the algorithm we are using might be so sensitive to rounding errors that we would have to use more than 100 bits of precision. But if a problem requires more than this precision, it may require substantially more, so it does not seem necessary to support any specific precision beyond about 100 bits. Instead, we would like to have subroutines to perform N-fold-precision arithmetic, where the user can specify N.

Suppose that we want to perform 16-fold-precision arithmetic on the IBM System/360. That is, we want to perform arithmetic in a system FP(16, 96, a). If our representation of these numbers were a generalization of the one we used for double-precision numbers in Sections 5.3–5.7, each part of the number would have its own characteristic and mantissa. Suppose that X is on the order of 1. Since the smallest normalized positive number on the IBM System/360 is 16^{-65}, the low-order parts of the representation of X would underflow. Therefore, it is not reasonable to require that each part of the representation of the N-fold precision number be a valid floating-point number. Also, consider the case in which we want to compute $X \oplus Y$, where X is on the order of 1 and $|Y| < 16^{-65}$. Then replacing Y by zero would damage about one-third of the digits in $X \oplus Y$. Thus, as our precision increases, we also want to increase the range of the characteristic.

It is common for N-fold-precision subroutines to use fixed-point operations instead of floating-point operations to produce arithmetic in FP(r, Np, a). Then the parts of the N-fold-precision number need not be valid floating-point numbers, so we can use, say, a full word to represent the characteristic.

Suppose that the machine we are using provides fixed-point arithmetic operations which handle signed p-digit integers in the radix r. Except for the pre- and postshifts, we can think of each word as a digit and perform the arithmetic in the radix r^p. [If N is extremely large, it might suffice to normalize in the radix r^p. That is, we might be willing to perform the arithmetic in the system FP(r^p, N, a).]

EXERCISES

1. Show that the number $1/(r-1)$ has the representation .111111 $\cdots$ in the base r. Use this result to find the hexadecimal representation for the decimal numbers .2 and .1.

2. Write a FORTRAN program for the double-precision version of the quadrature problem in Section 4.1. For the computation of X use $X = I*H$ and $X = (I + 1)*H$. Program SDD addition to accumulate the sum S in twice-double-precision.

3. Assume that we are given SSD operations for addition and multiplication.
   a. Program coarse DDD multiplication.
   b. Program clean DDD multiplication.
   c. Program coarse triple-precision multiplication.
   d. Program clean triple-precision multiplication.

4. To test the programs written in Exercise 3, we need SSD operations. If we have a version of FORTRAN which provides clean double-precision arithmetic, the following FORTRAN programming will produce SSD multiplication. Let A, B, C1, and C2 be typed single-precision, and let D be typed double-precision. Write

$$D = DBLE(A)*B$$
$$C1 = D$$
$$C2 = D-C1$$

Then C1, C2 is the SSD product AB. For SSD addition, we may use either this approach or the coding described in Section 4.2.

Using these SSD operations, test the programs written for Exercise 3. What assumptions do you have to make about the representation of double- and triple-precision numbers?

5. In Section 5.4 we saw that clean DDD multiplication could be produced by combining the four products in (5.4.5). Here the first three products are computed using SSD multiplication, but we may use SSS multiplication to form $A_2 * B_2$. If single-precision multiplication takes a long time, we can use a different approach which requires only three SSD multiplications. We assume that our representation of double-precision numbers requires that if the low-order part of the number does not vanish, it has the same sign as the high-order part and its characteristic is exactly p less than that of the high-order part. (Then we cannot require that the low-order part be normalized.) Let A and B be given by (5.4.1), and set

$$C = r^{-p}A_1 - A_2$$
$$D = r^{-p}B_1 - B_2.$$

We assume that multiplication by a power of r can be performed rapidly by adjusting the characteristic. Now

$$AB = A_1B_1 - (r^pCD - r^{-p}A_1B_1 - r^pA_2B_2) + A_2B_2.$$

Then we need only three SSD multiplications to form A_1B_1, CD, and A_2B_2.
a. Show that C and D can be represented exactly as p-digit floating-point numbers.
b. Assume that we have SSD operations for addition and multiplication. How should the terms in the above formula for AB be combined to produce the clean DDD product $\overline{AB}$ of A and B?
c. Program the approach devised in part b and use the method described in Exercise 4 to test the program.

6. What changes must be made in Procedures 1, 2, and 3 of Section 5.5 to produce subtraction instead of addition?

7. Suppose that we have an SDD add operation available and that we want to program DDD addition. Let A and B be given by (5.4.1). We consider coding the DDD addition as either

$$B_1 \oplus (B_2 \oplus A)$$

or

$$B_2 \oplus (B_1 \oplus A),$$

where $\oplus$ denotes SDD addition. Which of these formulas is better in the add magnitude case? Which is better in the subtract magnitude case? How much difference does it make which formula we use?

8. Consider Procedure 3 in Section 5.5. We observed that the SSD addition in step 7 was unnecessary in the add magnitude case, since it would suffice to set $S_1 = F_1$ and $S_2 = H$. Show by an example that F_1 and H might overlap in the subtract magnitude case, so step 7 is needed to clean up the answer.

9. What simplifications can be made in Procedure 3 of Section 5.5 if we are willing to accept coarse DDD addition?

10. Write a program to compare two double-precision numbers. The program should determine whether or not the numbers are equal, and if they are unequal, it should determine which number is larger. You may assume that the representation of double-precision numbers requires that the two parts of the number do not overlap. However, the program should work even if the high- and low-order parts of the double-precision number have opposite signs. It should also work when the high-order part of the number is not normalized. You may assume that underflow does not occur during the execution of the program.

11. Suppose that we use the method of coarse DDD division based on the power series. Let $C_1 + C_2$ be the result obtained when $A_1 + A_2$ is divided by B_1 using DSD division, and let D be the SSS quotient $B_2 \div B_1$. Let E be the SSS product $C_1 * D$, and let $Q_1 + Q_2$ be the SDD sum of $-E$ and $C_1 + C_2$. How accurate is the approximation $Q_1 + Q_2 \approx A/B$?

12. Suppose that we use the approach based on Newton's method to compute the reciprocal of a double-precision number B. Let x_0 be the SSS quotient $1 + B_1$.
   a. Show that (5.6.10) holds.
   b. Suppose that x_1 is computed exactly as $x_0(2 - Bx_0)$. How accurate is the approximation $x_1 \approx 1/B$?
   c. To produce a coarse reciprocal, we let C be the single-precision number x_0 and let $E_1 + E_2$ be the SDD product of B and C. We then compute D using the SSS operations

$$D = (1 \ominus E_1) \ominus E_2,$$

   and let F be the SSS product $C * D$. Our approximation for $1/B$ is the SDD sum of C and F. How accurate is this approximation for $1/B$?

13. Consider the procedure for clean DDD division given in Section 5.6. Let A and B be positive numbers.
   a. Show by an example that it is possible for $A_1 \div B_1$ to overflow even though $A/B < \Omega$.
   b. Show by an example that it is possible for R_1 to underflow even though $A/B > 1$.
   c. Devise a strategy to avoid spurious overflows and underflows in this procedure by testing and scaling. If $A/B < \Omega$, the program should not overflow. There should be no underflows unless one or both parts of the answer underflow.

14. Consider the procedure for clean DDD division described in Section 5.6.
   a. Show by an example that there may be a two-digit overlap of C_2 with C_1.
   b. What is the minimum number of digits spanned by $C_1 + C_2 + C_3$?
   c. How accurate is the approximation $C_1 + C_2 + C_3 \approx A/B$?

15. Write a program to perform clean DDD division and test it.

16. How can the procedure for clean DDD division in Section 5.6 be shortened to produce coarse DDD division? Estimate the maximum error for your procedure.

6 ROUNDING

6.1. GENERAL CONSIDERATIONS

For any real number x, the values $\bar{x}$ and $\overset{-\circ}{x}$ obtained by chopping or rounding x to p digits in the base r were defined in Section 1.5. In Section 3.2, we saw that

(6.1.1)
$$\bar{x} = (1 - \rho)x, \qquad 0 \le \rho < r^{-(p-1)}$$
$$\overset{-\circ}{x} = (1 + \rho)x, \qquad |\rho| \le \tfrac{1}{2}r^{-(p-1)}.$$

Thus, with rounding we have a smaller bound for the relative error, but we do not know its sign. For binary systems FP(2, p, a), we note that $\tfrac{1}{2}r^{-(p-1)} = r^{-p}$, so the bound for the relative error introduced by a single operation is the same in FP(2, p, R) as it is in FP(2, $p + 1$, c). From this result we might be led to think of rounding as having the same value as adding a bit to the precision and using chopping. However, there are distinctions between FP(2, p, R) and FP(2, $p + 1$, c), and depending on what we are doing we may have a decided preference for one system or the other.

Let $a = r^e m$ be a positive, normalized number in $S(r, p)$, and suppose that a is not a power of r. Then $\bar{x} = a$ if and only if x lies in the interval I_c: $a \le x < a + r^{e-p}$, while $\overset{-\circ}{x} = a$ when x is in the interval I_R: $a - \tfrac{1}{2}r^{e-p} \le x < a + \tfrac{1}{2}r^{e-p}$. Thus, either chopping or rounding maps an interval of length r^{e-p} into a number in $S(r, p)$. (A slight modification of I_R is necessary if a is a power of r, and the obvious modification of I_R and I_C must be made if a is negative.) We may think of $\overset{-\circ}{x}$ as representing the interval I_R by its midpoint, while $\bar{x}$ represents I_C by the end point closest to the origin. Thus, if we are given either $\bar{x}$ or $\overset{-\circ}{x}$, we know only that x lies in a given interval of length r^{e-p}, so either $\bar{x}$ or $\overline{x}^{\circ}$ gives us the same amount of information about x. Similarly, the bounds for the relative error given in (6.1.1) for either rounding or chopping restrict p to an interval of length $r^{-(p-1)}$.

The distinction between rounding and chopping is twofold. First, rounding gives us a smaller bound for the absolute value of the relative error introduced at a given stage in the calculation. This, in turn, may lead to a smaller bound for the error in the answer. The other distinction is that chopping always decreases the absolute value of x, while rounding may either increase or decrease $|x|$. In some situations, consistently chopping numbers may introduce a bias in the results which rounding would eliminate. (See, for example, the quadrature problem in Section 4.1.) In other situations, the errors introduced by chopping may compensate.

However, even in the case in which the errors introduced by chopping tend to compensate, chopping usually will not produce a smaller bound for the error than rounding would. For example, suppose that we have a and b in $S(10, 6)$, and that

$$x \approx a = 6.54321$$
$$y \approx b = 1.11111.$$

Suppose that we want to compute $x - y$, so we form

$$a \ominus b = a - b = 5.43210.$$

If a and b are $\bar{x}$ and $\bar{y}$, respectively, we find that

$$(6.1.2) \qquad a \ominus b - 10^{-5} < x - y < a \ominus b + 10^{-5}.$$

But we find that (6.1.2) still holds if a and b are $\overline{x}^{\circ}$ and $\overline{y}^{\circ}$. Thus, even though the errors introduced by chopping tend to compensate while the errors introduced by rounding might add, we still obtain the same bounds. On the other hand, if we had been interested in $x + y$ instead of $x - y$, we would have formed $a \oplus b = 7.65432$. In place of (6.1.2) we would find

$$a \oplus b \leq x + y < a \oplus b + 2 \cdot 10^{-5}$$

if $a = \bar{x}$ and $b = \bar{y}$, while

$$a \oplus b - 10^{-5} \leq x + y < a \oplus b + 10^{-5}$$

if $a = \overline{x}^{\circ}$ and $b = \overline{y}^{\circ}$. In each case $x + y$ is known to lie in an interval of length $2 \cdot 10^{-5}$, but rounding produces a smaller bound for the absolute value of the error.

6.2. USES OF ROUNDING

There are several ways in which rounding might be used in a program. In some cases, we might perform the entire calculation in FP(r, p, R). In other cases, we might perform most of the calculation in some other system FP(r, p, a), with an occasional arithmetic operation performed in FP(r, p, R). Finally, we might use rounding when we shorten a number from double-precision to single-precision. The way in which rounding is used will depend to a large extent on what is provided by the hardware and software.

We are likely to perform almost all our floating-point arithmetic in whatever system FP(r, p, a) is supplied by the hardware or software of the machine we are using. This system probably will not be exactly FP(r, p, c) or FP(r, p, R), but it is more likely to be a variant of chopping than rounding—particularly when the floating-point arithmetic is performed by the hardware. There have, however, been a few machines which provided a form of rounded arithmetic. For example, the CDC 6600 has operation codes which preround the operands. This produces a system FP($2, 48, a$) which is different from FP($2, 48, R$) but which does provide some of the effects of rounding.

Rounded arithmetic is somewhat more common when the floating-point arithmetic is performed by software instead of hardware. Again, there are many variations in the details of the implementation of rounding but in some cases software has actually provided arithmetic in FP(r, p, R).

We have seen that the advantages of rounded arithmetic are that it tends to produce smaller error bounds and that it tends to reduce bias. Although we shall use whatever system FP(r, p, a) is provided by the machine we are using, we would probably prefer FP(r, p, R) to FP(r, p, c) if we were given our choice. The principal reason for the prevalence of chopped arithmetic is that it is easier to implement and it will probably be a little faster. Of course, in considering a given implementation of either rounding or chopping, there are other questions which must be considered. For example, we would have to explore the question of whether there are any anomalies in the arithmetic and the question of how it can be used for operations such as FLOAT to FIXED conversion and programmed double-precision arithmetic.

We shall now consider using rounding selectively. Suppose that we perform almost all our arithmetic in FP(r, p, clq) but that we want to use rounded arithmetic in a few operations in the program. An example of this is the quadrature problem discussed in Section 4.1. In that section we discussed the idea of performing the entire calculation in FP($16, 6, cl1$) except for the accumulation of the sum S. The only difference between the programs which produced the output shown in Figures 4.1.1 and 4.1.6 was that in the latter program the addition of terms to S was performed in FP($16, 6, R$). From these results we can see the advantage of inserting rounding at a few crucial points in the program. It is particularly advantageous on a machine on which it is easier to produce rounded results than to perform double-precision arithmetic. (This is not the case with the IBM System/360.)

Finally, we consider the problem of shortening a number from double-precision to single-precision. This situation arises when double-precision arithmetic is used selectively at a crucial point in the program. Suppose that X is typed single-precision, and we write

(6.2.1) $X = \cdots,$

where the expression on the right-hand side of (6.2.1) produces a double-precision result D. With many FORTRANs, including those for the IBM System/360, this result D is chopped to single-precision and stored in X. That is, $X = \bar{D}$. But our use of double-precision in (6.2.1) suggests that we were concerned about accuracy, so we might prefer to have the slightly more accurate value $X = \bar{D}^{\circ}$. In the next section we shall discuss how this could be coded.

6.3. IMPLEMENTATION OF ROUNDING

We shall now address the question of how one can incorporate rounded arithmetic in his program. This will depend on the type of arithmetic provided by the hardware, and it will also depend on whether or not the language in which the program is being written provides a way to request rounding.

We have mentioned that the commonest situation is for the hardware to provide some variant of chopped arithmetic. But some machines are more versatile. For example, the CDC 6600 has one set of operation codes which produce chopped arithmetic and another set of operation codes which produce a version of rounded arithmetic. With such a machine, the programmer who is using Assembler language has his choice of which arithmetic to use, although this is not necessarily true for the programmer who writes his program in FORTRAN.

To produce a rounded result, we must look at the digits discarded. Suppose that we are using a machine which performs arithmetic in a system $FP(r, p, clq)$ and produces only the high-order p digits of the result. Unfortunately, there is no way we can use this number to produce the correctly rounded result. Somehow we must gain access to the digits discarded to find out whether or not what we discarded is as large as half a unit in the last place retained. We shall assume that r is even, since this is true on all machines currently in use. Then we need look at only the first digit discarded to see whether it is as large as $r/2$. Thus, if the hardware makes available one or more extra digits, we can program rounding—at least in Assembler language.

Indeed, arithmetic in FP(r, p, R) can be produced by performing arithmetic in FP(r, $p + 1$, c) and then rounding the result to p digits.

Thus, if we want to perform rounded arithmetic on a machine which provides only chopped arithmetic, we are faced with the requirement of developing at least one extra digit in the result and then rounding the result to p digits. Usually this means that we have to develop the result in double-precision and then round it to single-precision. For example, suppose that we have a machine which performs arithmetic in FP(r, p, c) and that we want to compute

$$(6.3.1) \qquad\qquad X = A \oplus B$$

in FP(r, p, R). We first use an SSD add to produce the double-precision sum of the two single-precision numbers A and B, and then we round this sum to single-precision. There are some machines, for example, the IBM 7090, on which many of the single-precision operations are really SSD operations. Then rounding can be programmed quite cheaply—at least in Assembler language. But on other machines, for example, the IBM System/360, most single-precision operations produce only single-precision results, so rounded arithmetic takes longer than double-precision arithmetic. In that case, arithmetic in FP(r, p, R) usually would not be used for computational purposes, although it might be used for study purposes, as in our study of the quadrature program in Section 4.1.

As we saw in Chapter 5, the FORTRAN coding to produce the SSD add needed in (6.3.1) would be

$$(6.3.2) \qquad\qquad D = DBLE(A) + B$$

where D is typed double-precision. To complete the operation in (6.3.1), D must be rounded to single-precision and stored in X. We saw in Section 6.2 that rounding a number from double-precision to single-precision is of interest in itself.

Suppose that S and D are typed single-precision and double-precision, respectively, and that we want to round the value of D to single-precision and store it in S. First, we write the FORTRAN statement

$$(6.3.3) \qquad\qquad\qquad S = D$$

and we assume that this statement will store the high-order digits of D in S. (This will usually be the case, but it may not be true if the double-precision arithmetic is performed by software and the routines do not guarantee sign agreement between the high-order and low-order parts of the double-precision number.) Then the FORTRAN expression D−S produces the remaining digits of D, and if it does not vanish, it has the same sign as D and S. If

$|D-S|$ is less than one-half a unit in the last place of $|S|$, S is the correctly rounded result; otherwise the absolute value of S must be increased by 1 in the last place. Thus, we want to increase the absolute value of S whenever $2 \cdot |D-S|$ is at least 1 in the last place of $|S|$. This may be accomplished by the following FORTRAN statement:

(6.3.4) $S = S+2.*(D-S)$

Since S contains the high-order digits of D, in place of (6.3.4) we could use the FORTRAN statement

(6.3.5) $S = D+(D-S)$

If we are writing the program in Assembler language, we might use this same approach. On the other hand, some machines, such as the IBM 7090 and the IBM System/360 models 85 and 195, have an operation code which rounds a double-precision number to single-precision.

We shall now consider the question of how rounding can be supported in higher-level languages when we are working on a machine which does not perform its arithmetic in FP(r, p, R). First, we should have in the language a function ROUND(D) whose argument D is a double-precision number and whose value is D rounded to single-precision. The result should be typed as single-precision. Since the coding for this function is so simple, it should be incorporated as an in-line subroutine rather than as a call to a subprogram. PL/I has the syntax for such a function, but instead of rounding D it uses the bias removal operation B(X) discussed in Section 6.4.

Performing all arithmetic in FP(r, p, R) is more difficult. If the hardware provides only chopped arithmetic and delivers only the high-order p digits of the result, there is really no way to produce rounded arithmetic short of producing the double-precision result and rounding it to single-precision. On the other hand, suppose that the hardware operations for single-precision arithmetic are really SSD operations. We would like to be able to write

(6.3.6)   ROUND(expression)

and have the compiler round the double-precision number produced by the expression in the parentheses in (6.3.7) whenever the last operation used in producing this expression was an SSD operation. This facility was provided by the modifications to the FORTRAN compiler for the IBM 7090 made at the University of Toronto. [See Kahan (1965a, 1966).]

6.4. BIAS REMOVAL

We shall now turn to a procedure which we shall call *bias removal*. Another name for this procedure is the *statistical round*. When we compared rounding with chopping in Section 6.1, we noted that rounding had two effects—reducing the size of the maximum error and eliminating bias. The bias removal feature which we shall discuss here does not reduce the size of the maximum error, but it does tend to reduce or eliminate bias.

We shall assume that the radix r of our floating-point number system is even. The bias removal procedure performs an operation which replaces a floating-point number x by a number which we shall designate by $B(x)$. If $x = 0$, $B(x) = 0$. For $x \neq 0$, $B(x)$ forces the low-order digit of x to be odd. Thus, $B(x) = x$ if x is zero or if the low-order digit of x is odd. Otherwise, $B(x)$ is obtained from x by increasing the absolute value of x by 1 in the pth digit of the mantissa. If the radix r is a power of 2, this simply means that the low-order bit of the mantissa is set to 1 whenever $x \neq 0$. Similarly, if we used the binary coded decimal representation for decimal digits in FP(10, p, a), then whenever $x \neq 0$ we produce $B(x)$ by setting the one bit of the low-order digit of the mantissa to 1.

When the computation of $B(x)$ is performed by hardware, the operation usually turns out to be very fast because it does not propagate any carries. Also, the operation is sometimes quite easy to perform in software. For example, suppose that r is a power of 2 and that negative numbers are stored as *sign and true magnitude*. To compute $B(x)$ we would first perform a test to see that x is not zero; then we would OR x with a word which is all zeros except for the low-order bit which is 1—at least if this operation is available on the machine.

In using bias removal, the basic approach is that instead of doing arithmetic in either FP(r, p, c) or FP(r, p, R), one uses bias removal in conjunction with chopping. That is, if we are given a real number x, then instead of forming either $\bar{x}$ or $\bar{x}^{\circ}$, we form $B(\bar{x})$. Now there are two different ways in which bias removal may be used. First, we might decide to use it in every arithmetic operation. Thus, we might define a system FP(r, p, B) where the arithmetic is defined by

$$a \oplus b = B(\overline{a + b})$$
$$a \ominus b = B(\overline{a - b})$$
$$a * b = B(\overline{ab})$$
$$a \div b = B(\overline{a/b})$$

and perform all arithmetic in FP(r, p, B). On the other hand, we might want to use bias removal selectively. That is, we might decide to perform most of the arithmetic in, say, FP(r, p, clq) but perform a few specific operations in FP(r, p, B).

Unless bias removal is used selectively, some care must be taken in its implementation. If every number x which arises in the problem is replaced by $B(\bar{x})$, it is impossible to represent small integers exactly. (See Exercise 10.) This is extremely annoying. An alternative is to use a function $B'(x)$ defined for all real numbers x by

$$B'(x) = x \qquad \text{if } x \text{ is in } S(r, p)$$
$$B'(x) = B(\bar{x}) \qquad \text{if } x \text{ is not in } S(r, p).$$

That is, we shall represent x exactly if possible; otherwise we shall use $B(\bar{x})$. We could then define a system $FP(r, p, B')$ by setting $x \oplus y = B'(x + y)$, etc. Such a system was implemented in hardware on the NORC (Naval Ordnance Research Calculator) built in the early 1950s. [See Eckert and Jones (1955).] NORC performed arithmetic in roughly the system $FP(10, 13, B')$.

It is far more common for bias removal to be used selectively, performing an occasional operation in $FP(r, p, B)$. We have noted that if we want to produce $\overset{-\circ}{x}$, we have to see the first digit discarded. If the hardware provides some variant of chopped arithmetic and produces only the high-order p digits of the result, we cannot use this number to program arithmetic in $FP(r, p, R)$. The advantage of bias removal lies in the fact that we do not have to look at the digits discarded. Thus, if the hardware performs an arithmetic operation in $FP(r, p, c)$ and produces a result $\bar{x}$, we can perform the operation in $FP(r, p, B)$ by forming $B(\bar{x})$.

The implementation of the ROUND function in the PL/I compilers for the IBM System/360 uses bias removal whenever the argument is a floating-point number. Thus, ROUND$(\bar{x})$ will form $B(\bar{x})$. When this is used in conjunction with the chopped arithmetic of the IBM System/360, it allows us to perform arithmetic in (approximately) the system $FP(r, p, B)$. Since bias removal will be performed only when we write ROUND explicitly, there is no danger of changing numbers we wanted to be exact.

Let x be any real number and set $\hat{x} = B(\bar{x})$. If we write

$$x = r^e m, \qquad r^{-1} \le |m| < 1,$$

then we also have

$$\hat{x} = r^e \hat{m}, \qquad r^{-1} \le |\hat{m}| < 1.$$

Here $|\hat{m}|$ is either $|\bar{m}|$ or $|\bar{m}| + r^{-p}$, so $|\hat{m} - m| \le r^{-p}$. Then we may write

$$\hat{x} = (1 + \rho)x, \qquad |\rho| \le r^{-(p-1)}.$$

We see that this provides the same bound for $|\rho|$ that we would have had with chopping, but we no longer know the sign of ρ. Unlike rounding, bias removal does not reduce the size of the error produced by shortening x to p digits. In fact, the representation of x by $B(\bar{x})$ gives us less information about x than $\bar{x}$ does, because it only restricts x to an interval of length $2r^{e-p}$.

The advantage of $B(\bar{x})$ is that its absolute value may be either too large or too small, so we tend to reduce bias. As an illustration of the effectiveness of this approach, we refer to the quadrature problem discussed in Section 4.1. We encountered difficulties there because we were adding a large number of terms of the same sign, so the sum got to be much larger than the terms. We saw that chopped arithmetic introduced a bias in the sum, so the sum consistently drifted below the correct value. In Figure 4.1.6 we saw that rounding produces a significant improvement in the answer. Similarly, in Figure 4.1.7 we saw that bias removal produces about the same improvement. We conclude that the improvement produced by rounding was primarily due to the tendency to reduce bias. The fact that it introduced smaller errors was of less importance. Thus, this is an example of a problem in which we would have a decided preference for performing arithmetic in $FP(2, p, R)$ instead of $FP(2, p + 1, c)$, even though the two systems produce the same bound for the absolute value of the error introduced in any operation.

The improvement produced by bias removal in the quadrature problem of Section 4.1 is typical of many quadrature problems and of some differential equation problems. In problems of this sort, the tendency to reduce bias is the most important aspect of rounding, and forming $B(\tilde{x})$ may be a satisfactory substitute for rounding. On the other hand, there are problems in which the smaller error bounds in $FP(r, p, R)$ are used to produce highly accurate results. For example, the program SQRT written by W. Kahan for the 7090 at the University of Toronto claims that the error is never more than .50000163 units in the last place. Such accuracy could not be achieved without judicious use of rounding, and bias removal would be no substitute.

6.5. OTHER "ROUNDING" PROCEDURES

When we define the floating-point arithmetic operations, we are faced with the problem of approximating a real number x by a number $\tilde{x}$ in $S(r, p)$. By far the commonest approaches in digital computing are forms of chopping, rounding, or bias removal. That is, the commonest choices for $\tilde{x}$ are $\tilde{x}$, $\overset{\_\circ}{x}$, or $B(\tilde{x})$. However, there are other approaches that could be used, and they might be useful in special situations.

First, we note that $\overset{\_\circ}{x}$ has a slight bias because we round the magnitude upward whenever the digits discarded are exactly one-half a unit in the last place kept. There is a slight modification of our rule for rounding which is quite popular for hand computation. This approach defines $\tilde{x}$ to be $\overset{\_\circ}{x}$ unless there are two numbers in $S(r, p)$ which are equally close to x. In that case, we let $\tilde{x}$ be the one in which the low-order digit of the mantissa is even. This eliminates the bias introduced in our definition of $\overset{\_\circ}{x}$. Although this rule is well known, it has seldom, if ever, been used for machine computation.

For any real number x, let x_L and x_R be the left and right *neighbors* of x in $S(r, p)$. That is, x_L is the largest number in $S(r, p)$ which is $\leq x$, and x_R is the smallest number in $S(r, p)$ which is $\geq x$. Then $x_L \leq x \leq x_R$, and $x_L < x_R$ unless x is in $S(r, p)$. We almost always want our approximation $\tilde{x}$ to be one of the numbers x_L, x_R. Our definition of rounding specifies that $\bar{x}^\circ$ is the neighbor closest to x, with a special convention to handle the case in which x_L and x_R are equally close to x. Similarly, our definition of chopping specifies that $\tilde{x}$ is the neighbor with smaller absolute value. This suggests other rules for selecting the neighbor we want. For example, we could define a rule *antichopping* which would always select the neighbor with larger absolute value. Similarly, we could define rules *chop left* and *chop right* which would always select the neighbor x_L and x_R, respectively. (x_L and x_R are often referred to as the *floor* and *ceiling*, respectively.) These rules could be useful in certain cases in which we want the error (or the relative error) to have a specific sign. Another example of their use arises in the definition of *interval* arithmetic in Section 7.4. There, we shall want to round an interval outward.

In Section 6.4 we discussed the possibility of performing bias removal after chopping by forming $B(\tilde{x})$. Similarly, we can perform these other operations after chopping, although this will introduce additional error when x is in $S(r, p)$. To facilitate these operations, it would be convenient to have the compilers support functions such as AUGMENT, DECREMENT, AUGMENT ABSOLUTE VALUE, and DECREMENT ABSOLUTE VALUE. These operations can also be useful in testing programs.

EXERCISES

1. Write a program to shorten a double-precision number D to single-precision using the following rules. (In some cases you may have to use techniques discussed in Section 4.4.)
   a. Rounding.
   b. Antichopping.
   c. Chop left.
   d. Chop right.
   e. $B(\bar{D})$.
   f. $B'(D)$.

2. Let D and S be typed double-precision and single-precision, respectively, and suppose that we want to round D to single-precision and store it in S. In Section 6.3 we suggested the FORTRAN coding

$$S = D$$
$$S = S+2.*(D-S)$$

What can we say about the result stored in S if our representation of double-precision numbers does not require sign agreement between the high-order and low-order parts?

3. There are other ways to round a double-precision number to single-precision.

Suppose that D and S are typed double-precision and single-precision, respectively, and that we want to store $\overline{D}^{\circ}$ in S. If $D = 0$, set $S = D$. If D is not zero, construct a double-precision number X having the same sign and characteristic as D, having a mantissa which is all zeros except for the $(p + 1)$st digit, which is $r/2$. (Clearly X is unnormalized.) Form

$$S = \overline{D \oplus X},$$

where the operation $\oplus$ is performed in double-precision. If our representation of double-precision numbers requires sign agreement and requires that the two parts of the number be nonoverlapping, this coding will store $\overline{D}^{\circ}$ in S.
a. Will this coding store $\overline{D}^{\circ}$ in S on the machine you are using?
b. Why must zero be handled specially?
c. Write a program to perform these operations. If you wish to write the program in a higher-level language, use the techniques described in Section 4.4.

4. Suppose that we want to form $\overline{A + B}^{\circ}$, where A and B are in $S(r, p)$. We first perform a test and interchange A and B if $|B| > |A|$. Then, using the FORTRAN coding studied in Section 4.3, we write

$$S = A \oplus B$$
$$T = B \ominus (S - A)$$
$$S = S \oplus 2. * T.$$

We consider different systems in which the operations $\oplus$, $\ominus$, and $*$ may be performed. In which of the following systems will this coding produce $\overline{A + B}^{\circ}$?
a. $FP(r, p, cl1)$.
b. $FP(r, p, c)$.
c. $FP(r, p, cl0)$.
d. $FP(r, p, R)$.
e. The floating-point number system provided by the machine you are using.

5. Suppose that we want to perform addition using bias removal. Write a program to produce the following quantities:
a. $B(\overline{X + Y})$.
b. $B'(X + Y)$.

6. Write a program to solve the quadrature problem discussed in Section 4.1. For the computation of X use I*H and (I+1)*H. Perform the addition of terms to S in the following systems:
a. $FP(r, p, R)$.
b. $FP(r, p, B)$.

7. Convert the programs written for Exercise 6 to double-precision.

8. Write a program to solve the quadrature problem discussed in Section 4.1. Compute X by repetitively adding H. Perform both the addition of terms to S and the addition of H to X in the following systems:
a. $FP(r, p, R)$.
b. $FP(r, p, B)$.

9. Convert the programs written for Exercise 8 to double-precision.

10. Consider the system FP(r, p, B).

    a. If we represent the integer n by $B(n)$, show that

    $$1 \oplus 1 \neq 2.$$

    b. Suppose that $x = B(2)$. Is $x * x$ the same as $x \oplus x$?

11. Let x be a real number and let $\tilde{x}$ be the number in $S(r, p)$ obtained from x by one of the following rules: chopping, antichopping, chop left, chop right. Let

    $$\tilde{x} = x + \epsilon$$

    and

    $$\tilde{x} = (1 + p)x.$$

    Which of the rules will guarantee that
    a. $\epsilon \geq 0$.
    b. $\epsilon \leq 0$.
    c. $p \geq 0$.
    d. $p \leq 0$.

12. Write a program to use the Runge-Kutta method to solve the differential equation $y' = y$ for $0 \leq x \leq 1$. Take $y(0) = 1$ and print only $y(1)$. Run the program using 2^N steps, for $N = 0, 1, 2, \ldots, 12$. Run three versions of the program, performing the arithmetic in the following systems:
    a. FP(r, p, c).
    b. FP(r, p, R).
    c. FP(r, p, B).

7 AUTOMATIC ANALYSIS OF ERROR

7.1. INTRODUCTION

In this chapter we shall study several approaches which have been used to try to let the computer assist us in the analysis of error. With these approaches, we ask the computer to produce both an answer and an indication of how accurate the answer is. As we shall see, none of these approaches is a panacea. They have not yet reached the point at which we can recommend that they be used in place of floating-point arithmetic as the normal computing procedure. But each of these approaches has its advocates, and they have succeeded in producing reasonable error estimates.

Getting the computer to give us an indication of the accuracy of the answers it produces is a major problem facing the computer profession. There are a few subroutines for which detailed error analyses have been performed manually, but such programs are distressingly rare. Far too often, we have little idea of how accurate the answer is. Since the difficulty of performing an error analysis increases as the complexity of the problem increases, this situation is not likely to change. It appears that our only hope is to get the computer to produce a reliable error estimate.

At first glance it might appear that we are carrying so may extra digits in the calculation that we could ignore the loss of accuracy. But, as we saw in Section 5.9, we may produce bad results if we are using an algorithm which is sensitive to rounding error. In Section 4.2 we studied the use of the power series to compute e^{-x} and sin x, and we observed extreme loss of accuracy in a reasonably well-conditioned problem. We would like to be warned of this loss of accuracy by an error analysis.

Another reason for the importance of an error analysis is that we do not observe what is happening in the calculation. Suppose that our algorithm requires us to form $\tilde{x} - \tilde{y}$, and for one set of input data $\tilde{x}$ and $\tilde{y}$ are nearly equal. Then errors in $\tilde{x}$ and $\tilde{y}$ may produce large relative error in $\tilde{x} - \tilde{y}$. If we were performing the calculation by hand, we would observe this loss of accuracy and know that we have to worry about the accuracy of the answer. But when the calculation is performed by a machine, we do not see the intermediate result, so we do not get any warnings.

It seems likely that this problem will be magnified as time goes on. The use of formula manipulation languages, such as FORMAC, is growing. It is reasonable to expect that the ability to manipulate formulas will be incorporated into some future compilers. Then we might code a formula for a function $f(x, y)$ and ask the compiler to produce the code for, say, $\partial^3 f/(\partial x^2 \partial y)$. Since we would not see the formulas the computer was using, we could not be expected to know whether the formulas were sensitive to rounding errors. Our only hope would be to have the computer estimate the error.

Thus, the problem of producing an automatic error analysis is extremely important. We shall discuss some of the approaches which have been used in attempts to solve this problem, and we shall try to show the difficulties inherent in each approach.

None of these techniques can be used blindly. Rather, they are tools that we can use to try to write programs which will give us both numerical results and a reasonable estimate of their accuracy. In many cases, we shall have to modify the algorithms we are using in order to produce good results with these tools. Unfortunately, we are not yet able to produce reliable error estimates automatically and painlessly.

In Sections 7.2, 7.3, and 7.4, we shall discuss three different techniques which address this problem. Each of these techniques involves replacing the standard floating-point arithmetic by slightly different operations. The modified arithmetic is usually performed by subroutines, but there have been some machines on which it was performed by the hardware. For example, the IBM 7030 provided noisy mode, and the Maniac III provided significance arithmetic.

7.2. SIGNIFICANCE ARITHMETIC

Significance arithmetic is a technique for the automatic analysis of rounding error which has been studied by Metropolis, Ashenhurst, and others. Its best known implementation was in the Maniac III, where one had the option of using either significance arithmetic or normalized arithmetic. [See Ashenhurst (1962).] But it has also been provided on other machines. For example, it was implemented on an IBM 7090 at New York University by installing a special feature. [See Goldstein (1963).]

With significance arithmetic, we use unnormalized numbers throughout the calculation. The objective is to represent each number with enough leading zeros so that the remaining digits will be significant. Then the appearance of the number will tell us how many significant digits it has. The arithmetic operations must be modified so that they will produce an unnormalized result with the correct number of leading zeros. This is fairly easy to implement for addition and subtraction, since it merely means that we omit postnormalization. For multiplication and division, the rules are based on the idea that the answer should have as many leading zeros as the least significant operand. Naturally there are slight variations in the way these rules have been implemented on different machines, but they usually require only modest changes in the way floating-point arithmetic is performed.

Ashenhurst (1964) discusses the design of function subroutines to be used with significance arithmetic. The objective is to produce an answer whose significance is determined by the significance of the input and the condition of the problem. For example, suppose that we want to compute e^x. As we saw in Section 3.7, the relative error in e^x due to a relative error p in x is approximately px. We may use the number of significant digits in x to estimate p, and then use the value of px to determine how many significant digits e^x has. A similar approach may be used for other functions. We would also want special programs for radix conversion. They should produce an unnormalized result with the correct number of significant digits. [See Metropolis and Ashenhurst (1965).]

Thus, to use significance arithmetic, we want to change both the arithmetic and the library programs. Following this approach, the advocates of significance arithmetic have produced good results.

Unfortunately, significance arithmetic has several drawbacks. Indeed, the objections raised in Chapter 3 to the use of significant digits as a measure of accuracy apply here. First, as we saw in Section 3.1, the discreteness of the number of significant digits poses a problem, and this is particularly pronounced when the radix is large. As a consequence, most implementations of significance arithmetic have been on binary machines.

A second drawback is that there is no guarantee that the estimate of accuracy produced by using significance arithmetic is correct. Thus, if our answer is $.00054321 \times 10^6$, there is no guarantee that five (or even two) of the digits are significant. (Of course standard floating-point arithmetic gives us no indication at all of the accuracy of the answer. But since the objective of significance arithmetic is to tell us how accurate the answer is, it is disappointing that it cannot guarantee that the digits it produces are accurate.) However, experiments such as those described by Ashenhurst (1965b) show that when significance arithmetic is used carefully, it will often give a good indication of the accuracy of the answer.

The exact behavior of significance arithmetic will depend on the details of its implementation. But if it used chopped arithmetic, it would almost surely fail to assess the decay of the answer to the quadrature problem of Section 4.1. The answer would not be as accurate as significance arithmetic would lead us to believe.

A more serious disadvantage of significance arithmetic is that it causes us to introduce larger errors at each step in the calculation. Depending on whether we use chopped or rounded arithmetic, the error introduced by an arithmetic operation is bounded by either one unit or one-half a unit in the last place kept. Since we retain fewer digits of the answer when we use significance arithmetic, the errors are larger. Indeed, the new error introduced at each step is of more or less the same order of magnitude as the inherited error. Thus, significance arithmetic uses the rule mentioned in Chapter 3 which suggests that we should develop only those digits that we can guarantee are significant. In Section 3.5, we considered the computation of

(7.2.1) $$x = \prod_{i=1}^{n} x_i$$

for large n. We saw that if we shortened the precision as the propagated error increased, we would needlessly damage the answer. But significance arithmetic must shorten the precision in this way if it is to estimate the error in the computation of (7.2.1). Therefore, we may expect it to produce a less accurate answer. This is the price we pay for getting an estimate of the error.

Although we may be willing to sacrifice a little accuracy to get an indication of how accurate the answer is, we would like to limit this loss of accuracy. One way to do this was proposed by Gray and Harrison (1959).† They use normalized floating-point numbers, but along with each number they carry an *index of significance* which indicates how many of the digits are significant. With this approach, we would set aside a few bits in the word for the index of significance, and we would modify the floating-point arithmetic so that it would produce both the answer and its index of significance. Thus, if we were using a machine which performed arithmetic in FP(2, 48, a), we would set aside six bits for the index of significance and perform arithmetic in FP(2, 42, a). This would sacrifice six bits, but never more than six bits.

A final difficulty with significance arithmetic arises in situations such as those described in Section 3.11. In that section, we considered the transformation of a problem into another problem having the same answer. The crucial question was whether the answer to the new problem was close to the answer to the original problem. We discussed two problems in which it was important to compute the new coefficients to full precision, even though this involved the division of a number with eight significant digits by a number with two significant digits. We could compute a good answer to these problems with normalized arithmetic but not with the unnormalized form of significance

†This was implemented in FLIP at Argonne National Laboratory.

arithmetic. We could also compute a good answer by using the version of significance arithmetic based on an index of significance, but the index of significance would mislead us by indicating that the answer was rather poor.

Finally, we point out that significance arithmetic suffers from the *simultaneity problem*, which will be discussed in Section 7.4.

7.3. NOISY MODE

Some machines, for example, the IBM 7030, have provided noisy mode in addition to the normal floating-point arithmetic. Performing the floating-point arithmetic in noisy mode changes the digits which will be shifted into the answer when postnormalization is required. To use noisy mode, we perform the calculation twice, first in the normal mode and then in noisy mode. The extent to which the two answers agree is taken as an indication of the accuracy of the answer.

The details of the implementation of noisy mode may vary from machine to machine. Usually, noisy mode changes the answer only when postnormalization is required. To illustrate a typical implementation of noisy mode, we shall consider an example in FP(10, 8, a). Let

$$(7.3.1) \qquad x = 1234.5678$$
$$y = 1234.4321$$

and suppose that we want to form $z = x \ominus y$. In either FP(10, 8, c) or FP(10, 8, R), the value of z would be .13570000. Here the result had to be shifted four places to the left to postnormalize it, and we shifted in four zeros. If x and y are known only approximately, we really do not know what digits should be shifted in. In noisy mode, we shift in 9s instead of zeros, so the value computed for z would be .13579999. Thus, the idea of noisy mode is to insert "noise" when we are uncertain of what the digit should be. We hope that this noise will propagate in about the same way that the errors introduced by normal floating-point arithmetic do, so the change in the answer will give us an indication of how accurate our answer is.

If the subtraction described above had been performed in the base r, the digits shifted in would have been ($r - 1$)s instead of 9s. In general, when the arithmetic is performed in noisy mode, the digits shifted in during postnormalization are the ($r - 1$)'s complements of the digits that would have been shifted in by the normal floating-point arithmetic. This definition can also be used for multiplication and for addition and subtraction when the operands have different characteristics. The effect of noisy mode on division may vary considerably from one implementation to another. One approach is to extend the dividend by appending several digits of ($r - 1$)s to it before dividing.

When noisy mode is provided by the hardware, there must be a way to specify whether we want noisy arithmetic or normal arithmetic. If there were separate operation codes for noisy arithmetic, we would have to change the floating-point instructions in order to rerun the program in noisy mode. It would be much more convenient if we could simply specify that the machine should operate in noisy mode until we tell it to change back to normal mode. Thus, we would like to tell the computer to "enter noisy mode" and have it perform all floating-point arithmetic in noisy mode until we tell it to "leave noisy mode." (This is approximately the way noisy mode was handled in the IBM 7030.)

The cost that we pay for using noisy mode is running the problem twice. But the results produced by the normal run of the program have not been damaged.

Unfortunately, there is no guarantee that the true answer lies between the results produced in noisy arithmetic and normal arithmetic. As a simple example, consider the computation of $z = x \ominus y$, where x and y are given by (7.3.1). For these data, the result produced for z using noisy mode is $\tilde{x} - y$, where $\tilde{x} = 1234.56789999$. However, it is possible that the correct value for z is $x - \tilde{y}$, where $\tilde{y} = 1234.43219999$. But in this case, the correct value for z would be .13560001. Thus, we may have inserted noise in the wrong direction. Indeed, experiments show that noisy mode may either overestimate or underestimate the error. Nevertheless, it has been used successfully to produce an indication of the accuracy of the answer.

If noisy mode is supported by the hardware in the manner described above, it is quite easy to control the mode of the arithmetic. Suppose that we are coding in FORTRAN. We would want a subroutine NOISY which we could call when we wanted to enter or leave noisy mode. This subroutine would have to be written in Assembler language, but it could be called by FORTRAN programs. The call would have the form

$$\text{CALL NOISY (J)}$$

where J is an integer. Then NOISY would enter noisy mode if J is 1 and leave noisy mode if J is 0. If J is neither 0 nor 1, NOISY would not change the mode, but it would set J to 0 or 1 to indicate which mode the machine is in. With such a routine it would be easy to rerun the problem in noisy mode, and we could even write a DO loop to execute the program twice, first in normal mode and then in noisy mode. Then we could compare the answers before printing them.

If noisy mode is supported in this way, it is easy to incorporate it in a program. But we may have to exercise care in using it. The difficulties which can arise depend on the details of the floating-point arithmetic and on how the compiler handles various operations. As an illustration, consider the FORTRAN statement

$$X = I$$

This requires that the integer I be converted into a floating-point number and stored in X. There are various ways to do this, but a common approach is to begin by constructing an unnormalized floating-point number whose exponent is p and whose mantissa is $r^{-p}I$. (On many machines this can be accomplished by inserting the appropriate characteristic in the high-order digits of the word.) Then this number is normalized by adding zero to it. Suppose that this approach is used on an eight-digit decimal machine. If I has the value 2, we first construct the unnormalized number $.00000002 \times 10^8$, and then we add zero to it. But when the addition of zero is performed in noisy mode, it will produce the number 2.9999999. Since the integer I is usually exact, this value for X is unacceptable. It can lead us to produce a ridiculous answer when the problem is run in noisy mode. Thus, noisy mode does not always give us a good indication of the accuracy of our answer.

Noisy mode cannot be used blindly. We must examine the library to see whether the library subroutines will produce acceptable results when they are run in noisy mode, and we must understand when the compiler will compile floating-point instructions. It is quite likely that we shall encounter difficulties unless the compiler was specifically designed to compile programs that will run in both the normal mode and noisy mode.

7.4. INTERVAL ARITHMETIC

The use of interval arithmetic has been studied extensively by R. Moore, E. Hansen, and others. [See Moore (1966).] Although it has been used on many different machines, it has been implemented by calling subroutines instead of by hardware operation codes. The basic idea is that each number x in the calculation will be represented by an interval (x_1, x_2), where x_1 and x_2 are chosen in such a way that we can guarantee that $x_1 \leq x \leq x_2$. Thus, if we have an approximation $\tilde{x}$ for x with $|\tilde{x} - x| \leq \epsilon$, we would represent x by the interval $(\tilde{x} - \epsilon, \tilde{x} + \epsilon)$ instead of by the number $\tilde{x}$. We shall require that $x_2 \geq x_1$, but we shall allow the use of the degenerate interval (x, x) to represent a number which is known exactly.

Throughout the calculation we deal with intervals instead of numbers. Our objective is to represent the answer y by an interval (y_1, y_2) with $y_1 \leq y \leq y_2$. If the interval (y_1, y_2) is small enough, the midpoint of the interval provides a good approximation for y. For example, if y were represented by the interval $(1.2345612, 1.2345678)$, we could approximate y by $\tilde{y} = 1.2345645$. Then we would have $|\tilde{y} - y| \leq .33 \times 10^{-6}$. If the interval were large, say $(7, 29)$, we would get little information about y, but we would know that we did not have a good approximation for y. By contrast, if we were using normal floating-point arithmetic, we would produce a number, such as 12.345678, and have no indication of its accuracy.

In Sections 7.2 and 7.3 we saw that both significance arithmetic and noisy mode might overestimate the accuracy of the answer. But when interval arithmetic is carefully implemented and carefully used, it is possible to guarantee the result. That is, we can guarantee that if the answer y is represented by the interval (y_1, y_2), then $y_1 \leq y \leq y_2$. In this respect it is superior to either significance arithmetic or noisy mode, and this is one of the reasons for its receiving wider use.

When we use interval arithmetic, we perform arithmetic operations on intervals instead of on numbers. For example, instead of adding two numbers x and y, we "add" the intervals (x_1, x_2) and (y_1, y_2) to produce an interval (z_1, z_2) such that $z_1 \leq x + y \leq z_2$ holds if $x_1 \leq x \leq x_2$ and $y_1 \leq y \leq y_2$. The natural definitions for addition and subtraction are

$$(x_1, x_2) + (y_1, y_2) = (x_1 + y_1, x_2 + y_2)$$
(7.4.1)
$$-(x_1, x_2) = (-x_2, -x_1)$$
$$(x_1, x_2) - (y_1, y_2) = (x_1 - y_2, x_2 - y_1).$$

Similarly, the natural definition for multiplication is

$$(x_1, x_2) \cdot (y_1, y_2) = (z_1, z_2)$$
(7.4.2)
$$z_1 = \min(x_1 y_1, x_1 y_2, x_2 y_1, x_2 y_2)$$
$$z_2 = \max(x_1 y_1, x_1 y_2, x_2 y_1, x_2 y_2).$$

For example, if x_1 and y_1 are positive, we have

$$(x_1, x_2) \cdot (y_1, y_2) = (x_1 y_1, x_2 y_2),$$

and if $x_1 < 0 < x_2$ but $0 \leq y_1$, we have

$$(x_1, x_2) \cdot (y_1, y_2) = (x_1 y_2, x_2 y_2).$$

Similar definitions hold for the other cases. We do not define the division of (x_1, x_2) by (y_1, y_2) if zero lies in the interval (y_1, y_2). Otherwise, we would like to define division by

$$\frac{(x_1, x_2)}{(y_1, y_2)} = (z_1, z_2)$$
(7.4.3)
$$z_1 = \min\left(\frac{x_1}{y_1}, \frac{x_1}{y_2}, \frac{x_2}{y_1}, \frac{x_2}{y_2}\right)$$
$$z_2 = \max\left(\frac{x_1}{y_1}, \frac{x_1}{y_2}, \frac{x_2}{y_1}, \frac{x_2}{y_2}\right).$$

But we cannot use the definitions (7.4.1)–(7.4.3) directly, because the number of digits required to represent the end points of the intervals would grow too rapidly. Instead, we shall use only intervals whose end points are floating-point numbers. Since we want to be able to guarantee that the answer lies in the interval we have produced, the arithmetic operations will be defined by rounding the intervals on the right-hand side of (7.4.1)–(7.4.3) outward. For example, in place of (z_1, z_2) in (7.4.2) we use the interval (w_1, w_2), where w_1 is the largest number in $S(r, p)$ which is $\leq z_1$, and w_2 is the smallest number in $S(r, p)$ which is $\geq z_2$. The other operations are defined similarly.

We shall now consider the library programs which compute functions such as $\sin x$ or e^x. When we are using interval arithmetic, the argument will be an interval (x_1, x_2), and the answer will be an interval (y_1, y_2). For the function $f(x)$, we would like to have y_1 and y_2 defined by

$$
\begin{aligned}
y_1 &= \min_{x_1 \leq x \leq x_2} f(x) \\
y_2 &= \max_{x_1 \leq x \leq x_2} f(x).
\end{aligned}
$$
(7.4.4)

As above, the interval (y_1, y_2) is rounded outward to make the end points floating-point numbers. If the function $f(x)$ is monotonic, the numbers in (7.4.4) are quite easy to compute. For example, for the function e^x the result is the interval (e^{x_1}, e^{x_2}). But when the function $f(x)$ is not monotonic, the calculation of y_1 and y_2 in (7.4.4) may be more complicated. If the argument for the sine function is the interval (1.5, 1.6), we would like to produce the interval $(\sin 1.5, 1)$ as the answer. Thus, even for simple functions we may have to perform some tests to compute the values in (7.4.4). We would like to have an *interval library* which would contain function subroutines which would produce these values.

The principal problem with interval arithmetic is that we may produce intervals which are so large that the midpoint is not a good approximation for the answer. If we want four decimal digit accuracy in the answer, we have to produce an interval (y_1, y_2) whose end points y_1 and y_2 differ by at most 1 in the fourth digit. But as we shall see below, there are situations in which we may produce a very large interval, say (.5, 2), even though normal floating-point arithmetic would produce a good answer.

Interval arithmetic suffers from some of the same problems that arise in a manual error analysis. The bounds we obtain for the errors are based on the assumption that we have incurred the worst possible error at every step. But in practice we are usually more fortunate, so the bounds tend to be larger than the errors we produce in a typical calculation. This is an inherent property of a rigorous error analysis.

A much more serious problem is that we may produce a bound which is larger than it need be. The objective of an error analysis is to produce a bound which is reasonably close to the smallest rigorous bound. Unfortunately, interval arithmetic has a tendency to produce intervals which are larger than necessary. Suppose that a quantity x enters the calculation in more than one place and that the errors it introduces tend to compensate. Interval arithmetic usually does not recognize the fact that x must have the same value at every point in the calculation, and this makes it produce too large an interval for the answer. This is known as the *simultaneity problem*.

To illustrate this problem we shall consider an example. Suppose that we want to compute

$$(7.4.5) \qquad\qquad w = \frac{x+y}{x+z},$$

and that x, y, and z are represented by the intervals (x_1, x_2), (y_1, y_2), and (z_1, z_2), respectively.

Then (7.4.5) will be replaced by

$$(7.4.6) \qquad\qquad (w_1, w_2) = \frac{(x_1, x_2) + (y_1, y_2)}{(x_1, x_2) + (z_1, z_2)},$$

Suppose that the data are

$$(7.4.7) \qquad\qquad (x_1, x_2) = (1, 2)$$
$$(y_1, y_2) = (.01, .02)$$
$$(z_1, z_2) = (.001, .002).$$

Then the numerator of (7.4.6) is the interval $(1.01, 2.02)$, and the denominator is $(1.001, 2.002)$, so

$$(7.4.8) \qquad\qquad (w_1, w_2) = \frac{(1.01, 2.02)}{(1.001, 2.002)}.$$

Then (w_1, w_2) is the interval

$$\left(\frac{1.01}{2.002}, \frac{2.02}{1.001} \right)$$

rounded outward. If we use eight-digit decimal arithmetic, this yields

$$(7.4.9) \qquad\qquad (w_1, w_2) = (.50449550, 2.0179821).$$

But if we write (7.4.5) in the form

$$w = \frac{1 + y/x}{1 + z/x},$$

our interval formulation becomes

$$(7.4.10) \qquad\qquad (w_1, w_2) = \frac{(1, 1) + (y_1, y_2)/(x_1, x_2)}{(1, 1) + (z_1, z_2)/(x_1, x_2)}.$$

Here $(y_1, y_2)/(x_1, x_2) = (.005, .02)$, so the numerator is the interval $(1.005, 1.02)$. Similarly, the denominator is $(1.0005, 1.002)$, so (w_1, w_2) is the interval

$$\left(\frac{1.005}{1.002}, \frac{1.02}{1.0005}\right)$$

rounded outward. This yields

(7.4.11) $(w_1, w_2) = (1.0029940, 1.0194903)$,

which is a much better answer than (7.4.9).

When we used (7.4.6), we first reduced the problem to the calculation (7.4.8). The upper bound for w was obtained by dividing the largest numerator, 2.02, by the smallest denominator, 1.001. If the numerator in (7.4.8) is 2.02, x must be 2. But if the denominator is 1.001, x must be 1. Clearly x cannot have these two values simultaneously, so it is impossible for the worst case in the division in (7.4.8) to arise. Since our interval formulation of the problem (7.4.5) discarded the fact that x must have the same value in the numerator as it does in the denominator, we really computed

$$(w_1, w_2) = \frac{(x_1, x_2) + (y_1, y_2)}{(u_1, u_2) + (z_1, z_2)},$$

where $(u_1, u_2) = (1, 2)$ and the other intervals are given by (7.4.7). Even the formulation in (7.4.10) suffers slightly from the simultaneity problem, because we do not use the fact that x must have the same value in the two places it appears. Therefore, (7.4.11) does not yield the best bounds for w.

Unfortunately, it is not simply a matter of finding the best formula to use. Suppose that we want to solve the same problem, but with the data

(7.4.12)
$$(x_1, x_2) = (.001, .002)$$
$$(y_1, y_2) = (1.001, 1.002)$$
$$(z_1, z_2) = (1.001, 1.002).$$

Then both the numerator and the denominator of (7.4.6) are represented by the interval $(1.002, 1.004)$, so (w_1, w_2) is the interval

$$\left(\frac{1.002}{1.004}, \frac{1.004}{1.002}\right)$$

rounded outward. This yields

$$(w_1, w_2) = (.99800796, 1.0019961).$$

But if we use the formulation (7.4.10) with the data in (7.4.12), we find that

$(y_1, y_2)/(x_1, x_2) = (500.5, 1002)$, so the numerator is the interval $(501.5, 1003)$. We get the same interval for the denominator, so

$$(w_1, w_2) = \left(\frac{501.5}{1003}, \frac{1003}{501.5}\right) = (.5, 2).$$

For these data, (7.4.6) gives a much better answer than (7.4.10) does.

Thus, we cannot reject (7.4.6) and always use (7.4.10). Which of these formulations we should use depends on the data, and if we select the wrong one, we may produce an interval which is very much larger than necessary.

If the calculation of w in (7.4.5) were the crucial point in the program, we might use the approach we discussed for the function subroutines in the library. This would be more work, but it would overcome the simultaneity problem. We would set

$$f(x, y, z) = \frac{x + y}{x + z}$$

and write a program to compute w_1 and w_2, where

(7.4.13) $w_1 = \min f(x, y, z),$ $x_1 \leq x \leq x_2, y_1 \leq y \leq y_2, z_1 \leq z \leq z_2$

(7.4.14) $w_2 = \max f(x, y, z),$ $x_1 \leq x \leq x_2, y_1 \leq y \leq y_2, z_1 \leq z \leq z_2.$

As an illustration, we shall consider the case in which x_1, y_1, and z_1 are all positive. (Programming the general case forms Exercise 3.) Then

$$w_1 = \min_{x_1 \leq x \leq x_2} \frac{x + y_1}{x + z_2}$$

$$w_2 = \max_{x_1 \leq x \leq x_2} \frac{x + y_2}{x + z_1}.$$

Since the derivative of $(x + a)/(x + b)$ is $(b - a)/(x + b)^2$, it follows that

$$w_2 = \frac{x_2 + y_2}{x_2 + z_1} \qquad \text{if } y_2 \leq z_1$$

$$w_2 = \frac{x_1 + y_2}{x_1 + z_1} \qquad \text{if } y_2 > z_1.$$

We may use a similar approach to find w_1. For the data in (7.4.7), this yields the interval

$$(w_1, w_2) = (1.0039960, 1.0189811).$$

This shows that even the interval in (7.4.11) was larger than necessary.

Thus, although interval arithmetic gives us guaranteed bounds, it has a tendency to produce intervals which are too large. When it is used carefully, it can be quite effective. [See Moore (1965a).] However, it does not appear to be a panacea which will replace floating-point arithmetic as the standard computing procedure.

We need two floating-point numbers to represent an interval, so interval arithmetic doubles the storage required to hold the data. It also makes the arithmetic more complicated. In these respects it is quite similar to double-precision arithmetic. Indeed, when we use interval arithmetic we are faced with the same programming problems that face us when we use double-precision arithmetic with a compiler which does not support the double-precision data type. Therefore, we can use the techniques discussed in Section 5.7. Arithmetic statements must be replaced by subroutine calls, and we must use some technique to allocate two words for every variable. It is often convenient to use an additional subscript for this purpose, or we may be able to use the COMPLEX data type. A few compilers have supported the data type OTHER, which has proved to be very effective for interval arithmetic.

7.5. RERUNNING THE PROGRAM IN HIGHER-PRECISION

Each of the approaches to automatic error analysis discussed in Sections 7.2, 7.3, and 7.4 has been used successfully, but none of them has acquired wide usage as yet. As we saw, they had to be used with care. It is to be hoped that we shall learn more about the use of techniques such as these, so that it will be easier to use the computer for the automatic analysis of error.

Presently, the commonest way to use the computer to study the accuracy of our answers is simply to rerun the program in higher-precision. Suppose that our original answer was y_1, and that when we reran the program in higher-precision we produced an answer y_2. Our objective is to determine the accuracy of y_1, and for this purpose we need to know only the first two or three decimal digits of the error. Therefore, if y_2 is accurate to at least two or three more decimal places than y_1 is, we can use $y_1 - y_2$ and $(y_1 - y_2)/y_2$ to estimate the error and the relative error in y_1.

The use of this approach is based on the assumption that running the program in higher-precision will produce a more accurate answer. While this will generally be true, there are several situations in which it may fail to hold. First, the error in the answer may be due primarily to errors in the data. Then the answer would not be improved by increasing the precision of the calculation. By contrast, both significance arithmetic and interval arithmetic offer us a way to study the effect that noise in the data has on the answer. Another reason that increasing the precision might not increase the accuracy of the answer is that our program might have used approximations which are not accurate to higher-precision. Increasing the precision cannot correct errors due to such approximations. Rather, it is a technique for studying the effect that rounding error has on the answer. The question of whether this approach will provide a good measure of this effect will be discussed below.

A second assumption is that it is easy to rerun the program in higher-precision. This may or may not be true. The conversion of a FORTRAN or PL/I program from single-precision to double-precision was discussed in Section 5.1. We saw that it is easy to overlook one variable or constant in the program and produce a result which is good to only single-precision accuracy. A more difficult problem arises if the compiler does not support the high-precision data type. Then we have to use the techniques discussed in Section 5.7.

We shall now turn to the question of whether rerunning the program in higher-precision will produce a more accurate answer. To avoid the problems described above, we shall assume that the data are known to as many places as we need and that any approximations used in the program are accurate to the highest precision we use. Then increasing the precision of the arithmetic will usually increase the accuracy of the answer, but we cannot guarantee that it will do so. Indeed, we shall give examples of situations in which it does not. Fortunately, such situations are quite rare. In most instances, rerunning the program in higher-precision will provide a good estimate of the effect of rounding error.

To study the effect of precision on the accuracy of the answer, we shall select a problem whose answer is known and solve it in $FP(r, p, c)$ for several values of p. The accuracy of the result $\tilde{y}$ produced by the program will be determined by comparing it with the correct answer y. We shall plot d versus p, where d is the number of correct digits in $\tilde{y}$.

There are several ways to perform such experiments. On some variable word length machines the number of digits in the floating-point numbers can be specified at the beginning of the compilation. Then we can simply recompile the program specifying different values for p. But more often we have to use subroutines to produce results we want. For example, we can code subroutines to perform addition, subtraction, multiplication, and division in $FP(r, p, c)$, where p is specified by the user.

We shall adopt a somewhat different approach. Suppose that $FP(r, p_{max}, clq)$ is the highest-precision arithmetic supported by the machine. We shall perform the arithmetic operations in $FP(r, p_{max}, clq)$ and then chop the results to p digits. Then we are really performing the arithmetic in $FP(r, p, clq')$, where $q' = q + p_{max} - p$. All variables will be declared to have the precision p_{max}, and we shall have to write a special subroutine CHOP which will chop a number X to p digits, replacing the low-order $p_{max} - p$ digits by zeros. Instead of writing a FORTRAN statement such as X = A + B + C, we must write

$$X = A+B$$
$$\text{CALL CHOP (X)}$$
$$X = X+C$$
$$\text{CALL CHOP (X)}$$

CHOP should be written in such a way that the value of p can be specified by calling it with a different entry point.

The subroutine CHOP can also be used to handle input data. We may enter numbers with an accuracy of p_{max} and then chop them to p digits. We can even use this approach for library programs such as SIN(X) and SQRT(X). To avoid having to write p-digit versions of these subroutines, we shall simply call the version of the subroutine for the precision p_{max} and then chop the result to p digits. Since the subroutine used arithmetic with the precision p_{max}, this may produce a slightly better result than we would expect to produce on a p-digit machine. But it will be a reasonable result, and with this approach we do not have to rewrite the library for each precision p.

Our objective is to plot d versus p, where d is the number of correct digits in the answer. We would expect to produce a graph such as the one shown in Figure 7.5.1. For each digit we add to the precision of the arithmetic, we expect to get one more digit of accuracy in the answer, so we expect the points to lie on a straight line with a slope of $45°$. If the x intercept of this line is α, we have lost α digits of accuracy in the calculation. This is a measure of the condition of the problem.

However, the points will not fall exactly on a straight line. One reason for this is the discreteness of the number of correct digits in the answer. To overcome this difficulty, we shall use a continuous measure of accuracy.

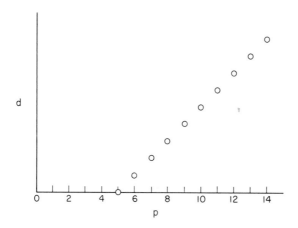

Figure 7.5.1

Suppose that the correct answer is y and that we have produced a value $\tilde{y}$. We let

(7.5.1) $e = |\tilde{y} - y|,$

and note that $-\log_r \epsilon$ has the property that it is increased by 1 when ϵ is divided by r. To produce a measure of the number of correct digits in $\tilde{y}$, we determine e such that

$$r^e > |y| \geq r^{e-1}$$

and set

(7.5.2) $d = e + \log_r .5 - \log_r \epsilon.$

If ϵ is one-half a unit in the ith position of y, we have $\epsilon = \frac{1}{2}r^{e-i}$, so $d = i$. Then the integer part of d is the number of correct digits in $\tilde{y}$, and (7.5.2) gives us a continuous measure of the accuracy of $\tilde{y}$.

The question of how to measure the number of digits of accuracy in the answer becomes more complicated when we consider a program whose output is several numbers instead of one. For example, a matrix inversion program produces n^2 numbers, and some of these numbers will be more accurate than others. Usually it is best to use an appropriate norm to measure the accuracy of the answer. In our discussion here, we shall consider only programs that produce a single number as the answer.

We first consider two simple problems to illustrate the way our results may depart from the behavior shown in Figure 7.5.1. Let $y = \frac{1}{47}$ and $\tilde{y} = 1 \div 47$. When the arithmetic is performed in FP(16, p, c), we produce the results shown in Figure 7.5.2. This is about the sort of behavior we would expect. The points do indeed follow the line quite well, but they do not lie exactly on the line. The variations from the line are due to variations in the digits chopped. The sixth hexadecimal digit of $\frac{1}{47}$ is zero, so we produced the same results using six digits as we did using five digits. Thus, we cannot guarantee that increasing the precision of the arithmetic will increase the accuracy of the result.

We next consider a problem in which the behavior is a little more erratic. Let

(7.5.3) $y = (\frac{1}{7})^2 - (\frac{68}{483})(\frac{70}{483}).$

To compute $\tilde{y}$, we set

$$A = 1 \div 7$$
$$B = 68 \div 483$$
$$C = 70 \div 483$$
$$\tilde{y} = (A * A) \ominus (B * C).$$

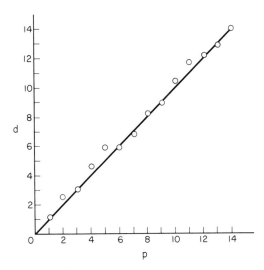

Figure 7.5.2

Figure 7.5.3 shows the results produced when this calculation was performed in FP(16, p, c). The numerical values for d were

p	d
4	.42
5	2.64
6	2.64
7	4.00
8	6.37
9	5.38
10	6.47
11	8.11
12	8.48
13	9.96
14	12.46

Of particular interest is the fact that we lost one digit of accuracy by increasing the precision from eight digits to nine digits.

To see how this loss of accuracy can happen, we examine the calculation of $\tilde{y}$ more closely. Let

$$u = (\tfrac{1}{7})^2$$
$$v = (\tfrac{68}{483})(\tfrac{70}{483}),$$

so $y = u - v$. Also, let

$$\tilde{u} = A * A$$
$$\tilde{v} = B * C,$$

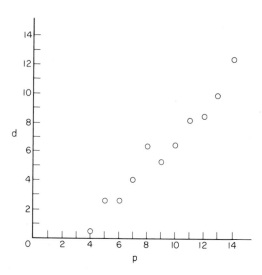

Figure 7.5.3

so $\tilde{y} = \tilde{u} \ominus \tilde{v}$. We have used chopped arithmetic in the calculation of $\tilde{u}$ and $\tilde{v}$, so $\tilde{u} \leq u$ and $\tilde{v} \leq v$. Write

$$u = \tilde{u} + \epsilon_1, \qquad \epsilon_1 \geq 0$$
$$v = \tilde{v} + \epsilon_2, \qquad \epsilon_2 \geq 0.$$

Then

(7.5.4) $\epsilon = |\tilde{y} - y| = |\epsilon_1 - \epsilon_2|.$

We expect the number of digits accuracy in $\tilde{u}$ and $\tilde{v}$ to behave more or less like the graph in Figure 7.5.2. When p is increased by 1, we expect ϵ_1 and ϵ_2 to be decreased by about a factor of 16. This does indeed happen when p is increased from 8 to 9. But the errors ϵ_1 and ϵ_2 are both nonnegative, so they tend to compensate. For $p = 8$, ϵ_1 and ϵ_2 are nearly equal, so $|\epsilon_1 - \epsilon_2|$ is about $\frac{1}{256}$ as large as either ϵ_1 or ϵ_2. When p is increased to 9, ϵ_1 is decreased by more than a factor of 16, so $\epsilon \approx |\epsilon_2|$. Even though ϵ_2 has been decreased by a factor of almost 16, it is still about 16 times as large as ϵ was for $p = 8$. Thus, the decrease in accuracy when p is increased from 8 to 9 is due to the fact that the errors ϵ_1 and ϵ_2 in (7.5.4) almost compensate when p is 8, so the answer is more accurate than we would expect it to be.

In spite of the unexpected behavior when we increased p from 8 to 9, the results shown in Figure 7.5.3 do not invalidate the approach of increasing the precision to test the accuracy of the answer. But they suggest that increasing p by one or two digits will not suffice. Since we normally increase the precision from single-precision to double-precision or from double-precision to triple-precision, it is typical to increase p by several digits rather than one or two digits. We hope that this is enough to overcome such local anomalies. Increasing p by five or six digits would allow us to produce a good estimate of the error in the computation in (7.5.3).

The results shown in Figure 7.5.3 displayed a slight irregularity because the answer produced when $p = 8$ was more accurate than we expected it to be. But this discrepancy can be much larger. Suppose that we want to compute $y = A - B$, where

$$A = 2.3456110000000234567$$

and

$$B = 1.1111109999999111111.$$

Figure 7.5.4 shows the results produced when this computation is performed in FP(10, p, c). The answer produced using five- or six-digit arithmetic is far more accurate than we expect it to be. In fact, it is more accurate than the answer produced using 13-digit arithmetic. Thus, we cannot guarantee that increasing the precision by five or six digits will increase the accuracy of the answer.

The anomalies we observed in Figures 7.5.3 and 7.5.4 were due to the fact that we produced an unusually good answer for certain values of p. Our next example shows a different way in which increasing p can fail to produce a more

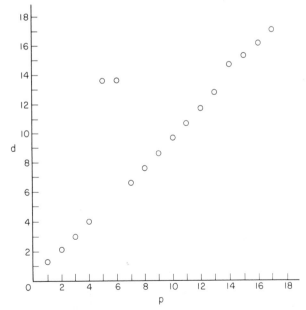

Figure 7.5.4

accurate answer. Let

(7.5.5) $y = x + A \sin Bx,$

where

$$x = \tfrac{1}{7}$$
$$A = 10^{-8}$$
$$B = 10^{16}.$$

Figure 7.5.5 shows the results produced when the computation is performed in FP(16, p, c). The numerial values are shown in the following table:

p	d
1	1.20
2	1.59
3	2.71
4	4.20
5	4.58
6	5.66
7	6.13
8	6.20
9	6.21
10	6.30
11	6.54
12	6.67
13	6.79
14	7.40
15	8.54
16	9.47
17	10.36
18	11.54
19	12.47
20	13.36

The flat part of this graph ranging from $p = 7$ through $p = 13$ differs dramatically from the behavior we expect.

To see what happens in the calculation of (7.5.5), write $z = Bx$. If $Bx > 16^p$, our approximation $\bar{z}$ for z is likely to have an error of several radians. Then the result computed for sin z is pure noise. Now $16^{13} > 10^{16}/7 > 16^{12}$, so when $p = 13$ the error in $\bar{z}$ is less than 1 radian. Increasing p beyond 13 produces the sort of behavior we expect. For $p < 7$, the contribution of $A \sin Bx$ is negligible, so this part of the graph is also normal.

Figures 7.5.4 and 7.5.5 show that we cannot guarantee that increasing the precision will improve the answer. These examples may appear to be rather

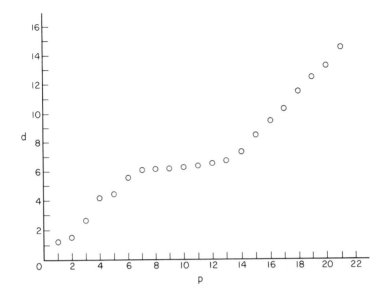

Figure 7.5.5

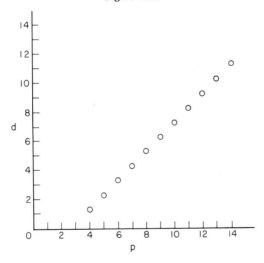

Figure 7.5.6

pathological, but that is because increasing the precision by five or six digits usually does increase the accuracy of the answer. The behavior shown in Figures 7.5.2 and 7.5.3 is much more typical.

To illustrate the effect of precision on a less trivial computation, we consider the two problems discussed in Sections 4.1 and 4.2. Figure 7.5.6 shows the result produced when the quadrature problem of Section 4.1 was

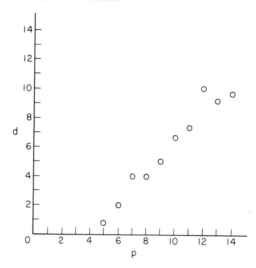

Figure 7.5.7

run in FP(16, p, c) with $N = 2^{13}$. Similarly, Figure 7.5.7 shows the results produced when we use the power series to compute e^{-7} and perform the arithmetic in FP(16, p, clq'), where $q' = 15 - p$.

EXERCISES

1. Write a subroutine to perform the operation of multiplication in interval arithmetic. That is, the subroutine should compute the values of z_1 and z_2 in (7.4.2) and round the interval (z_1, z_2) outward.

2. Write a sine routine to be used with interval arithmetic. The input is the interval (x_1, x_2), and the routine should compute

$$y_1 = \min_{x_1 \le x \le x_2} \sin x$$

$$y_2 = \max_{x_1 \le x \le x_2} \sin x$$

and round the interval (y_1, y_2) outward.

3. Let

$$f(x, y, z) = \frac{x + y}{x + z}.$$

Write a subroutine which can be used to compute this function when we are using interval arithmetic. The input will be the intervals (x_1, x_2), (y_1, y_2), and (z_1, z_2). Assume that the interval $(x_1, x_2) + (z_1, z_2)$ does not contain zero. The output will be the interval (w_1, w_2) rounded outward, where w_1 and w_2 are given by (7.4.13) and (7.4.14).

4. Let X be in $S(r, p)$ and let N be a positive integer. Select several values of X close to 1 and compute X^{1023} using one of the algorithms described in Section 3.6. (The value of X must be close enough to 1 so that you do not encounter either overflow or underflow.) Perform the calculation using both interval arithmetic and normal floating-point arithmetic. Compare the interval produced with the error bound obtained in Section 3.6.

5. Compute $\frac{1}{47}$ in hexadecimal and show that the sixth hexadecimal digit is zero. Verify that the points in Figure 7.5.2 are above or below the line depending on the size of the first digit dropped.

6. Write the subroutine CHOP described in Section 7.5 and perform some experiments of the sort described there.

8 RADIX CONVERSION

Unless we are using a decimal machine, we are faced with the problem of radix conversion for both input and output. These conversion programs will be the subject of this chapter, but before discussing the conversion techniques themselves, we shall try to determine the number of decimal digits needed to produce an accuracy which is "equivalent" to the accuracy we are using inside the machine. We suppose that we are using a machine which performs arithmetic in a system FP(r, p, a), where $r \neq 10$. Then we want to know the number of decimal digits needed to produce the same accuracy as p-digit numbers in the radix r. This will not depend on the sign of the numbers, so we may restrict our attention to positive numbers.

Consider the set $S(r, p)$ of p-digit floating-point numbers in the radix r. The numbers in $S(r, p)$ which lie in the interval $r^{e-1} \leq x \leq r^e$ are uniformly spaced at a distance r^{e-p} apart. We shall designate this distance by δ and we note that δ changes as we go past a power of r.

We shall now consider two systems FP(r_1, p_1, c) and FP(r_2, p_2, c) with $r_1 \neq r_2$. Unless one of the radices is a power of the other, the spacing of the numbers in the sets $S(r_1, p_1)$ and $S(r_2, p_2)$ will change at different points. Consider an interval I: $a \leq x \leq b$ which is large enough to contain at least two points of each of the sets $S(r_1, p_1)$ and $S(r_2, p_2)$. Suppose that

(8.1.1)
$$r_1^{e_1-1} \leq a < b \leq r_1^{e_1}$$
$$r_2^{e_2-1} \leq a < b \leq r_2^{e_2},$$

and let $\delta_i = r_i^{e_i-p_i}$ be the distance between consecutive points of $S(r_i, p_i)$ in I.

For any real number x in I, the bound for the error incurred by chopping x to $S(r_i, p_i)$ is δ_i. Then the system with the smaller δ_i is the more accurate in I in the sense that the maximum error incurred by chopping a number to $S(r_i, p_i)$ is smaller. This will be our criterion for comparing the systems, so we shall say that $FP(r_1, p_1, c)$ is more accurate than $FP(r_2, p_2, c)$ if $\delta_1 < \delta_2$, that is, if

$$(8.1.2) \qquad r_1^{e_1 - p_1} < r_2^{e_2 - p_2}.$$

[We would also arrive at the criterion (8.1.2) if we compared $FP(r_1, p_1, R)$ with $FP(r_2, p_2, R)$.]

This criterion depends on I, so it suggests that we might prefer one of the systems $FP(r_i, p_i, c)$ in some intervals and the other system in other intervals. This is often true. As an illustration, we shall compare $FP(16, 6, c)$ with $FP(10, p, c)$. In the interval $.1 \leq x \leq 1$, the value of δ is $16^{-6} = 2^{-24}$ for $FP(16, 6, c)$, and it is 10^{-p} for $FP(10, p, c)$. Now $10^{-7} > 2^{-24} > 10^{-8}$, so in this interval we find that $FP(16, 6, c)$ is more accurate than $FP(10, 7, c)$ but not as accurate as $FP(10, 8, c)$. Indeed, $2^{-24} \approx .6 \times 10^{-7}$, so for $.1 \leq x < 1$ the six-digit hexadecimal machine is more accurate than a seven-digit decimal machine by about one bit.

Now suppose that $1 < x < 10$. The values of δ for $FP(16, 6, c)$ and $FP(10, p, c)$ are 2^{-20} and $10^{-(p-1)}$, respectively. Since $2^{-20} \approx .954 \times 10^{-6}$, $FP(16, 6, c)$ is only slightly more accurate than $FP(10, 7, c)$ in this interval. Next, suppose that $10 \leq x \leq 16$. The value of δ for $FP(16, 6, c)$ is still 2^{-20}, but the decimal representation of x requires two digits to the left of the decimal point, so δ is $10^{-(p-2)}$ for $FP(10, p, c)$. Then $FP(16, 6, c)$ is slightly more accurate than $FP(10, 8, c)$ in the interval $10 \leq x \leq 16$.

Finally, suppose that $16 \leq x \leq 100$. The hexadecimal representation of x requires two digits to the left of the radix point, so the value of δ for $FP(16, 6, c)$ is 2^{-16}. But δ is still $10^{-(p-2)}$ for $FP(10, p, c)$, so $FP(10, 7, c)$ is more accurate than $FP(16, 6, c)$ for numbers in this interval.

Thus, the number of decimal digits needed to produce the same accuracy as $FP(16, 6, c)$ depends on the size of the numbers considered. The accuracy of $FP(16, 6, c)$ may lie between that of $FP(10, 6, c)$ and $FP(10, 7, c)$, between that of $FP(10, 7, c)$ and $FP(10, 8, c)$ or between that of $FP(10, 8, c)$ and $FP(10, 9, c)$, because the spacing of the numbers in $S(r, p)$ changes when we go past a power of r, so the spacing of the numbers in $S(16, 6)$ and $S(10, p)$ changes at different points.

To establish bounds for the variation in the equivalent number of digits, we shall consider the relative error. Let $x = r^e m$, $r^{-1} \leq m < 1$, and let $\bar{x}$ be x chopped to $S(r, p)$, so $\bar{x} = r^e \bar{m}$. Then $\bar{x} = (1 - \rho)x$, where

$$\rho = \frac{m - \bar{m}}{m}.$$

Write $\epsilon = m - \bar{m}$, so

(8.1.3)
$$p = \frac{\epsilon}{\bar{m} + \epsilon}, \qquad 0 \le \epsilon < r^{-p}.$$

Let $a = r^e \bar{m}$ and $b = a + r^{e-p}$. Then (8.1.3) holds for any x in the interval $a \le x < b$. Let β be the least upper bound for the relative error p introduced by chopping a number x in the interval $a \le x < b$ to $S(r, p)$. Since $dp/d\epsilon > 0$,

$$\beta = \frac{r^{-p}}{\bar{m} + r^{-p}},$$

and since $r^{-1} \le \bar{m} \le 1 - r^{-p}$,

(8.1.4)
$$r^{-p} \le \beta \le \frac{r^{-(p-1)}}{1 + r^{-(p-1)}}.$$

Thus, if I is any interval containing at least two points of $S(r, p)$, the least upper bound for the relative error introduced by chopping numbers in I to $S(r, p)$ is at least r^{-p} and at most $r^{-(p-1)}/(1 + r^{-(p-1)})$.

Instead of using the bound for absolute error as the criterion for deciding which of the systems $\mathrm{FP}(r_i, p_i, c)$ is the more accurate, we could use the bound for the relative error. Let I be an interval which contains at least two points of each of the sets $S(r_1, p_1)$ and $S(r_2, p_2)$, so the bound for the relative error introduced by chopping a number in I to $S(r_i, p_i)$ is at least $r_i^{-p_i}$ and at most

$$\frac{r_i^{-(p_i-1)}}{1 + r_i^{-(p_i-1)}}.$$

If

(8.1.5)
$$\frac{r_1^{-(p_1-1)}}{1 + r_1^{-(p_1-1)}} < r_2^{-p_2},$$

we always get a smaller bound for the relative error by chopping the numbers in I to $S(r_1, p_1)$ instead of $S(r_2, p_2)$. Since (8.1.5) does not depend on I, this would lead us to say that $\mathrm{FP}(r_1, p_1, c)$ is always more accurate than $\mathrm{FP}(r_2, p_2, c)$ when (8.1.5) holds. It is sometimes convenient to write the criterion (8.1.5) in the equivalent form

(8.1.6)
$$r_2^{p_2} < r_1^{p_1-1} + 1.$$

We shall now show that when (8.1.5) holds we would also say that $\mathrm{FP}(r_1, p_1, c)$ is always more accurate than $\mathrm{FP}(r_2, p_2, c)$ if we used the criterion based on absolute error. Suppose that (8.1.5) holds and consider an interval I which contains at least two points of each of the sets $S(r_i, p_i)$ and whose end points a and b satisfy (8.1.1). Since I contains at least two points of $S(r_1, p_1)$, it follows from (8.1.1) that

$$r_2^{e_2} \geq r_1^{e_1}(r^{-1} + r^{-p})$$

With (8.1.5), this yields

$$r_2^{e_2-p_2} = r_2^{e_2}r_2^{-p_2} > r_1^{e_1}(r^{-1} + r^{-p})\frac{r_1^{-(p_1-1)}}{1 + r_1^{-(p_1-1)}} = r_1^{e_1-p_1}.$$

Thus, (8.1.2) holds, so we conclude that $FP(r_1, p_1, c)$ is always more accurate than $FP(r_2, p_2, c)$ when (8.1.5) holds.

We again compare $FP(16, 6, c)$ with $FP(10, p, c)$. The values of r^p and r^{p-1} for $FP(16, 6, c)$ are 2^{24} and 2^{20}. Now $10^6 < 2^{20}$ and $2^{24} < 10^8$, so $FP(16, 6, c)$ is always more accurate than $FP(10, 6, c)$ but less accurate than $FP(10, 9, c)$.

When we say that a system $FP(r_1, p_1, c)$ is more accurate than a system $FP(r_2, p_2, c)$, we are referring to the bounds for the error due to chopping. This does not mean that for every number x, less error will be introduced if we chop x to $S(r_1, p_1)$ instead of $S(r_2, p_2)$. For example, the decimal number .1 is in $S(10, 3)$ but not in $S(2, 100)$, so we introduce more error by chopping it to $S(2, 100)$ than we do by chopping it to $S(10, 3)$. Nevertheless, the comparison based on error bounds appears to be a reasonable way to compare the accuracy of systems with different radices.

Some slight variations of the criterion (8.1.6) are useful. For example, (8.1.6) follows from

(8.1.7) $$r_1^{-(p_1-1)} \leq r_2^{-p_2},$$

and (8.1.7) is often easier to remember. We might also use the criterion

(8.1.8) $$r_2^{p_2} \leq r_1^{p_1-1} + 1.$$

If equality holds in (8.1.7), $FP(r_1, p_1, c)$ is always at least as accurate as $FP(r_2, p_2, c)$, and there are some intervals in which it is more accurate. The criterion (8.1.7) was obtained by Goldberg (1967), and (8.1.8) was obtained by Matula (1968b). Each of them arrived at the criterion by considering the effect of conversion and reconversion.

Comparisons of the accuracy of systems $FP(r_i, p_i, c)$ are fundamental to many aspects of our programs. For instance, they are often the basis for our decisions about the number of decimal digits we should specify in our input and output formats. They also affect the decisions made by the compilers about the typing of variables and constants. For example, in the implementation of PL/I for the IBM System/360, a floating-point constant will be treated as single-precision if it has at most six decimal digits, but it will be treated as double-precision if it has more than six decimal digits. This decision is based on the criterion (8.1.5).

When we want to compare the overall accuracy of two different machines, there is another viewpoint we may take. Suppose that we are asked whether we would prefer to have the arithmetic performed in $FP(r_1, p_1, c)$ or $FP(r_2, p_2, c)$. The question is easily answered if $r_2^{p_2} \leq r_1^{p_1-1} + 1$ or if $r_1^{p_1} \leq r_2^{p_2-1} + 1$. But if neither of these relations holds, which of the systems is the more accurate may depend on the size of the numbers we are working with. In that case, it is reasonable to base our decision on the bounds for the relative error. Then we would select $FP(r_1, p_1, c)$ in preference to $FP(r_2, p_2, c)$ if

(8.1.9)
$$\frac{r_1^{-(p_1-1)}}{1 + r_1^{-(p_1-1)}} < \frac{r_2^{-(p_2-1)}}{1 + r_2^{-(p_1-1)}}.$$

This criterion is equivalent to

$$r_2^{p_2-1} + 1 < r_1^{p_1-1} + 1$$

or

(8.1.10)
$$r_1^{-(p_1-1)} < r_2^{-(p_2-1)}.$$

When we use this criterion, we are comparing the worst case for the relative error in $FP(r_1, p_1, c)$ with the worst case for the relative error in $FP(r_2, p_2, c)$. This appears to be a reasonable basis for comparing the accuracy of the systems $FP(r_i, p_i, c)$ over a wide range of problems. On the basis of (8.1.10), we would prefer $FP(16, 6, c)$ to $FP(10, 7, c)$, but we would prefer $FP(10, 8, c)$ to $FP(16, 6, c)$.

8.2. PROPERTIES OF CONVERSION TRANSFORMATIONS

When we write input or output statements in a language such as FORTRAN or PL/I, there are various formats we may use for floating-point numbers. But it is usually quite easy to change a number from one of these formats to another, so we shall consider only the E format. Then the problem becomes one of converting a floating-point number in $S(r_1, p_1)$ to a floating-point number in $S(r_2, p_2)$. One of the sets $S(r_i, p_i)$ is the set of decimal floating-point numbers, and the other is the set of floating-point numbers handled by the machine we are using. By considering the general case, we may discuss input and output conversion simultaneously.

Mathematically, the conversion program is a *transformation* τ of the set $S(r_1, p_1)$ into the set $S(r_2, p_2)$. To each element a in $S(r_1, p_1)$, τ assigns an element $\tau(a)$ in $S(r_2, p_2)$, called the *image* of a. It is convenient to use the standard terminology for transformations to describe properties of conversion programs. The transformation τ is said to be *well defined* if the image $\tau(a)$ is independent of the representation of a. We note that there are many different ways to represent an input number in an E format, for example,

$$27.E0, \quad 2700.E-2, \quad .0027E4, \quad 2.70E1,$$

etc. If τ is well defined, all these representations of the number 27 will be converted into the same number. For some conversion programs, the transformation τ is not well defined.

Suppose that τ is a transformation of $S(r_1, p_1)$ into $S(r_2, p_2)$. We say that τ is a transformation of $S(r_1, p_1)$ *onto* $S(r_2, p_2)$ if for every element b in $S(r_2, p_2)$ there is at least one element a in $S(r_1, p_1)$ with $b = \tau(a)$. Thus, the statement that τ is onto $S(r_2, p_2)$ makes the additional assertion that every element in $S(r_2, p_2)$ is the image of some element in $S(r_1, p_1)$ under τ.

The transformation τ is said to be *one to one* if distinct elements of $S(r_1, p_1)$ have distinct images. Thus, the assertion that the transformation τ is one to one means that for a and b in $S(r_1, p_1)$, $\tau(a) = \tau(b)$ if and only if $a = b$. The statement that τ is one to one makes no assertion about whether or not τ is a transformation onto $S(r_2, p_2)$.

We shall say that the transformation τ is *monotone* provided that $\tau(a) \leq \tau(b)$ holds for every pair of elements a, b in $S(r_1, p_1)$ with $a < b$. If $\tau(a) < \tau(b)$ holds for every pair of elements in $S(r_1, p_1)$ with $a < b$, τ is said to be *strictly monotone*. It follows at once that τ is strictly monotone if and only if it is both monotone and one to one.

Since we have both input and output conversion programs, we also have a transformation σ of $S(r_2, p_2)$ into $S(r_1, p_1)$. We define a transformation $\sigma\tau$ of $S(r_1, p_1)$ into itself by setting

$$\sigma\tau(a) = \sigma(\tau(a))$$

for all a in $S(r_1, p_1)$. It is natural to ask whether $\sigma\tau(a) = a$ holds for all a in $S(r_1, p_1)$.

We may also define a transformation $(\sigma\tau)^2$ of $S(r_1, p_1)$ into itself by setting

$$(\sigma\tau)^2(a) = \sigma\tau[\sigma\tau(a)]$$

for all a in $S(r_1, p_1)$. This definition is easily extended to $(\sigma\tau)^n$ for every positive integer n. We shall be interested in the behavior of $(\sigma\tau)^n$, that is, in the effect of converting and reconverting a number n times.

The terminology we have introduced is the standard mathematical terminology for properties of transformations. But a conversion program has the additional requirement that for each a in $S(r_1, p_1)$, $\tau(a)$ must be approximately a. For any real number a, let a_L and a_R be the left and right neighbors of a in $S(r_2, p_2)$. That is, a_L is the largest number in $S(r_2, p_2)$ which is $\leq a$, and

a_R is the smallest number in $S(r_2, p_2)$ which is $\geq a$. If a is in $S(r_2, p_2)$, $a_L = a_R = a$. We shall say that τ is a *neighbor transformation* if τ converts each a in $S(r_1, p_1)$ into one of its neighbors a_L, a_R. Any conversion transformation which is not a neighbor transformation introduces unnecessarily large error. There are two neighbor transformations which are of particular interest. We say that τ is a *truncation conversion transformation* if $\tau(a)$ is the number $\bar{a}$ obtained by chopping a to $S(r_2, p_2)$. Similarly, τ is a *rounding conversion transformation* if $\tau(a)$ is the number $\overset{\circ}{a}$ obtained by rounding a to $S(r_2, p_2)$.

There are some numbers which belong to both $S(r_1, p_1)$ and $S(r_2, p_2)$. For example, this is true of small integers. It is extremely desirable for τ to have the property that $\tau(a) = a$ whenever a is in both $S(r_1, p_1)$ and $S(r_2, p_2)$. This always holds for a neighbor transformation.

Consider an interval I which does not contain a power of either r_1 or r_2. For $i = 1, 2$, let S_i be the set consisting of those points in $S(r_i, p_i)$ which lie in I. The spacing of the numbers in $S(r_i, p_i)$ is uniform in any interval which does not contain a power of r_i, so the points in each S_i are uniformly spaced. Let τ be a transformation of $S(r_1, p_1)$ into $S(r_2, p_2)$ such that $\tau(a) \approx a$ holds for all a in $S(r_1, p_1)$. If a is in S_1, we would expect $\tau(a)$ to be in S_2. This need not always be true, because points near the ends of I might have images outside of I, but it should hold for almost all a in S_1. Similarly, if an element b in S_2 is the image $\tau(a)$ of some element a in $S(r_1, p_1)$, we would expect a to be in S_1. Again, this may not always be true, but it will be true for points that are not too close to the ends of I.

Now suppose that the transformation τ is onto. Then every point in S_2 is the image of at least one point in $S(r_1, p_1)$, and most of the points in S_2 are images of points in S_1. If every point in S_2 were the image of some point in S_1, we would conclude that S_1 must have at least as many points as S_2 does. Clearly S_2 cannot have very many more points than S_1 does. Therefore, we expect the points of S_1 to be closer together than the points of S_2 are. We saw in Section 8.1 that this will be true for all such intervals I if (8.1.8) holds. Thus, if τ is onto, we expect that

(8.2.1) $r_2^{p_2} \leq r_1^{p_1 - 1} + 1.$

Similarly, if τ is one to one, the images of the points in S_1 are all distinct, and almost all of them lie in S_2. Therefore, we expect S_2 to have at least as many points as S_1. This will hold for all such intervals I if

(8.2.2) $r_1^{p_1} \leq r_2^{p_2 - 1} + 1.$

The statements we have made are rather vague. We have said that we expect (8.2.1) to hold if τ is onto and we expect (8.2.2) to hold if τ is one to one, but we have not stated that they must hold, because we have considered the general case in which we know only that $\tau(a) \approx a$. Much stronger statements can be made if we are more specific about τ. Matula (1968b) considers the case in which $r_1^i \neq r_2^j$ for any positive integers i and j, and he shows that if τ is either a truncation conversion transformation or a rounding conversion transformation, then (8.2.1) is necessary and sufficient for τ to be onto and (8.2.2) is necessary and sufficient for τ to be one to one.

Consider a machine which performs arithmetic in FP(r, p, c), and suppose that we use E formats for input and output with p'-digit decimal numbers. Then we are working with the sets $S(r, p)$ and $S(10, p')$. If

$$(8.2.3) \qquad 10^{p'} \leq r^{p-1} + 1,$$

the points in $S(r, p)$ are always at least as close together as the points in $S(10, p')$. Then the input conversion can be one to one, and the output conversion can be onto. On the other hand, if

$$(8.2.4) \qquad r^p \leq 10^{p'-1} + 1,$$

the input conversion can be onto and the output conversion can be one to one.

If neither (8.2.3) nor (8.2.4) holds, we would expect that neither the input conversion nor the output conversion would be onto and that neither of them would be one to one. (Matula's results show that this is the case for either a truncation conversion transformation or a rounding conversion transformation.) Of course, if one of the relations (8.2.3) or (8.2.4) does hold, it does not guarantee that the corresponding conversion transformation will be onto or one to one. This depends on the quality of the conversion program.

We are generally given the system FP(r, p, a) in which the machine performs arithmetic, so we are given the set $S(r, p)$ of floating-point numbers handled by the machine. We must decide how many digits to use for input and output, so we must select the length p' of the numbers in $S(10, p')$. In some cases, such as the PUT DATA statement in PL/I, the compiler selects p' for us. PL/I bases its decision on (8.2.3), so it guarantees that the floating-point numbers in the machine will be more closely spaced than the decimal numbers are. The use of the criterion (8.2.3) is based on the view that the programmer will think of his computation as being performed in FP(10, p', a), so the arithmetic in the machine should be at least this accurate. When two decimal numbers are different, they should still be different after they are converted to $S(r, p)$. That is, the input conversion should be one to one. The use of the criterion (8.2.3) for determining the number of digits to be printed on output is based on the idea that we should print only those digits that are significant. We would not expect the answer to have p' significant decimal digits if the numbers in $S(10, p')$ are closer together than the numbers in $S(r, p)$ are. This suggests that the selection of p' should be based on the criterion (8.2.3). Also, if we expect p' decimal digits of the answer to be significant, we would want to be able to produce every number in $S(10, p')$, so we would want the output conversion transformation to be onto.

But there is another point of view we may adopt in selecting p'. In many carefully written programs it is essential that we write the program with the realization that the computation will be performed in FP(r, p, a). Suppose that we want to enter a constant such as $\pi/2$. We would like to produce the number in $S(r, p)$ which is closest to $\pi/2$, so we want the input conversion transformation to be onto. This suggests that we use the criterion (8.2.4) for the selection of p'. With this criterion, we can also hope that the output conversion transformation will be one to one. Then when the same value is printed for two numbers A and B, we would know that A and B were equal before they were converted. This is quite helpful in debugging programs, because it helps us determine which branches were taken in the program.

Thus, we would use the criterion (8.2.3) for the selection of p' if we want to think of the calculation as being performed in FP(10, p', a). But if we program with the realization that the calculation will be performed in FP(r, p, a), then (8.2.4) is the natural criterion to use for the selection of p'.

Finally, we shall consider the effect of conversion and reconversion. This was studied in detail by Matula (1968a). First, suppose that our conversion programs produce a truncation conversion transformation for both input and output. For any positive number a in $S(r_1, p_1)$, we have

$$\sigma\tau(a) \leq \tau(a) \leq a.$$

Then each conversion and reconversion decreases the number until we reach a number which belongs to both $S(r_1, p_1)$ and $S(r_2, p_2)$. If the number is converted and reconverted many times, this downward drift can be quite annoying.

Matula (1968a) considers the case in which $r_1^i \neq r_2^j$ for any positive integers i and j. He shows that if σ is a rounding conversion transformation and τ is a truncation conversion transformation, then $\sigma\tau(a) = a$ holds for all a in $S(r_1, p_1)$ if and only if

$$r_2^{p_2-1} \geq 2r_1^{p_1} - 1.$$

He also shows that if both σ and τ are rounding conversion transformations, then $\sigma\tau(a) = a$ holds for all a in $S(r_1, p_1)$ if and only if

$$r_2^{p_2-1} > r_1^{p_1}.$$

8.3. CONVERSION TECHNIQUES

We shall now consider several techniques which can be used to convert a number from one radix to another. To discuss input and output conversion simultaneously, we shall consider the general case of converting a number from $S(r_1, p_1)$ to a number in $S(r_2, p_2)$. Since the sign of the number is easy to handle, it suffices to consider the conversion of positive numbers.

It is easy to change a number in $S(r, p)$ from the form

(8.3.1) $$x = r^e m, \qquad r^{-1} \leq m < 1,$$

to

(8.3.2) $$x = r^f I,$$

where I is an integer less than r^p. We shall discuss the conversion of numbers written in each of these forms.

Many machines are able to perform arithmetic in only one radix. Then for the input conversion we want to perform the arithmetic in the radix r_2 which we are converting to, and for the output conversion we want to perform the arithmetic in the radix r_1 which we are converting from. The techniques we shall use will depend on whether we perform the arithmetic in the radix r_1 or r_2. But we shall still discuss the general case of converting a number from $S(r_1, p_1)$ to $S(r_2, p_2)$, because there are some machines, the IBM System/360, for example, which can perform both decimal and binary arithmetic.

Some aspects of the conversion programs are dependent on the representation of the decimal numbers. We shall assume that the decimal numbers are represented as a string of decimal digits.† We shall also assume that it is easy to convert one-digit integers from one radix to the other. For example, suppose that we are using a four-bit binary-coded decimal representation for decimal digits. Then it is easy to convert a decimal digit to binary or to convert a binary or octal digit to decimal, but converting a hexadecimal digit to decimal requires a little more work.

We shall first consider the conversion of a positive integer I from the radix r_1 to the radix r_2. This is part of the problem of converting a number in the form (8.3.2), and it is also of interest in its own right for handling I formats. We shall assume that $I < r_i^{p_i}$ for $i = 1, 2$, so that I can be represented exactly in both systems. We may write

(8.3.3) $$I = a_{p_1-1} r_1^{p_1-1} + a_{p_1-2} r_1^{p_1-2} + \cdots + a_1 r_1 + a_0$$
(8.3.4) $$I = b_{p_2-1} r_2^{p_2-1} + b_{p_2-2} r_2^{p_2-2} + \cdots + b_1 r_2 + b_0,$$

where $0 \leq a_i < r_1$ and $0 \leq b_i < r_2$. Then we are given the a_i and we want to find the b_i.

We begin by considering techniques which allow us to perform the arithmetic in the radix r_1. One approach is to divide I by $r_2^{p_2-1}$ to get a quotient Q_1 and a remainder R_1 less than $r_2^{p_2-1}$. Then $b_{p_2-1} = Q_1$, and we may continue

---

†Occasionally other representations have been used. An example was the *card image* used on the IBM 704. We shall not discuss the conversion of data represented in this way.

the process by dividing R_1 by $r_2^{p_2-2}$ to get the next digit. This procedure requires $p_2 - 1$ divisions, each division producing one more digit b_i. The final remainder is b_0. There are two disadvantages to this approach. First, we still do about the same amount of work if I is small. (We may save a little time by comparing the powers of r_2 with I and not performing the division unless $r_2^i \leq I$.) A second disadvantage is that we must store the representation of each of the $p_2 - 1$ numbers r_2^i in the radix r_1.

We may overcome both of these disadvantages by developing the digits b_i in the opposite order. Divide I by r_2 to get a quotient Q_1 and a remainder R_1 less than r_2. Then $b_0 = R_1$. Then divide Q_1 by r_2 to get a quotient Q_2 and a remainder $R_2 = b_1$. Continuing in this way, we may develop the rest of the digits b_i. We terminate the procedure whenever the quotient Q_i is zero, so less work is required when I is small. With this procedure we need to store only the representation of r_2 in the radix r_1. With either of these procedures we develop the b_i in the radix r_1, so each digit must be converted to the radix r_2.

Next, suppose that we want to use arithmetic in the radix r_2 to convert I. Let $f(x)$ be the polynomial

$$f(x) = \sum_{i=0}^{p_1-1} a_i x^i,$$

so $I = f(r_1)$. We first convert each digit a_i to the radix r_2. Then we compute $f(r_1)$ from the formula

$$f(r_1) = a_0 + r_1(a_1 + r_1(a_2 + \cdots + r_1(a_{p_1-2} + r_1 a_{p_1-1})) \cdots),$$

where the arithmetic is performed in the radix r_2. This requires at most $p_1 - 1$ multiplications and additions, and it is easy to shorten the procedure when the representation of I in the radix r_1 has fewer than p_1 digits.

With any of these techniques for the conversion of I, it is often convenient to perform the radix conversion in fixed-point arithmetic and then convert the result to floating-point. In particular, this is the natural way to proceed if the machine has a convert instruction for the conversion of fixed-point integers. But the word length for fixed-point numbers may be shorter than the precision of the floating-point numbers we are using, so we may have to partition I. In this case, we select an integer k such that integers less than r_1^k can be converted directly, and then we write I as a polynomial in r_1^k. For example, if $I < r_1^{3k}$, we can write

$$I = c_0 + c_1 r_1^k + c_2 r_1^{2k},$$

where the c_i are less than r_1^k. It is easy to obtain the c_i from the representation of I in (8.3.3), and we can convert the c_i to floating-point numbers in $S(r_2, p_2)$.

Then we form the floating-point representation of I by computing

$$I = c_0 + \{r_1^k * [c_1 + (r_1^k * c_2)]\}$$

in $FP(r_2, p_2, a)$.

Next, we shall consider the conversion of a fraction F. We are given the representation

$$F = \sum_{i=1}^{p_1} a_i r_1^{-i},$$

and we want to convert F to

$$F' = \sum_{i=1}^{p_2} b_i r_2^{-i}.$$

It may be impossible to represent F in the radix r_2 with p_2 digits, so we want $F' \approx F$. We shall try to produce the value $\bar{F}$ of F chopped to $S(r_2, p_2)$. First suppose that we can perform arithmetic in the radix r_1. We multiply F by r_2 and note that b_1 is the integer part of $r_2 F$. Let F_1 be the fractional part of $r_2 F$. Then we multiply F_1 by r_2, and let b_2 be the integer part of $r_2 F_1$. Continuing in this way, we obtain the first p_2 digits of F, so $F' = \bar{F}$.

Suppose that we want to use arithmetic in the radix r_2 to convert the fraction F. Let

$$f(x) = \sum_{i=1}^{p_1} a_i x^i,$$

so $F = f(r_1^{-1})$. Then we could compute F' by evaluating $f(r_1^{-1})$ in the radix r_2. Unfortunately, it may be impossible to represent r_1^{-1} exactly in the radix r_2 with p_2 digits, and the error in the representation of r_1^{-1} may introduce more error in F' than we would like. One way to overcome this difficulty is to use more than p_2 digits in the evaluation of $f(r_1^{-1})$. But it is usually easier to convert F by converting the integer $r_1^{p_1} F$ exactly and then dividing the result by $r_1^{p_1}$ in $FP(r_2, p_2, a)$.

We shall now consider the conversion of a floating-point number $x = r_1^e m$ from $S(r_1, p_1)$ to $S(r_2, p_2)$. First, suppose that we can perform arithmetic in the radix r_2. We adjust the exponent e so that m is either an integer or a fraction, and then we convert m to the radix r_2. Call this value m'. Then we complete the conversion of x by multiplying m' by r_1^e in the radix r_2, and this is often done by performing the arithmetic in the system $FP(r_2, p_2, a)$. To do this, we need a representation of r_1^e in the radix r_2 for all values e which can arise without producing overflow or underflow. We can reduce the number of different powers of r_1 that have to be stored by dividing by $r_1^{|e|}$ when e is negative. But if the range of the exponents is large, it may require too much storage to store even the positive powers of r_1. One way to save storage is to compute $r_1^{|e|}$ using one of the techniques described in Section 3.6 for the computation of X**N. Alternatively, we might store a few powers of r_1 and compute the others. For example, we could store r_1^k and r_1^{10k} for $k = 1, 2, 3,$ $\ldots, 9$. Then it would be very easy to compute r_1^e for any positive integer e less than 100.

The accuracy of the result produced by this technique depends on the accuracy of the multiplication and division and on the accuracy of the representation of r_1^e. Unfortunately, we may need more than p_2 digits to represent r_1^e when e is large. (See Exercise 9.) By using only p_2 digits in the representation of r_1^e, we may introduce an error in the answer which is larger than the error introduced by the multiplication or division. If $r_1 = 10$ and r_2 is a power of 2, $S(r_2, p_2)$ does not contain r_1^{-1} but it does contain r_1^e for small positive integers e. This provides an additional motivation for dividing by $r_1^{|e|}$ when e is negative.

Finally, suppose that we want to perform the arithmetic in the radix r_1. We first find the integer e for which

$$r_2^e > |x| \geq r_2^{e-1},$$

and then we divide x by r_2^e to produce a number y. Clearly $r_2^{-1} \leq y < 1$, so e is the exponent of the answer. To obtain the mantissa of the answer, we convert the fraction y using the technique described earlier.

With either of these techniques, we need to store or to compute the representation of the powers of one radix in the other radix. We are concerned with both the conversion of decimal numbers to the radix r and the conversion of numbers in $S(r, p)$ to decimal. If we store the representation of powers of 10 in the radix r, then we can use arithmetic in the radix r for both of these conversions. Alternatively, we could store the decimal representation of the powers of r and use decimal arithmetic for both conversions.

If we are using chopped arithmetic and we want the conversion program to produce a rounding conversion transformation, we usually try to produce the first $p_2 + 1$ digits of $\bar{x}$ and then round the result to p_2 digits. The round operation requires us to perform an addition in the radix r_2.

The various techniques described above will differ in speed, accuracy, and the storage required. We shall not discuss the speed or the storage required, because they depend on the machine we are using. But we shall consider a few of the problems related to accuracy. We shall assume that the machine performs arithmetic in the system $FP(r, p, c)$, where r is a power of 2.

We shall first consider input conversion. Suppose that we begin by changing the decimal number to the form (8.3.1) and that we then program the conversion using chopped arithmetic. Let

$$x = 10^e m, \qquad .1 \leq m < 1,$$

and let $m' = \bar{m}$ be the fraction m converted to the radix r. To complete the conversion of x, we multiply or divide m' by $10^{|e|}$ in $FP(r, p, c)$. Suppose that

$x = 2$, so x will be changed to the form $10^1 \times .2$. Since r is a power of 2, .2 is not in $S(r, p)$. Then m' will be less than .2, so $10 * m' < 2$. Thus, 2 will be converted into a number less than 2, so we do not have a neighbor transformation. The fact that small integers are not converted exactly is quite annoying. A possible remedy is to use higher-precision arithmetic in the conversion and then round the result to $S(r, p)$.

Another approach is to begin by changing the decimal number to the form (8.3.2), so we convert the integer I instead of the fraction m. If I does not have too many digits, it can be converted exactly. Similarly, if $|e|$ is not too large, $10^{|e|}$ is in $S(r, p)$. If both I and $10^{|e|}$ are in $S(r, p)$, then the multiplication or division of I by $10^{|e|}$ in FP(r, p, c) will produce the correctly chopped answer $\bar{x}$. This means that small integers will be converted exactly. In fact, we can find a wider class of numbers for which this procedure will produce the correctly chopped answer. Let k be the integer for which

$$(8.3.5) \qquad\qquad 10^k < r^p < 10^{k+1},$$

and let $\hat{e}$ be the largest integer such that 10^e is in $S(r, p)$ for all integers e with $1 \le e \le \hat{e}$. Then this conversion procedure will produce the correctly chopped answer $\bar{x}$ whenever the decimal representation of x in the form (8.3.2) satisfies both

$$(8.3.6) \qquad\qquad |e| \le \hat{e}$$

and

$$(8.3.7) \qquad\qquad |I| < 10^k.$$

The condition (8.3.7) means that I has at most k decimal digits. Once we have determined k and $\hat{e}$, it is easy to see whether the conditions (8.3.6) and (8.3.7) hold. These conditions are satisfied by many of the numbers we use in our programs, so there is a significant class of numbers for which this conversion procedure will produce the correctly chopped answer. But if either (8.3.6) or (8.3.7) fails to hold, we may not even have a neighbor transformation. In fact, the transformation might not even be monotone.

We may encounter the same sort of problems with the output conversion program. Suppose that we begin by determining e such that

$$(8.3.8) \qquad\qquad 10^e > |x| \ge 10^{e-1}.$$

Then we divide x by 10^e in FP(r, p, c) to produce a fraction y. (If e is negative, we can multiply x by $10^{|e|}$.) Here e is the exponent of the answer, and the mantissa of the answer is obtained by converting the fraction y to decimal. We assume that the conversion of y produces the value $\bar{y}$ of y chopped to $S(10, p')$. Consider the case in which $x = 2$. Here $e = 1$, so $y = 2 \div 10$.

Since .2 is not in $S(r, p)$, $y < .2$. Then when y is converted to decimal we produce the value $\bar{y} \leq y < .2$. We again have the problem that small integers are not converted exactly.

One remedy for this problem is to change the approach so that we convert integers instead of fractions. We begin by finding e satisfying (8.3.8) and k satisfying (8.3.5). Instead of dividing or multiplying x by $10^{|e|}$, we divide or multiply x by $10^{|e-k|}$. This produces a number z in $S(r, p)$ with

$$10^k > |z| \geq 10^{k-1}.$$

The integer part of z may be converted exactly. We convert the fractional part of z independently, producing the chopped result z. With this approach, we can guarantee that small integers are converted exactly.

EXERCISES

1. Consider the conversion of decimal numbers into numbers in $S(16, 14)$. Suppose that the conversion program transforms the decimal number x into the value $\tau(x)$ of x chopped to $S(16, 14)$. Answer the following questions for $x = \pi/4, \pi/2, \pi, 2\pi,$ and 4π.
    a. What is the value $\overset{\_\circ}{x}$ of x rounded to $S(16, 14)$?
    b. Find a decimal number y for which $\tau(y) = \overset{\_\circ}{x}$. Use as few decimal digits as possible in the representation of y.
    c. In some implementations of PL/I for the IBM System/360, we were allowed a maximum of 16 decimal digits to specify a floating-point constant. Find the 16-digit decimal number y for which $\tau(y)$ is closest to $\overset{\_\circ}{x}$. By how many digits in the last place do $\tau(y)$ and $\overset{\_\circ}{x}$ differ?

2. Let $S(r, p)$ be the set of floating-point numbers handled by the machine you are using. Find the largest integer p' which satisfies (8.2.3), and find the smallest integer p' which satisfies (8.2.4). What are these values for p' if $S(r, p)$ is the set of double-precision floating-point numbers on the machine you are using?

3. Let τ be a transformation of $S(r_1, p_1)$ into $S(r_2, p_2)$, and let σ be a transformation of $S(r_2, p_2)$ into $S(r_1, p_1)$. Find a bound for the relative error in the approximation $\sigma\tau(x) \approx x$ if both σ and τ are neighbor transformations.

4. Let τ be a rounding conversion transformation of $S(r_1, p_1)$ into $S(r_2, p_2)$ and let σ be a rounding conversion transformation of $S(r_2, p_2)$ into $S(r_1, p_1)$. Prove that if $r_1^{-p_1} > r_2^{-(p_2-1)}$, then $\sigma\tau(x) = x$ holds for all x in $S(r_1, p_1)$.

5. Let τ be a truncation conversion transformation of $S(r_1, p_1)$ into $S(r_2, p_2)$, and let σ be a rounding conversion transformation of $S(r_2, p_2)$ into $S(r_1, p_1)$. Prove that if $r_2^{p_2-1} \geq 2r_1^{p_1} - 1$, then $\sigma\tau(x) = x$ holds for all x in $S(r_1, p_1)$.

6. Let τ be a rounding conversion transformation of $S(16, 6)$ into $S(10, 8)$.
    a. Prove that τ cannot be a transformation onto $S(10, 8)$.
    b. Find a number in $S(10, 8)$ which is not the image of any number in $S(16, 6)$ under τ.

7. Let τ be a truncation conversion transformation of $S(16, 6)$ into $S(10, 9)$, and let σ be a truncation conversion transformation of $S(10, 9)$ into $S(16, 6)$.
   a. Write a program which will produce the transformations σ and τ for numbers in the interval $16 \le x < 100$.
   b. Select a number x which is in $S(16, 6)$ and lies in the interval $16 \le x < 100$. Convert and reconvert x several times. That is, form $(\sigma\tau)^n(x)$. How many times must the conversion and reconversion be repeated before you reach a number y in $S(16, 6)$ with $(\sigma\tau)(y) = y$?
   c. Repeat part b for several other values of x.

8. Let τ be a truncation conversion transformation of $S(16, 6)$ into $S(10, 9)$, and let σ be a truncation conversion transformation of $S(10, 9)$ into $S(16, 6)$. Let x be a positive number in $S(16, 6)$. Suppose that we convert and reconvert x many times. That is, we form $(\sigma\tau)^n(x)$ for large n.
   a. If $16 \le x < 100$, what is the maximum number of units in the last place by which x and $(\sigma\tau)^n(x)$ may differ?
   b. If x lies between $\frac{1}{2}$ and 1, what is the maximum number of units in the last place by which x and $(\sigma\tau)^n(x)$ may differ?
   c. Find a value of x such that the downward drift of $(\sigma\tau)^n(x)$ will never stop. That is, we shall never reach a number y with $(\sigma\tau)(y) = y$.

9. Let $S(r, p)$ be the set of floating-point numbers on the machine you are using. Find the largest integer $\hat{e}$ such that 10^e is in $S(r, p)$ for every integer e with $1 \le e \le \hat{e}$. Is it possible for 10^e to be in $S(r, p)$ for some integer $e > \hat{e}$?

10. Let $S(r, p)$ be the set of floating-point numbers on the machine you are using. Suppose that you have an input conversion program which begins by changing the decimal number to the form (8.3.2). Then it converts the integer I to $S(r, p)$ and multiplies or divides I by $10^{|e|}$ in the system $FP(r, p, c)$. If I is not in $S(r, p)$, assume that I is converted into the properly chopped value $\bar{I}$. If $10^{|e|}$ is not in $S(r, p)$, then I is multiplied or divided by the number $P = \overline{10^{|e|}}$ obtained by chopping $10^{|e|}$ to $S(r, p)$.

   a. When can you guarantee that this program will produce the correctly chopped answer. [That is, find the values of k and $\hat{e}$ to use in (8.3.6) and (8.3.7).]
   b. Is the input conversion transformation τ well defined?
   c. Is the input conversion transformation monotone?
   d. Suppose that the input conversion program converts x into $\tilde{x}$. Find a bound for the relative error in the approximation $\tilde{x} \approx x$.

11. Change the specifications of the conversion program described in Exercise 10 by changing the number P used as an approximation for $10^{|e|}$. Let P be the number $\overline{10^{|e|}}^{\circ}$ obtained by rounding $10^{|e|}$ to $S(r, p)$. Answer all the questions in Exercise 9 for this conversion program.

12. Suppose that you can perform arithmetic only in the radix r_1. How would you program a rounding conversion transformation of $S(r_1, p_1)$ into $S(r_2, p_2)$?

9 CAREFULLY WRITTEN PROGRAMS

9.1. INTRODUCTION

In this chapter we shall illustrate some of the problems that face us when we try to produce high-quality programs. We shall refer to these programs as *carefully written programs* in contrast to "quick and dirty" programs. Carefully written programs are typified by library routines, although not all library routines have achieved high quality. Since a library program will be used extensively, its author is usually willing to devote a great deal of time and effort to the program. His objective is to produce a high-quality program rather than a quick and dirty program.

To write a good program, we must have a thorough understanding of the problem we are solving. In fact, writing the program often forces us to study aspects of the problem which we would not have considered otherwise. This suggests that our illustrations of carefully written programs should deal with problems which are extremely familliar. We shall discuss the solution of a quadratic equation and computing the average of two numbers. By considering these simple problems, we hope to show why it takes so long to write a high-quality program.

We shall not discuss either the speed of execution of the program or the amount of storage it requires, although both of these affect the quality of the program. Instead, we shall restrict our attention to the quality of the answers produced by the program.

9.2. AVERAGE PROBLEM

As our first illustration of a carefully written program, we shall consider a program which computes the average of two normalized floating-point numbers. We could begin by specifying the accuracy we expect in the answer—say an error of less than one unit in the last place. But it might turn out that the accuracy we have specified is extremely difficult to achieve without resorting to higher-precision arithmetic and that we could produce a much more efficient program if we relaxed the accuracy criterion slightly. Instead of specifying the accuracy of the program in advance, we shall try to produce as accurate an answer as we can without degrading the speed of the program too much. However, we shall require the write-up of the program to provide the user with a bound for the error.

The accuracy of our answer may depend on the precision of the arithmetic used in the program. If we performed the entire calculation in higher-precision arithmetic and then rounded the result to single-precision, we would expect to produce a very accurate answer. But this seems to be a very high price to pay, so we would not be willing to use this approach if it produced only a slight improvement over a single-precision version of the program. We usually start with the hope that we can write the entire program in single-precision. As we saw in the discussion of the quadrature problem in Section 4.1, it is sometimes highly advantageous to perform a few operations in a precision which is higher than that used in the rest of the program. Whether this approach is reasonable often depends on whether the hardware and software support double-precision arithmetic. For the average problem, we shall assume that double-precision arithmetic is extremely expensive, so we shall require that only single-precision arithmetic be used.

When we use a library program to compute a familiar function, we expect the program to preserve many well-known properties of the function. For example, we would be quite annoyed if a cosine routine produced an answer greater than 1, or if it failed to produce the value 1 for the cosine of zero. Similarly, the average of two numbers has several familiar properties, and we shall demand that the program preserve them. For example, the answer should be independent of the order of the arguments. Another requirement is that the average of A and B should be zero if and only if $B = -A$. Also, we would be annoyed if the average failed to satisfy

$$(9.2.1) \qquad \min(A, B) \leq \text{average}(A, B) \leq \max(A, B).$$

We note that equality can hold in (9.2.1) if $A = B$. When $A \neq B$, we would like to have

$$(9.2.2) \qquad \min(A, B) < \text{average}(A, B) < \max(A, B).$$

But if A and B were consecutive floating-point numbers, we could not satisfy (9.2.2), so we shall require only that our program satisfy (9.2.1).

Finally, consider the problem of overflow and underflow. The input numbers A and B are floating-point numbers, and their average cannot exceed the larger of them. Therefore, the correct answer does not exceed the largest floating-point number Ω. If overflow occurs, it is the fault of the algorithm, not the problem. We shall require that the program never overflow.

The situation with underflow is different. If we are asked to compute the average of zero and the smallest normalized positive number ω, the answer underflows. We shall require that no underflows occur unless the answer has an absolute value less than ω.

Based on these ideas, we give the following specifications for the program.

Specifications for the Average Problem

1. The program may be written as either a function subprogram or as a subroutine subprogram. The input A and B may be assumed to be normalized floating-point numbers, and the result produced for the average must be a normalized floating-point number. Only single-precision arithmetic may be used.

2. *Accuracy*: The value produced for the answer must be approximately $(A + B)/2$, and it must have the correct sign. The write-up must contain a reasonable bound for the error. The error bound may be given either in terms of units in the last place or as a bound for the relative error.

3. *Properties*: The program must produce a value for the average which has the following properties:
a. $\min(A, B) \leq \text{average}(A, B) \leq \max(A, B)$.
b. $\text{average}(A, B) = \text{average}(B, A)$.
c. $\text{average}(A, B) = 0$ if and only if $B = -A$.
d. $\text{average}(-A, -B) = -\text{average}(A, B)$.
(Property c may be modified if the average underflows.)

4. *Overflow/Underflow*: The program should never produce an overflow, and it should not underflow unless

$$(9.2.3) \qquad\qquad 0 < |(A + B)/2| < \omega.$$

There should be a reasonable strategy for handling the case in which (9.2.3) holds. The write-up should tell the user what happens in this case and what number will be returned as the answer. It should also tell him how to find out whether this case arose.

The reader is invited to stop here and write a program meeting these specifications. We shall discuss some aspects of the problem, but part of the problem will be left as exercises. Since the details of the problems we encounter will depend on the floating-point arithmetic we are using, we shall assume throughout that the arithmetic is performed in the system $FP(r, p, c)$.

The natural formulas to consider using for the average of A and B are

(9.2.4) $$\mathrm{AV} = (A \oplus B) \div 2$$

(9.2.5) $$\mathrm{AV} = (A \div 2) \oplus (B \div 2)$$

(9.2.6) $$\mathrm{AV} = A \oplus [(B - A) \div 2].$$

We shall discuss some of the problems that arise with the use of each of these formulas.

If A and B have the same sign, then formula (9.2.4) might produce overflow. Similarly, formula (9.2.6) can produce overflow if A and B have opposite signs. Then one way to avoid overflow is to test A and B for sign agreement and use one or the other of these formulas depending on whether they have like signs or unlike signs. Indeed, we shall see that many aspects of the behavior of formulas (9.2.4)–(9.2.6) depend on whether or not A and B have the same sign, so it is quite natural to use one formula when they have the same sign and another formula when they have different signs.

It is easy to show that (9.2.4) cannot produce underflow unless (9.2.3) holds. The division by 2 in formula (9.2.5) will not underflow unless A or B has an absolute value which is positive but less than 2ω. Even if $|A|$ and $|B|$ are both $\geq 2\omega$, the addition in (9.2.5) can underflow if A and B have opposite signs. (See Exercise 6.) With formula (9.2.6), we have to worry about underflow both in the computation of

(9.2.7) $$C = (B \ominus A) \div 2$$

and in the addition of C to A.

One way to avoid underflow with any of these formulas is to scale the problem when A and B are both small. We shall illustrate this approach by considering the computation of C in (9.2.7). Suppose that $A = r^e m$, where $r^{-1} \leq |m| < 1$. If $|B \ominus A| \neq 0$, we have $|B \ominus A| \geq r^{e-p-1} > |A|/r^{p+1}$. It follows that (9.2.7) cannot underflow unless both

(9.2.8) $$|A| < r^{p+2}\omega$$

and

(9.2.9) $$|B| < r^{p+2}\omega.$$

When both (9.2.8) and (9.2.9) hold, we can scale the problem by multiplying A and B by r^{p+2}. This will not produce overflow so long as the machine provides a reasonable range of exponents. The advantage of scaling by a power of r is that it does not introduce any error. We compute a value D for the average of $r^{p+2}A$ and $r^{p+2}B$, and then we divide D by r^{p+2}. Before dividing D

by r^{p+2}, we can perform a test to see whether $|D| \geq r^{p+2}\omega$. If $0 < |D| < r^{p+2}\omega$, we can omit the division by r^{p+2} and provide whatever treatment we have decided on for the case in which the answer underflows.

Another problem with formula (9.2.6) is that it is not symmetric in A and B, so it might produce a different result if the arguments A and B were interchanged. To overcome this difficulty, we could require that, say,

$$(9.2.10) \qquad\qquad |A| \leq |B|,$$

and interchange A and B if (9.2.10) does not hold.

Next, we shall consider the properties listed under heading 3 in the specifications. To see how these properties can fail to hold, we shall consider some examples in FP(10, 6, c). It is easy to modify these examples so that they apply to FP(16, p, c), but they are not applicable to a binary machine.

First, suppose that

$$A = .500001$$
$$B = .500003.$$

Then $A + B = 1.000004$, so $A \oplus B = \overline{A + B} = 1$. Then $(A \oplus B) \div 2 = .5$, so formula (9.2.4) can produce a result which is smaller than min(A, B). It can be shown that the value produced by formula (9.2.4) in FP(r, p, c) satisfies (3a) if and only if $r = 2$.

Similarly, if

$$A = .500001$$
$$B = -.500003,$$

then $B \ominus A = -1$, so formula (9.2.6) produces the answer .000001. But the correct answer is $-.000001$, so formula (9.2.6) failed to produce the correct sign.

We even find difficulties with formula (9.2.5). Let

$$A = B = .500001.$$

Then $A \div 2 = B \div 2 = .25$, so formula (9.2.5) produces the value .5, which is smaller than min(A, B). Similarly, if

$$A = .500001$$
$$B = -.500000,$$

then (9.2.5) produces the value zero even though $B \neq -A$. It can be shown that these two examples represent the only cases in which the result produced by this formula fails to have the properties listed under heading 3 in the specifications. That is, it fails to satisfy (3a) if and only if $A = B$ and $A \div 2 \neq A/2$, and it fails to satisfy (3c) if and only if $A \div 2 = -B \div 2$ and $A \neq -B$. If we want to use formula (9.2.5), we can provide special treatment for these cases.

It can be shown that formula (9.2.4) produces a result which has all the properties listed under heading 3 in the specifications if A and B have opposite signs. Similarly, it can be shown that the result produced by (9.2.6) satisfies (3a), (3c), and (3d) when A and B have the same sign. We may use these observations to devise a strategy for selecting the appropriate formula based on the signs of A and B.

We shall now turn to the question of the accuracy of the answers produced by these formulas. To see how the error behaves, we shall consider some examples in FP(10, 6, c). The first example uses formula (9.2.5) for the data

$$A = 2.00001$$
$$B = .0000199999.$$

Here

$$\frac{A}{2} = 1.000005$$

$$\frac{B}{2} = .00000999995.$$

Then (9.2.5) produces the answer 1.00000 instead of 1.00001499995, so the error is almost $1\frac{1}{2}$ units in the last place. It can be shown that if A and B have the same sign, then (9.2.5) always produces an error of less than $1\frac{1}{2}$ units in the last place.

Now suppose that A and B have opposite signs. We have seen that (9.2.5) can produce the answer zero even though $B \neq -A$. Even if we provided special treatment for the case in which $A \div 2 = B \div 2$, the formula can produce large relative error if $|A|$ is close to $|B|$. Suppose that

$$A = 2.00004$$
$$B = -2.00001.$$

Then (9.2.5) would produce the answer .00002 instead of .000015. Since this is a relative error of $\frac{1}{3}$, we would not use this formula for the case in which A and B have opposite signs.

Next, we shall consider formula (9.2.4). Let

$$A = 1.00002$$
$$B = -.000001.$$

Then $A \oplus B = 1.00001$, so $(A \oplus B) \div 2 = .500005$. The correct answer is

.5000095, so the error is $4\frac{1}{2}$ units in the last place. It can be shown that if r is even and A and B have opposite signs, then the error produced by (9.2.4) is less than $r/2$ units in the last place. We shall not consider the error produced by (9.2.4) in the case in which A and B have the same sign, because the formula can produce a result which does not satisfy (3a).

For formula (9.2.6), we consider the example

$$A = .0000008$$
$$B = 1.00002.$$

Here $B \ominus A = 1.00001$, so (9.2.6) produces the answer .500005. But the correct answer is .50000104, so the error is 5.4 units in the last place. It can be shown that the error produced by (9.2.6) is less than $r/2 + 1$ units in the last place if r is even and A and B have the same sign. We have seen that when A and B have opposite signs the value produced by (9.2.6) might have the wrong sign, which would produce a relative error greater than 1.

9.3. QUADRATIC EQUATION

Consider the problem of writing a subroutine to solve the quadratic equation

(9.3.1) $Ax^2 + Bx + C = 0.$

The input will be the coefficients A, B, and C, and the output will be the roots R1 and R2.[†] We shall assume that A, B, and C are real and that they are normalized floating-point numbers. Three aspects of the problem will be discussed: the form of the CALL, producing good accuracy, and avoiding overflow and underflow.

First, consider the CALL. It would be natural to use

(9.3.2)   CALL QUAD(A,B,C,R1,R2,I)

where I is an error indicator which would be set to indicate whether or not we have been able to solve the equation. But what should the subroutine do if the roots of (9.3.1) are complex? One approach would be to treat the case in which the roots are complex as an error. On the other hand, it could always return the roots as complex numbers, even when they are real. The CALL could be written as

(9.3.3)   CALL QUAD(A,B,C,R1R,R1I,R2R,R2I,I)

---

†Forsythe (1970) discusses the specifications for a good quadratic equation solver, and he describes a high-quality program produced by Kahan.

where, for example, R1R and R1I are the real and imaginary parts of R1. Another approach would be to use the CALL (9.3.2) and require that R1 and R2 be typed COMPLEX. But there are many cases in which we know that the roots of (9.3.1) are real, and in these cases it would be annoying to have to type the roots COMPLEX in the calling program. If we forgot to do so, we would change other variables in the calling program, producing disastrous results. Including the extra variables in the CALL statement (9.3.3) is a small price to pay, since we do not have to look at the imaginary parts if we know that the roots are real.

There is still another way to handle complex roots of (9.3.1). Since the coefficients of (9.3.1) are real, when the roots are not real they are complex conjugates $a \pm bi$. We could use the CALL (9.3.2) and store a and b in R1 and R2, using I to indicate that the roots are not real. From the point of view of the author of the subroutine, this would appear to be an ideal solution. It retains the simplest form of the CALL, and yet it provides complete information about the answer. Unfortunately, experience shows that many users of a subroutine such as this do not bother to test the error indicator. If the user expected the roots to be real and did not test I, he would get a bad answer when the roots are not real. For example, a slight change in the coefficients of (9.3.1) could change a double root at 2 into a pair of complex roots $2 \pm 10^{-8}i$. If we stored 2 in R1 and 10^{-8} in R2 and the user thought that the roots were real, he would be very dissatisfied with the value 10^{-8} for a root of the equation. Exercise 12 gives an example of an equation in which a double root is changed into a pair of complex conjugates by the error introduced by radix conversion.

Based on these observations, it seems reasonable to accept the form (9.3.3) for the CALL.

Another question related to the calling sequence is what the subroutine should do when $A = 0$. If we were solving (9.3.1) by hand, we would never consider using the quadratic formula

$$(9.3.4) \qquad\qquad x = \frac{-B \pm \sqrt{B^2 - 4AC}}{2A}$$

when $A = 0$. If we do try to use (9.3.4), we encounter division by zero and we cannot find the root of the equation $Bx + C = 0$. One approach is to use the error indicator I to indicate that $A = 0$, and set R1 $= -C \div B$ and R2 $= \Omega$.

The following example suggests that this might be a reasonable approach to use when $A = 0$. Suppose that we want to find the maximum value of the function

$$f(x) = ax^3 + bx^2 + cx + d$$

in the interval $0 \le x \le 1$. We would find the roots of the quadratic equation

$f'(x) = 0$ and test the value of $f(x)$ at 0, 1, and any root of $f'(x) = 0$ lying between 0 and 1. Now suppose that for one set of data we have $a = 0$, so $f'(x)$ is linear. We want the root of $f'(x) = 0$, and any number the quadratic equation solver produced for the second root would be acceptable. We would reject the number if it did not lie in the interval $0 \leq x \leq 1$, and if it did lie in this interval, we would test it. But if the quadratic equation solver did not produce the root of the linear equation $f'(x) = 0$, we would have to provide special treatment for the case $a = 0$.

Similarly, if $A = B = C = 0$, the subroutine could return any values for R1 and R2 and set I to indicate that every complex number satisfies the equation. If $A = B = 0$ but $C \neq 0$, I should indicate that the equation has no solution.

The way a subroutine treats degenerate cases such as these can be quite important, because they arise in computing far more often than one might expect. One reason for this is that we write the program to handle the general case, but we often test the program on simple cases that can be handled analytically. To find a problem which can be solved easily, we may simplify the formulation, and this can easily result in a degenerate case for some subroutine.

We shall now turn to the question of the accuracy of the answers. We shall assume that the arithmetic is performed in FP(r, p, c) and that $A \neq 0$. We shall also assume that the relative error introduced by the square root program we are using is small. As we saw in Chapter 3, the result produced by multiplication or division in FP(r, p, c) will have small relative error if the operands do. This is also true of addition and subtraction when we have the add magnitude case, so our primary concern is the subtract magnitude case. The quadratic formula (9.3.4) will produce a small relative error if we do not encounter this case.

First, suppose that $B^2 - 4AC > 0$ and consider the addition of $\pm\sqrt{B^2 - 4AC}$ to $-B$. For one of the roots we shall have the add magnitude case, and for the other root we shall have the subtract magnitude case. We shall test the sign of $-B$ and select the $+$ or $-$ sign in front of the square root so that we first compute the root R1 for which we have the add magnitude case. Since the product of the roots of (9.3.1) is C/A, we may compute R2 from

$$(9.3.5) \qquad\qquad R2 = \frac{C}{A \cdot R1}.$$

If we can compute $B^2 - 4AC$ accurately, this gives us a way to compute both roots of (9.3.1) with small relative error. [Another way to compute R2 is to *rationalize the numerator* in (9.3.4).]

Now consider the computation of the discriminant $B^2 - 4AC$. In Exercise 6 of Chapter 3 we saw that there is no reasonable bound for the relative error produced by coding $x^2 - y^2$ as $(x * x) \ominus (y * y)$. Similarly, we may produce a large relative error in the computation of $B^2 - 4AC$ if $B^2 \approx 4AC$. To see how this error will affect the answer, suppose that B is slightly less than 1 and that $B^2 - 4AC \approx 10^{-8}$. If we were using eight-digit decimal arithmetic, we might obtain the value 2×10^{-8} for $B^2 - 4AC$. Then $\sqrt{B^2 - 4AC}$ would be $\sqrt{2} \times 10^{-4}$ instead of 10^{-4}, so the error introduced in the calculation of $B^2 - 4AC$ has produced an error in digits near the middle of the answer.

When $B^2 \approx 4AC$, the roots of Eq. (9.3.1) are nearly equal. Consider the equation

$$(9.3.6) \qquad x^2 - 2ax + a^2 = 0,$$

which has a double root at a. If we change the constant term in (9.3.6) to $a^2 - \epsilon^2$, we have the equation

$$x^2 - 2ax + a^2 - \epsilon^2 = 0,$$

whose roots are $a \pm \epsilon$. Suppose that a is slightly less than 1 and that $\epsilon = 10^{-4}$. Then a change of 1 in the eighth digit of the constant term in (9.3.6) produces a change of 1 in the fourth digit of the roots, so the problem is not well conditioned. By means of a backward error analysis, it can be shown that the value we compute for $\sqrt{B^2 - 4AC}$ is exactly $\sqrt{B^2 - 4AC'}$, where $C' \approx C$. But when the roots are nearly equal, this change in C can produce a significant change in the answers.

Thus, we want the value computed for $\sqrt{B^2 - 4AC}$ to have a small relative error, and in the subtract magnitude case this requires us to use higher-precision arithmetic. We shall use double-precision arithmetic to compute $B^2 - 4AC$, and then shorten the result to single-precision and use a single-precision square root program. The rest of the calculation can be performed in single-precision.

If we compute the discriminant in this way and use the approach suggested earlier for the calculation of R2, we can compute both roots with small relative error.

Finally, we shall consider the problem of overflow and underflow. Our objective is to compute any root of (9.3.1) whose absolute value lies between ω and Ω. To simplify the discussion, we shall assume that we are using the IBM System/360, where $r = 16$, $\Omega \approx 16^{63}$, and $\omega = 16^{-65}$.

Since the quadratic formula requires us to compute B^2 and $4AC$, we can encounter overflow even when the coefficients and the roots are substantially less than Ω. An example of this is provided by the equation

$$(9.3.7) \qquad 16^{40}x^2 - 3 \cdot 16^{40}x + 2 \cdot 16^{40} = 0,$$

where B^2 and $4AC$ overflow, but the roots are 1 and 2. It is clear that we can avoid this overflow by dividing (9.3.7) by 16^{40}. This suggests that we should scale the coefficients of (9.3.1) so that the coefficient of x^2 is close to 1.

But consider the equation

$$(9.3.8) \qquad 16^{-40}x^2 - 3x + 2 \cdot 16^{40} = 0,$$

whose roots are 16^{40} and $2 \cdot 16^{40}$. If we tried to scale the coefficients by multiplying (9.3.8) by 16^{40}, the constant term would overflow. For this problem, it is convenient to introduce the scale factor in x instead of in the coefficients. If we substitute $x = 16^{40}y$, (9.3.8) becomes

$$(9.3.9) \qquad 16^{40}y^2 - 3 \cdot 16^{40}y + 2 \cdot 16^{40} = 0.$$

As above, we can solve (9.3.9) if we divide the coefficients by 16^{40}. Then the roots of (9.3.9) must be multiplied by 16^{40} to produce the roots of (9.3.8).

Our approach will be a combination of these two types of scaling. It will require us to extract the exponents of the floating-point numbers we are using and to change them. These operations can be performed by using the techniques discussed in Section 4.4. All the scaling will be done by adjusting the exponents, so it will not introduce any errors. We shall assume that neither A nor C is zero, since the other cases are easy to handle.

We shall scale the problem by substituting $x = 16^K y$ and multiplying the coefficients by 16^L. Then (9.3.1) becomes

$$(9.3.10) \qquad 16^{2K+L}Ay + 16^{K+L}By + 16^L C = 0,$$

which we write as

$$(9.3.11) \qquad A'y^2 + B'y + C = 0.$$

We shall select K and L so that the exponent of A' is zero and either $B = 0$ or the exponent of B' is zero. First, extract the exponents e_A, e_B, and e_C of A, B, and C. If $B = 0$, set $K = 0$ and $L = -e_A$. If $B \neq 0$, the exponents $e_{A'}$, $e_{B'}$, and $e_{C'}$ of A', B' and C' are given by

$$e_{A'} = e_A + 2K + L$$
$$e_{B'} = e_B + K + L$$
$$e_{C'} = e_C + L.$$

We want

$$2K + L = -e_A$$
$$K + L = -e_B,$$

so we set $K = e_B - e_A$ and $L = e_A - 2e_B$.

We now construct the floating-point numbers A' and B' by changing the exponents of A and B to zero. We would also like to change the exponent of C to $e_{C'} = e_C + L$, but $e_{C'}$ may be outside the range of the exponents handled by the machine. For example, this happens with the equation

$$16^{-40}x^2 + 16^{-40}x - 16^{40} = 0.$$

Even if $|C'| < \Omega$, the calculation of $4A'C'$ might produce overflow. But with our scaling of A' and B', it is easy to see that we shall not encounter either overflow or underflow in the solution of (9.3.11) if

(9.3.12) $-63 \leq e_{C'} \leq 62.$

We shall first consider the case in which (9.3.12) holds. We can change the exponent of C to $e_{C'}$ and solve (9.3.11). Then the roots of (9.3.11) must be multiplied by 16^K to produce the roots of (9.3.1). Since these multiplications can produce overflow or underflow, we might prefer to perform the multiplications by changing the exponents. This would allow us to provide any treatment we wanted to for the cases in which the answer overflows or underflows.

Now suppose that (9.3.12) does not hold. We may write

(9.3.13) $C' = 16^{2M}C'',$

where the exponent of C'' is 0 or 1 according to whether the exponent of C' is even or odd. We form C'' by changing the exponent of C, and we compute

$$S = \sqrt{\left|\frac{C''}{A'}\right|}.$$

First, suppose that $B = 0$. If $AC < 0$, the roots of (9.3.11) are $\pm 16^M S$. Then the roots of (9.3.1) may be obtained by adding $K + M$ to the exponents of S and $-S$. Similarly, if $AC > 0$, the roots of (9.3.1) are $16^{K+M}Si$. The real parts are zero and the imaginary parts are obtained by changing the exponents of S and $-S$.

Next, suppose that $e_{C'} < -63$ and that $B \neq 0$. Then $16^{-1} \leq |B'| < 1$ and $|4A'C'| < 16^{-62}$, so we may ignore the contribution of $4A'C'$ to the discriminant of (9.3.11). For the add magnitude case, we take $y_1 = -B' \div A'$, and to compute the other root we form $T = -C'' \div (A' * y_1)$. To get the roots of (9.3.1), we add K to the exponent of y_1 and $2M + K$ to the exponent of T.

Finally, suppose that $e_{C'} > 62$. Then we may ignore the contribution of $(B')^2$ to the discriminant, so our formula becomes

(9.3.14) $$y = \frac{-B' \pm \sqrt{-4A'C'}}{2A'}.$$

If $AC < 0$, the roots of (9.3.11) are real and we may ignore the contribution of B' in (9.3.14). This yields $\pm 16^M S$ for the roots of (9.3.11). But if $AC > 0$, the real parts of the roots are $-B'/(2A')$, and the imaginary parts are $\pm 16^M S$. As above, the roots of (9.3.1) are obtained by changing the exponents of the roots of (9.3.11).

EXERCISES

1. Find two floating-point numbers A and B such that

$$A \oplus [(B \ominus A) \div 2] \neq B \oplus [(A \ominus B) \div 2],$$

when the arithmetic is performed in the system
a. FP(10, 6, c).
b. FP(2, p, c).

2. Suppose that we use the formula

$$AV = (A \div 2) \oplus (B \div 2)$$

for the average problem. Assume that r is even and that the arithmetic is performed in the system FP(r, p, c). Show that

$$\min(A, B) \leq AV \leq \max(A, B)$$

fails to hold if and only if $A = B$ but $A \div 2 \neq A/2$.

3. Suppose that we use the formula $AV = (A \oplus B) \div 2$ for the average problem. Assume that the arithmetic is performed in the system FP(r, p, c), where $p \geq 2$. Show that

$$\min(A, B) \leq AV \leq \max(A, B)$$

holds for all A and B in $S(r, p)$ if and only if $r = 2$.

4. Show that if A and B have opposite signs, then formula (9.2.4) produces a result which has all the properties listed under heading 3 in the specifications for the average problem.

5. Suppose that we perform the arithmetic for the average problem in FP(r, p, c), where r is even and $p \geq 2$. For which of the formulas (9.2.4)–(9.2.6) can we assert that the result satisfies

$$\min(A, B) < AV < \max(A, B)$$

whenever there is a floating-point number between A and B?

6. Suppose that we are using a machine on which the arithmetic is performed in FP(16, 6, c) and that $\omega = 16^{-65}$. Find floating-point numbers A and B such

that

$$\left|\frac{A+B}{2}\right| \geq \omega$$

$$|A| \geq 2\omega$$

$$|B| \geq 2\omega$$

but $(A \div 2) \oplus (B \div 2)$ underflows.

7. In FP(10, 6, c), find an example of two floating-point numbers A and B for which $B = -A$ but

$$A \oplus [(B \ominus A) \div 2] \neq 0.$$

8. Write a program to solve the average problem on the machine you are using. It should meet all the specifications given in Section 9.2.

9. Suppose that we use formula (9.2.5) for the average problem. Assume that the arithmetic is performed in FP(r, p, c), where r is even. Show that if A and B have the same sign, then the error is less than $1\frac{1}{2}$ units in the last place.

10. Suppose that we use the formula (9.2.4) for the average problem. Assume that the arithmetic is performed in the system FP(r, p, c), where r is even. Show that the error is less than $r/2$ units in the last place.

11. Suppose that we use formula (9.2.6) for the average problem. Assume that the arithmetic is performed in the system FP(r, p, c), where r is even. Show that if A and B have the same sign, then the error is less than $r/2 + 1$ units in the last place.

12. Convert the coefficients of the equation

$$10^{30}x^2 - 2 \cdot 10^{30}x + 10^{30} = 0$$

to $S(16, 6)$ using a truncation conversion transformation. Show that the roots of the resulting equation

$$\tilde{A}x^2 + \tilde{B}x + \tilde{C} = 0$$

are not real.

13. Suppose that you want to solve the quadratic equation

$$Ax^2 + Bx + C = 0$$

without doing any scaling. Let e_A, e_B, and e_C be the exponents of the floating-point numbers A, B, and C. Find a number e such that you will not encounter either overflow or underflow on the machine you are using if $|e_A|$, $|e_B|$, and $|e_C|$ are all less than e.

14. Find an example of a quadratic equation

$$Ax^2 + Bx + C = 0$$

whose coefficients are normalized floating-point numbers on the machine you are using, and

a. One root overflows.

b. One root underflows.

c. One root overflows and one root underflows.

15. Let $Ax^2 + Bx + C = 0$ be a quadratic equation whose coefficients are normalized floating-point numbers. Suppose that we use the scaling described in Section 9.3 and that (9.3.12) holds. Find examples to show that we may encounter either overflow or underflow when we multiply the roots of (9.3.11) by 16^K.

16. Suppose that we use the quadratic formula to solve

$$Ax^2 + Bx - A = 0,$$

without providing any special treatment for the computation of R2. Also, assume that $B^2 + 4A^2$ is computed using single-precision arithmetic. Let $B = 1.23456$ and $A = .000123$. Describe what happens when the calculation is performed in FP(10, 6, c).

17. Consider the quadratic equation

$$Ax^2 + Bx = C = 0$$

where $A = .20000001$, $B = -.4$, and $C = .19999999$. Use the quadratic formula to find both roots of the equation, performing all operations in FP(10, 8, c). Assume that the square root routine produces the correctly chopped result. What is the error in the answers? What is the error in the answers if we perform the calculation of $B^2 - 4AC$ in FP(10, 16, c), but perform the rest of the calculation in FP(10, 8, c)?

18. Let A, B, and C be in $S(r, p)$, and suppose that we compute $B^2 - 4AC$ in FP(r, $2p$, c). Show that we obtain the correctly chopped value $\overline{B^2 - 4AC}$ if r is 2 or 4 but that we do not always obtain $\overline{B^2 - 4AC}$ if r is 10 or 2^k with $k > 2$. If $r \neq 3$, show that we always obtain $\overline{B^2 - 4AC}$ when the calculation is performed in FP(r, $2p + 1$, c).

10 CHECKING AND TESTING

10.1. RANGE CHECKING

We shall distinguish between the ideas of checking and testing. By *testing* we mean running test cases to test the behavior of the program. *Checking* refers to checks that are incorporated in the program to check the validity of the answers that are produced.

One of the simplest forms of checking is to check that the numbers lie in a prescribed range. For example, a statistical program can check that variances are nonnegative and that the absolute values of the correlation coefficients do not exceed 1. Then if one of these conditions is violated, the program can print an error message instead of producing a ridiculous answer.

Many programs check the input data and print a message if they detect an invalid character. In some cases, they also check the range of the numbers read in. When an error is detected, the error message should indicate where the error occurred, so we do not have to search through several hundred cards to find the one that is mispunched.

It is quite common for a library subroutine to check the range of its arguments. For example, square root programs usually check that the argument is nonnegative, an arcsine program may check that the absolute value of its argument is at most 1, and a matrix inversion program might check that the parameter specifying the order of the matrix is a positive integer. A subroutine to compute e^x may check that $x \leq \log_e \Omega$ in order to avoid the case in which the answer overflows. These checks are designed to detect cases in which the problem is incorrectly posed or in which the program is unable to produce a reasonable answer. We normally expect library programs to provide us with this protection.

Some compilers provide us with a similar sort of protection. Programs compiled by WATFOR check the range of the subscripts, and they print an error message if we try to use a subscript which is larger than the dimension of the array. PL/I provides a similar check on an optional basis. We can specify whether or not we want the program it produces to check the range of the subscripts.

10.2. MATHEMATICAL CHECKS

The most familiar form of checking pertains to the solution of an equation $f(x) = 0$. After we have solved the equation, we can check the answer by substituting it into the equation. This concept is extremely familiar from elementary courses in mathematics. Some algorithms for the solution of $f(x) = 0$ use an iterative procedure in which the stopping criterion is based on the size of $f(x)$, so the procedure automatically includes a check.

When automatic computers were first used, it was quite common to spend as much time and effort checking the answers as calculating them. But much less checking is done today, and this may be attributed to several factors. First, and probably most important, the hardware has become very much more reliable. Also, the problems we solve have become more complicated, so it is harder to find a satisfactory check. Finally, we have more confidence in our programs, because they are written in higher-level languages. (This confidence may not be justified, but it seems to exist.)

Our experience with solving problems analytically suggests that it is much easier to check the answer than it is to solve the problem. For example, consider the equation

$$x^6 + 2x^5 - 8x^3 - 4x^2 + 6x - 15 = 0.$$

It is more difficult to solve the equation than it is to check the fact that $\sqrt{3}$ is a root. Even in machine calculation, we would rather find the value of a polynomial $p(x)$ than find the roots of $p(x) = 0$. This is also true of solving a system of n linear equations. It requires about $\frac{1}{3}n^3$ multiplications and additions to solve the equations, but it requires only n^2 multiplications and additions to check the solution. For large values of n, checking the answer requires much less machine time than solving the equations.

Another type of problem in which checking has been used successfully is the solution of the partial differential equations that arise in hydrodynamics. Here we often find that conservation laws, such as the conservation of energy or the conservation of momentum, can be used as a check. But there are other problems in which it is quite difficult to find a satisfactory check. This is true of many problems referred to as *simulation*.

The commonest form of a check is an equation such as

(10.2.1) $$f(x) = 0$$

which is satisfied exactly by the correct answer x for our problem. Unfortunately, we usually have only an approximation $\tilde{x}$ for x, and we do not expect $f(\tilde{x})$ to be exactly zero. Moreover, we usually cannot calculate $f(\tilde{x})$ exactly, so we obtain an approximation $\bar{f}(\tilde{x})$ for $f(\tilde{x})$. We cannot expect $\bar{f}(\tilde{x})$ to be exactly zero, but we hope that $|\bar{f}(\tilde{x})|$ is small. We are immediately faced with the problem of deciding how large $|\bar{f}(\tilde{x})|$ must be before we should reject $\tilde{x}$.

Often we do not have a clear idea about where the cutoff point should be set, but the larger $|\bar{f}(\tilde{x})|$ is, the less confidence we have in the answer $\tilde{x}$. This suggests that we should select two criteria A and B. We would accept the answer $\tilde{x}$ if $|\bar{f}(\tilde{x})| \leq A$ and reject it if $|\bar{f}(\tilde{x})| \geq B$. If $A < |\bar{f}(\tilde{x})| < B$, then $\tilde{x}$ is questionable. For this case, we might print a warning message and allow the calculation to continue.

The criteria A and B cannot be chosen without some knowledge of the function $f(x)$. For example, suppose that we are working in FP(10, 8, c) and we set $A = 10^{-8}$ and $B = 10^6$. If

(10.2.2) $$f(x) = 10^{30}x^2 - 3 \cdot 10^{30}x + 2 \cdot 10^{30},$$

then either $\bar{f}(\tilde{x}) = 0$ or else $|\bar{f}(\tilde{x})| > 10^6$. Therefore we shall reject the answer unless $\bar{f}(\tilde{x})$ is exactly zero. On the other hand, if

(10.2.3) $$f(x) = 10^{-30}x^2 - 3 \cdot 10^{-30}x + 2 \cdot 10^{-30},$$

$|\bar{f}(\tilde{x})|$ will be less than 10^{-8} unless $|\tilde{x}|$ is extremely large. Thus, we need some idea of the size of the numbers that will arise in the computation of $\bar{f}(\tilde{x})$ before we select A and B.

The scaling of the coefficients of $f(x)$ is not the only source of difficulty in the selection of A and B. Consider the polynomial

(10.2.4) $$f(x) = x^n - \alpha x^{n-1} - x + \alpha,$$

where $n \geq 3$ and $\alpha = 8.76543_H$. Here $f(x) = (x - \alpha)(x^{n-1} - 1)$, so the roots of $f(x) = 0$ are α and the $(n-1)$st roots of unity. Suppose that we are working in FP(16, 6, c) and have obtained an approximation $\tilde{\alpha} = 8.76544_H$ for the root α. We would like to check $\tilde{\alpha}$ by computing $f(\tilde{\alpha})$, and since $\tilde{\alpha}$ differs from α by only one unit in the last place, we want to accept $\tilde{\alpha}$.

The following table shows the true values of $f(\tilde{\alpha})$ rounded to four decimal digits:

n	$f(\tilde{\alpha})$
3	$.6734 \times 10^{-4}$
5	$.4889 \times 10^{-2}$
10	$.2122 \times 10^{3}$
20	$.3996 \times 10^{12}$
40	$.1417 \times 10^{31}$
80	$.1781 \times 10^{68}$

When we evaluate $\tilde{f}(\tilde{\alpha})$ by performing the arithmetic in FP(16, 6, c), we shall produce approximately these results. Although the coefficients of (10.1.4) are of reasonable size, changing the root by 1 in the last place produces an extremely large value for $|\tilde{f}(\tilde{x})|$ when n is large. We clearly cannot set A large enough to accept the value $\tilde{\alpha}$ without accepting many very bad approximations for the roots of $f(x) = 0$. This behavior is not unusual for polynomials of high degree.

For some problems, we may be able to write the check in the form

$$(10.2.5) \qquad\qquad F(x) = G(x),$$

where $F(x)$ and $G(x)$ can be computed with small relative errors. Then we can base our criteria for accepting the answer $\tilde{x}$ on the relative difference

$$\frac{\tilde{F}(\tilde{x}) - \tilde{G}(\tilde{x})}{\tilde{G}(\tilde{x})}.$$

Although we still have to decide on the cutoff points, the influence of the scaling of the coefficients has been eliminated.

It may be advantageous to change the check from the form (10.2.1) to the form (10.2.5) by carrying some of the terms in $f(x)$ to the right-hand side of the equation. For example, if $f(x)$ is given by (10.2.4), we could write

$$x^n + \alpha = \alpha x^{n-1} + x.$$

We want to be able to compute $F(x)$ and $G(x)$ in (10.2.5) with small relative errors, so we would usually try to carry out this rearrangement in such a way that all the terms in $F(x)$ and $G(x)$ have the same sign. But the sign of a term may depend on the value of x, so we might want to let the rearrangement depend on x. We could compute each term in $\tilde{f}(\tilde{x})$ and let $\tilde{F}(\tilde{x})$ be the sum of the terms which are positive. While this approach is not foolproof, there are times when it can be quite effective.

Other strategies for checking may be devised. If $f(x)$ in (10.2.1) is a function of a single variable x, we may ask whether $f(x)$ changes sign in the neighborhood of $\tilde{x}$. That is, we could select a value p and compare the sign of

$\tilde{f}(\tilde{x})$ with the signs of $\tilde{f}(\tilde{x} \pm \rho \tilde{x})$. Since $f(x)$ may vanish without changing sign, this check cannot be used indiscriminately. But we could decide to accept the answer $\tilde{x}$ when either $|\tilde{f}(\tilde{x})| \leq A$ or $f(x)$ changes sign in the neighborhood of $\tilde{x}$.

10.3. TESTING

We shall now consider the problems involved in testing a program to determine the quality of the answers it produces.† These tests may be performed by either the author of the program or a user, but we shall assume that the objective of the test is program evaluation, not debugging. Sometimes the test merely verifies that the program produces reasonably good answers for a few problems whose answers are known, but we would prefer to have a more extensive test which would test the behavior of the program on a large number of different cases. We have seen that it is often difficult to perform an error analysis and find a good bound for the error produced by a program. Instead, we often try to obtain an estimate of the maximum error by testing the program on a large number of cases. Results of tests of this sort are often included in the documentation of the program.

Some guide lines for testing various classes of programs are beginning to appear in the literature.‡ Both the formulation of a good set of test cases and the evaluation of the results of the test require a detailed knowledge of the problem being solved, so we shall make only a few general comments about testing.

Since any test of the accuracy of the results produced by a program is based on the comparison of these results with the correct answers, one of the major difficulties we face is finding the correct answer to the problem. Sometimes this leads us to test the program on problems which can be solved analytically. Unfortunately, these cases may not be sufficiently general to provide a good test.

Suppose that we are testing a subroutine rather than a complete program. The data supplied to the subroutine will be floating-point numbers in $S(r, p)$, but we often find that the problems for which we know the answers have data which are not in $S(r, p)$. For example, consider a subroutine which computes the value of a function $f(x)$, where we know that $f(\frac{1}{3}) = \alpha$. The number $\frac{1}{3}$ will not be in $S(r, p)$ unless r is divisible by 3. Suppose that we call the subroutine with an argument $\tilde{x} \approx \frac{1}{3}$ and the subroutine returns the answer $\tilde{\alpha}$.

---

†The algorithms section of the Communications of the ACM and the SHARE SSDs publish reviews containing the results of such tests. An extensive review of the FORTRAN library for the IBM System/360 was published by Clark et al. (1967).

‡An example is the collection of guide lines published by Kuki et al. (1966). Usow (1969, 1970) published a bibliography of papers on the testing of programs.

The error introduced by the subroutine is the difference between $\tilde{\alpha}$ and $f(\tilde{x})$. Probably the commonest mistake in testing programs is to compare $\tilde{\alpha}$ with α instead of with $f(\tilde{x})$. The discrepancy between $\tilde{\alpha}$ and α is due partly to the error introduced by the subroutine and partly to the change in the argument. We cannot fault the subroutine for failing to guess the argument we had in mind.

Thus, we have to be able to find the value of $f(x)$ for x in $S(r, p)$, and this often requires us to write a program to compute $f(x)$ using higher-precision arithmetic. Since we are interested in only the first two or three decimal digits of the error, the answer produced by the higher-precision program has to be accurate to only a few more than p digits. Even so, writing a program which will produce answers that are this accurate may be a major undertaking. For example, we may have to develop new approximations which are more accurate than the ones used in the program we are testing.

When our test is based on a comparison of the results produced by a subroutine with the results produced by a higher-precision program, we must be careful about the way the arguments are generated. We should generate the p-digit arguments for the subroutine and then extend them to higher-precision by appending zeros. Then both programs are trying to compute $f(x)$ for the same value of x. A common mistake is to generate a higher-precision number x for the argument of the higher-precision program and use $\tilde{x}$ as the argument of the subroutine. This contaminates the test because the two programs will receive different arguments.

There are several ways in which we might compare the answer $\tilde{y}$ computed by the subroutine with the correct answer y. For instance, we can compute either the absolute error or the relative error. (Which of these is the appropriate measure of the accuracy of the result will depend on the nature of the problem.) Instead of computing the relative error, we sometimes express the error in terms of units in the last place. When the answer is a vector instead of a single number, it is often appropriate to compute the norm of the error.

Instead of comparing $\tilde{y}$ with the correct answer y, some tests have compared $\tilde{y}$ with the number $\bar{y}^{\circ}$ obtained by rounding y to $S(r, p)$. Then the program is considered to be perfect if it always produces the answer $\bar{y}^{\circ}$, and the maximum number of units in the last place by which $\tilde{y}$ and $\bar{y}^{\circ}$ differ can be used as a measure of the accuracy of the program. This leads to a description of the test results which is easy to understand. But suppose that we are using an eight-digit decimal machine and that

$$y = 1.234567850000001.$$

Then $\bar{y}^{\circ} = 1.2345679$, but $\tilde{y}$ is almost as good an approximation for y as $\bar{y}^{\circ}$ is. If we compare $\tilde{y}$ with $\bar{y}^{\circ}$, we shall consider 1.2345680 to be as good an answer as $\tilde{y}$ is, but it has about three times as large an error. Thus, we obtain a better estimate of the error by comparing $\tilde{y}$ with y instead of with $\bar{y}^{\circ}$.

Our test of a program is often based on random arguments selected from a suitable distribution. Then we can compute the sample mean and variance for the error as well as the maximum error observed. As we saw in Chapter 3, there are problems in which the average error is more important than the maximum error.

In addition to testing the program on a large number of typical problems, we may want to devote part of our test to trying to determine the class of problems for which the program will produce reliable answers. This leads us to test the program on some problems which are known to be difficult to handle. For example, it is common for a test of a program for the solution of simultaneous equations to include a few problems which are ill-conditioned, and the test cases for a differential equation solver usually include some "stiff" systems of differential equations. By testing the limits of the program in this way, we hope to determine the type of problem for which the program can be used with confidence. Also, tests of this sort will show us whether the program can recognize the cases in which it is unable to produce a good answer and provide us with a suitable warning. However, these should not be the only tests we perform, because our primary interest is in the behavior of the program on the sort of problem for which it was designed. Most of our test cases should be more typical of the use we expect to make of the program.

EXERCISES

1. Let $\alpha = 8.76543_H$, $\tilde{\alpha} = 8.76544_H$, and $\beta = \alpha - 1$. Let

$$p(x) = (x - \alpha)(x^{n-1} + x^{n-2} + \cdots + 1),$$

so

$$p(x) = x^n - \beta x^{n-1} - \beta x^{n-2} - \cdots - \beta x - \alpha.$$

Suppose that we obtain the approximation $\tilde{\alpha}$ for the root $x = \alpha$ of $p(x) = 0$. Check $\tilde{\alpha}$ by computing the value of $p(\tilde{\alpha})$ for $n = 3, 5, 10, 20, 40, 80$.

2. Let $\tilde{\alpha}$ be an approximation for the root α of

$$x^n - \alpha x^{n-1} - x + \alpha = 0,$$

where $\alpha = 8.76543_H$. Take several values of $\tilde{\alpha}$ close to α and check them using

$$x^n + \alpha = \alpha x^{n-1} + x$$

for $n = 3, 5, 10, 20, 40, 80$.

3. Show that the root $x = \alpha$ of

$$x^n - \alpha x^{n-1} - x + \alpha = 0$$

is well conditioned with respect to changes in the nonzero coefficients of the
equation.

4. Let $\tilde{\alpha}$ be an approximation for the root α of the equation

$$p(x) = x^n - \alpha x^{n-1} - x + \alpha = 0,$$

where $\alpha = 8.76543_H$ and $\tilde{\alpha} = 8.76544_H$. To check $\tilde{\alpha}$, we select a value of p
and compute $p(\tilde{\alpha} - p\tilde{\alpha})$ and $p(\tilde{\alpha} + p\tilde{\alpha})$. We shall accept $\tilde{\alpha}$ if these quantities
have opposite signs. Perform this check for $n = 3, 5, 10, 20, 40, 80$.

5. Let $\tilde{\alpha}$ be an approximation for the root α of

$$p(x) = x^n - \alpha x^{n-1} - x + \alpha = 0,$$

where $\alpha = 8.76543_H$ and $\tilde{\alpha} = \alpha + 16^{-13}$. Then $\tilde{\alpha}$ differs from α by 1 in the
fourteenth hexadecimal digit. Check $\tilde{\alpha}$ by computing $p(\tilde{\alpha})$ in FP(16, 14, $cl1$) for
$n = 3, 5, 10, 20, 40, 80$.

6. Let $x = \pi/6$ and let $\tilde{\pi}$ be a floating-point number which is an approximation for
π. To obtain an approximation $\tilde{x}$ for x, we form $\tilde{x} = \tilde{\pi} \div 6$ using floating-point
division. Let $\bar{y}$ be the value produced for sin $\tilde{x}$ by the sine routine. How much
of the error in the approximation $\bar{y} \approx \sin x$ is due to the change in the argument?

7. Test the sine routine on the machine you are using.

11 LANGUAGE FEATURES FOR FLOATING-POINT COMPUTATION

11.1. INTRODUCTION

We shall now turn to the question of the characteristics a higher-level language should have to enable us to write programs for floating-point computation. Some aspects of this question were considered in earlier chapters. The treatment of overflow and underflow was discussed in Chapter 2, and the language support for double-precision arithmetic was discussed in Chapter 5. Consequently, these subjects will not be addressed in this chapter.

Many discussions of compilers stress fast compilation, fast execution, machine independence, and ease of writing programs in the language. These objectives are desirable and widely recognized. Our intent is to discuss some other objectives which are less often mentioned. We shall consider the language from the point of view of the author of a carefully written program. The principal question we shall address is whether the language gives us suifficient control over the calculation so that we can produce a high-quality program.

The first properties we shall look for in the language are predictability, controllability, and observability. Then in Section 11.3, we shall discuss some operations that are often difficult to program in higher-level languages. Finally, machine independence will be discussed in Section 11.4. Many of our comments will refer to the implementation of the language rather than the language specifications themselves.

11.2. PREDICTABILITY, CONTROLLABILITY, OBSERVABILITY

Floating-point arithmetic usually produces only an approximation for the correct answer, and rearrangements of the calculation which would be mathematically equivalent if the arithmetic were performed in the real number system may have a significant effect on the accuracy of the answer. Therefore, we may want to know the sequence of operations which will be used to evaluate the arithmetic expressions we have written.

In Chapter 1, we saw that the associative laws of addition and multiplication fail to hold in FP(r, p, c). Consequently, instead of writing

(11.2.1) $X = A * B * C,$

we should use parentheses to indicate the order in which the operations are to be performed. But the compilers allow us to write expressions such as (11.2.1), and this can be justified by the fact that the associative law of multiplication holds approximately. (See Section 3.4.) Indeed, there are many cases in which we do not care whether the compiler treats (11.2.1) as $(A * B) * C$ or $A * (B * C)$. But there are other times when the distinction is important, so we would like to know which form will be used.

Thus, one aspect of predictability is that we want to know what arithmetic operations will be performed when we write a statement such as (11.2.1). This is even more important with a statement such as

(11.2.2) $X = A + B - C,$

because the associative law of addition does not even hold approximately. The number $(A \oplus B) \ominus C$ need not be close to $A \oplus (B \ominus C)$.

Almost any compiler is deterministic in the sense that a given program will always produce the same object code. But predictability means that the user can predict the arithmetic operations that will be performed. This requires simple rules, such as "the terms in a sum will be added from left to right." It also requires that these rules be communicated to the user.

There are times when we are uncertain about the way the compiler treats certain statements, so we want to find out what arithmetic operations have been compiled. We shall refer to this as observability. Many compilers provide this capability by allowing us to request an Assembler listing of the object code. This listing shows us the sequence of operations that will be performed, and it provides a way to resolve any ambiguities in the description of the language.

While predictability is desirable, controllability is essential. When we know exactly the way we want the calculation to be performed, we must be able to produce the desired sequence of arithmetic operations. This means that the compiler must honor our parentheses. While we would like to be able to predict the way (11.2.2) will be evaluated, it is essential that the statement

$$X = (A + B) - C$$

produce $(A \oplus B) \ominus C$.

A second aspect of predictability and controllability concerns the conversion of constants. When we write a statement such as $X = 2.1$, will the number stored in X be $\overline{2.1}, \overline{2.1}^{\circ}$, or merely some number close to 2.1? If the constant is a number C which can be represented exactly in $S(r, p)$, will the conversion program produce C exactly? Here the predictability of the conversion program depends on its documentation. Unfortunately, we are often given little or no information of this sort. Even if we cannot have complete predictability, it would be helpful to have a description of the cases in which the result can be described easily. (For example, see Exercise 10 of Chapter 8.)

Since constants will be converted to the radix r before they are used in the calculation, there are times when we want a constant to be converted to a specific number in $S(r, p)$. For example, we might want to produce the corrrectly rounded value of π. This is another aspect of controllability, and it means that the conversion transformation must be onto. That is, for any number x in $S(r, p)$, there must be a constant which will be converted into x.

Another way to produce a number we want in $S(r, p)$ is to enter the number in the radix r of the machine. This is the natural way to express numbers such as $1 - r^{-p}$, $1 + r^{-(p-1)}$, ω, Ω etc. It is also an appropriate way to enter constants which were computed by another program. For example, we may write a program to compute the coefficients of a polynomial approximation for a function $f(x)$ and then use these coefficients as constants in another program which computes $f(x)$. But converting these coefficients to decimal and then reconverting them to the radix r might introduce errors, so we might prefer to print them in the radix r and then enter them in that form in our program for $f(x)$.†

Finally, since our calculation will be performed in FP(r, p, a), there are times when we want to see the numbers in $S(r, p)$ that arise in the calculation. This is another aspect of observability, and it is another reason for our wanting the language to provide us the option of printing numbers in the radix of the machine.

11.3. EASE OF PROGRAMMING

The widespread use of higher-level languages for floating-point computation is due to the fact that they make it much easier to program many of the

---

†Some FORTRANs for the IBM System/360 support the Z format which provides this capability.

calculations we want to perform. Since the merits of these languages are widely recognized, we shall not dwell on them here. Instead, we shall indicate a few aspects of floating-point computation that are hard to program in many higher-level languages. The criterion we shall adopt is that operations which are easy to program in Assembler language should be easy to program in the higher-level language. Fortunately, this is true of most of the things we want to do with floating-point numbers.

However, in many higher-level languages it is difficult to perform operations which make explicit reference to the representation of the number. For example, in Section 4.4 we described ways to dismantle the floating-point number. We succeeded in coding this operation in both FORTRAN and PL/I, but the FORTRAN coding was quite devious and it obscured the intent of the code. It would be much nicer if the language had functions which would produce the parts of the number. We shall suggest four functions which would enable us to perform operations like this that are often difficult to program in higher-level languages. The names selected for these functions follow the FORTRAN conventions.

First, we shall consider functions which would allow us to dismantle and reassemble the floating-point number. We would like to have an integer valued function IEXP(X) whose value is the exponent of the floating-point number X. To reassemble the floating-point number, we would like to have a function ASSEMB(I,X). The value of ASSEMB(I,X) would be a floating-point number whose exponent is I and whose sign and mantissa are the same as those of X. This function would also allow us to extract the mantissa of X. If we set I = 0, then the value of ASSEMB(I,X) is a floating-point number which is equal to the mantissa of X.

Rounding is another operation that is often difficult to program in higher-level languages. We would like to have a function ROUND(D) whose argument is the double-precision number D and whose value is the single-precision number $\overline{D}^{\circ}$.

Finally, it would be helpful to have a function AUG(X,I) whose value is the floating-point number X augmented by I units in the last place. (When I is negative, X is decremented.) This function would allow us to perform operations such as rounding the intervals outward in interval arithmetic. (See Section 7.4.) It can also be useful in testing programs. In some tests we want to use consecutive floating-point numbers as arguments to see whether the answers are monotonic; in other tests we might want to step the argument by a fixed number of units in the last place. [See Turner (1969b).]

Since all these operations are easy to perform in Assembler language, we could produce the function subroutines described above and include them in the library for the machine we are using. Even if these subroutines were coded in Assembler language, they could have linkages which would allow them to be called by programs written in the higher-level language. But it would be much more convenient if they were provided by the higher-level language itself. Since they are so short, they could be compiled as in-line code instead of using subroutine calls. This would make the execution of these operations much faster, so we would be more inclined to use them. Moreover, they are especially useful when we are trying to produce a high-quality program, so we would like to use them in programs which will be distributed to others. But when we distribute a program, we would prefer not to have to distribute our library subroutines along with it. Thus, it would be much nicer if these functions were provided by the higher-level languages.

11.4. MACHINE INDEPENDENCE

We shall now ask whether a program can be written in a higher-level language and run on two (or more) different machines. We often find that this is not the case; the program must be modified to run on a second machine. The basic question is how difficult it is to make these modifications. In recognition of the fact that some modifications may be necessary, it is becoming common to speak of the *portability* of a program. Instead of asking whether a program is machine-independent, we ask how much work will be required to transport it to another machine.

We often find that the most serious problem we encounter in converting a program from one machine to another is the incompatibility of the compilers. Different compilers implement different language features, and in some cases they handle the same statement in different ways. This is even true of different FORTRAN compilers for the same machine. One way to enhance the portability of our programs is to restrict ourselves to the language features common to all the compilers we wish to use. But if we do this, we may have to sacrifice some of the power of the language, and in some cases the quality of the program will suffer. Since it is the high-quality programs that we are most interested in converting from one machine to another, we usually will not be willing to sacrifice quality to gain portability.

If a program must be modified to convert it to another machine, the first problem we face is identifying the statements that must be changed. Cases of obvious machine dependency, such as the use of UNSPEC in PL/I, are not nearly as troublesome as hidden dependencies are. Similarly, the functions IEXP and ASSEMB suggested in the previous section are easy to identify as possible sources of machine dependency. By making the machine dependence explicit, these functions would enhance the portability of our programs.

In Section 11.2 we mentioned the desirability of entering constants in the radix r of the machine. This is another language feature which makes the machine dependence clear. Since the constants which we would want to enter in the radix r are very likely to be machine-dependent, writing them in this form helps identify items that have to be modified. In addition to writing constants in the radix r, we sometimes want to use the radix r in input and output formats. The most common use of this format occurs in test programs and special-purpose programs which we are not interested in converting to another machine. But if we do want to convert a program that uses this format, the modifications that have to be made are usually quite clear.

After we find the items in the program that are machine-dependent, we have to decide how to change them. To do this correctly, we must understand the intent of the coding. But the intent of the coding is most easily understood when the language allows the programmer to specify the operations he wants to perform. For example, if the program uses the functions IEXP and ASSEMB described in Section 11.3, it is easy to see what the programmer was trying to do. By contrast, the FORTRAN coding for these operations described in Section 4.4 completely obscures the intent of the code. While the use of UNSPEC in PL/I indicates that the coding is machine-dependent, it does not describe the intent of the coding as well as the functions IEXP and ASSEMB do.

As we have seen, a carefully written program often uses constants which are chosen for the specific machine on which the program is to be run. There are several types of constants that are machine-dependent. First, there are constants such as r, p, ω, Ω, etc., which describe the characteristics of the machine. Second, there are mathematical constants, such as π, e, $\log_e 2$, etc. We would like to enter these constants in such a way that the number stored in the machine is the correctly rounded value of the constant. Finally, there are the coefficients for the approximations we shall use for various functions. Both the coefficients and the number of terms in the approximation will depend on the word length of the machine we are using. It has been proposed that the first two types of constants could be stored in a special subroutine which could be called by programs that need the constant. Thus, by calling this program and asking for the fourth constant we could obtain say, the correctly rounded value of $\pi/2$. Each machine would have such a subroutine with the constants arranged in the same order. To convert a program to a new machine, we would change this one subroutine instead of changing the constants in every program that used them. Here the price we pay for portability is some extra subroutine calls. A variation of this approach is to place the block of constants in COMMON, so that we can avoid the subroutine calls.

There are two different situations in which we are interested in the portability of programs, and they present slightly different problems. The first case is the one in which the machine we have been using is to be replaced by a new machine, so we want to convert all our programs. The second case arises when we have access to several different machines, and we want our programs to run on all these machines. We shall call the first case *conversion* and the second case the *multiple-machine* case. In the conversion case, once the conversion has been performed we are no longer interested in the original version of the program. But in the multiple-machine case, we want to avoid maintaining several different versions of the program, so we want to write the program in such a way that it will run on all the machines we are using.

Conversion appears to be the easier case, since we can modify the program in any way we want to. In fact, we can use a *sift* program to flag statements that are clearly machine-dependent. But the conversion case may also present some diffculties. We often find that the programs were written for the first machine without any consideration of the problems involved in transporting them to the new machine. (Some of the programs may have been written before we knew the specifications for the new machine and its compilers.) Also, we may have to convert the programs at a time when we have had little or no experience with the new machine, so we may be unaware of some idiosyncrasies of the machine that can affect the programs.

The difficulties in the multiple-machine case stem from the fact that we want genuine machine independence—that is, we want the same program to run on several different machines. However, we shall assume that at the time we write the program we are familiar with all the machines on which it is to be run. In principle, it should be possible to write such a program. If we want to use some coding that is machine-dependent, we can perform a test to determine which machine the program is being run on and then branch to the appropriate coding.† But the compilers sometimes make it quite difficult to do this. Some of the statements in the program we compile on machine A will be executed only on machine B, but the compiler for machine A may diagnose them as errors. It is quite acceptable for the compiler to print warning messages for these errors, but it must not consider them to be so serious that it fails to compile the rest of the program. Of course, it is also vital that these errors not contaminate the rest of the program.

If we use this approach, we must be able to perform a test to determine which machine the program is being run on. We shall call this the machine identification problem. We want to perform a simple computation which will produce different answers on two different machines. When the machines are specified, such a test is usually quite easy to devise. In Exercise 7 we shall address the more general problem of performing a series of tests to identify the system $FP(r, p, a)$ in which the floating-point calculation is being performed.

EXERCISES

1. Another way to obtain the exponent of a floating-point number is to use $\log_r |x|$, which we can compute by using

$$\log_r |x| = \frac{\log_e |x|}{\log_e r}.$$

---

†A slightly less elegant approach is to write the machine-dependent statements as comments. To convert the program to another machine, we only have to change a few characters to change the appropriate comments to executable statements.

Write a program to use $\log_r |x|$ to compute the exponent of x on the machine you are using. Does this computation produce the correct exponent for all floating-point numbers?

2. For mathematical constants, such as π and e, we want the number stored in the machine to be accurate to the precision of the machine. Then the number of digits used to represent the constant becomes a source of machine dependency in the program. At the expense of execution time, we can sometimes overcome this difficulty by using a standard library function. For example, we can compute $\pi/4$ by computing arctan(1). What other constants can we obtain in this way?

3. Consider the function ASSEMB(I,X) described in Section 11.3. What should this subroutine do when the value of I is outside of the range of the exponents of the floating-point numbers on our machine?

4. Write Assembler language subroutines for the functions IEXP, ASSEMB, ROUND, and AUG on the machine you are using.

5. How would you use the functions described in Section 11.3 to decide whether or not X differs from Y by less than K units in the last place of Y?
   a. Use AUG.
   b. Use IEXP and ASSEMB.

6. Suppose that you are writing a program which you intend to run on two different machines. Devise a simple test to identify the machine on which the program is being run, if the systems in which the two machines perform floating-point arithmetic are:
   a. FP(2, 40, a) and FP(2, 48, a)
   b. FP(16, 6, c) and FP(2, 24, c)
   c. FP(16, 6, c) and FP(10, 7, c)
   d. FP(2, 27, c) and FP(10, 8, c)
   e. FP(2, 27, R) and FP(2, 28, c)

7. Suppose that you are writing a program which is to be run on many different machines. Devise a collection of tests which will identify the system FP(r, p, a) in which the calculation is being performed. You may assume that all of the machines on which the program will be run have all of the following five properties:
   a. r is either 10 or 2^k with $1 \le k \le 6$.
   b. $r^p > 1000$.
   c. a is R, c, or clq. If a is clq, then q is 0, 1, or 2.
   d. $\Omega > r^{2p}$ and $\omega < r^{-2p}$.
   e. If your program contains a constant which is a positive integer less than 1,000, then the compiler does not introduce any error when it converts the constant to $S(r, p)$.

12 FLOATING-POINT HARDWARE

12.1. CHOICE OF RADIX

The choice of the radix is the most basic decision to be made in the design of the floating-point hardware. The criteria on which this decision is based are usually speed, cost, and ease of use. It is generally accepted that binary arithmetic is faster than decimal arithmetic—at least at the same cost—so the very fast machines have seldom used the decimal representation for floating-point numbers. (The NORC was an exception.) Since it is as easy to perform the basic arithmetic operations in the radix 2^k as it is in binary, several machines have been designed with a radix which is a power of 2.

We now have machines which are capable of performing floating-point arithmetic at remarkably high speeds. For example, the execution time for single-precision multiplication on the model 195 of the IBM System/360 is 162 nanoseconds, which is less than one quarter of a memory cycle. Indeed, a conditional branch may take longer than floating-point multiplication on this model of the IBM System/360. Thus, the speed of the floating-point arithmetic is no longer the major bottleneck, so the fact that binary arithmetic is faster than decimal arithmetic is not as decisive as it once was in the selection of the radix.

Another argument that is sometimes advanced for binary machines is that they make more efficient use of the storage. The decimal representation of numbers requires more bits than the binary representation does to produce the same accuracy, so a binary machine can have a shorter word length than a decimal machine with comparable accuracy. This can reduce the cost of the memory. Alternatively, if it uses the same word length, the binary machine can provide more accuracy.

There are several respects in which a machine with radix 2^k differs from a binary machine. One difference is that with larger values of k we may expect that less shifting will be required in the operations $\oplus$ and $\ominus$. [See Sweeney (1965).] Therefore, these operations might be slightly faster on an octal or a hexadecimal machine than on a binary machine. Another difference is in the size of the exponent. As an illustration, we shall compare a binary machine with a hexadecimal machine having the same word length and the same range of floating-point numbers. Since $16^e = 2^{4e}$, the binary machine will require two more bits for the characteristic. Then its mantissa will be two bits shorter. Thus, to produce the same range for the floating-point numbers, we need the same word length for the numbers in $S(16, p)$ as we do for the numbers in $S(2, 4p - 2)$.

A machine with the radix 2^k also differs from a binary machine in the accuracy of the computation. We shall illustrate this by comparing FP(16, 6, c) with FP(2, p, c). When we chop a number to $S(16, 6)$, we retain the high-order 21, 22, 23, or 24 bits, depending on the leading hexadecimal digit. It follows that the arithmetic operations in FP(16, 6, c) are at least as accurate as those in FP(2, 21, c), and they are never more accurate than the operations in FP(2, 24, c). The bounds for the relative error in a program are often the same in FP(16, 6, c) as they are in FP(2, 21, c), so we should think of FP(16, 6, c) as being roughly equivalent to FP(2, 21, c). [However, by exercising sufficient care, we can sometimes produce a slightly better result in FP(16, 6, c) than we can in FP(2, 21, c).] But since 24 bits are used to represent the mantissas of the numbers in $S(16, 6)$, it is easy to fall into the trap of thinking of FP(16, 6, c) as being about the same as FP(2, 24, c). This is a mistake, because we can seldom attain this accuracy.

The major advantage of decimal arithmetic is that it eliminates the need for radix conversion. This affects both the speed of computation and the ease of use of the machine. We may think of the radix conversion as the price we pay to use binary arithmetic instead of the slower decimal arithmetic. It is worth paying this price if the program does a lot of computation. But decimal arithmetic could be quite attractive in a program that reads in a lot of decimal input and performs relatively little computation on it. There might not be enough computation to offset the time required to convert the numbers to binary.

In many respects, decimal machines are easier to use than binary machines are. Many simple decimal numbers cannot be represented exactly in a binary machine, and this can lead to programming errors. For example, if we want to use the decimal number .1 in a double-precision FORTRAN program, we may have to write it as .1D0 to force the compiler to perform the conversion to double-precision accuracy. (See Section 5.1.) This problem does not arise on a decimal machine. Also, since binary numbers are not as familiar as decimal numbers are, many programmers find it much harder to understand what is happening in the calculation when they are using a binary machine. The calculation is further obscured in a binary machine by the fact that the output is decimal, so we do not see the numbers the machine is using. But it is often necessary for the programmer to understand the details of the arithmetic in order to write a high-quality program. It is usually easier to reach this level of understanding when the calculation is performed on a decimal machine.

Finally, we want the representation of the floating-point numbers to be compatible with the representation of the fixed-point numbers. Numbers may be converted from floating-point to fixed-point or from fixed-point to floating-point numbers in the inner loop of the calculation, so it is important that these conversions be easy to perform. We would not want them to involve radix conversion between decimal and binary, but we would not object to changing the radix from 2 to 2^k. Since fixed-point numbers are used for indexing and address calculations, their representation must be compatible with the addressing scheme of the machine. In many cases this produces a strong argument for making the fixed-point arithmetic binary. Another advantage of the binary representation of fixed-point numbers is that any bit configuration is a valid number, so we can accept any coding of the input data and use the fixed-point operations to convert it to the form we want to use.

12.2. THE REPRESENTATION OF FLOATING-POINT NUMBERS

We usually want to treat the floating-point number as a single entity, so it is convenient to have it stored in a single word of memory. Then a basic design decision is how to partition the word into the sign, characteristic, and mantissa of the number. Also, the designer must decide how negative numbers will be represented and how the sign of the exponent will be handled.

The signed exponent e will actually be stored as the characteristic $e + \gamma$. (See Section 1.4.) If $\gamma = 0$, the exponent is stored as a signed integer; otherwise the characteristic is nonnegative. Many decimal machines have allocated two decimal digits for the characteristic and used $\gamma = 50$. For binary machines, the common choice for γ is 2^{k-1}, where k is the number of bits used to hold the characteristic. But some binary machines have used $\gamma = 0$, and this approach has also been used on decimal machines when the representation of the decimal digits is such that we do not need an extra digit to hold the sign.

In addition to selecting γ, the designer must decide where to place the characteristic within the word. Insofar as the floating-point instructions are concerned, this decision can be made arbitrarily. But if we want to use non-floating-point instructions to manipulate floating-point numbers, the location of the characteristic may be important. For example, some machines have special instructions which manipulate subdivisions of the word called *bytes*. Then it is easy to extract the characteristic if it occupies a byte. Also, when the characteristic is placed in the high-order digits of the word, it may be possible to use fixed-point instructions to compare two normalized floating-point numbers.

Next, consider the representation of negative numbers. Throughout this book we have assumed that the machine stored the sign and true magnitude of the mantissa, but some machines have used either the r's complement or the $(r - 1)$'s complement to represent the mantissa of a negative number. Again, compatibility with the representation of the fixed-point numbers may be an important consideration in deciding which representation to use. Also, it may be a little harder for the programmer to understand what is happening in the calculation when the machine uses complements, because complements are not as familiar as signed numbers are.

If the machine uses the sign and true magnitude representation for negative numbers, then the subtract magnitude case for the operations $\oplus$ and $\ominus$ is handled by complementing one of the operands and adding. If the operand having the larger magnitude is complemented, the result will have to be recomplemented.† One advantage of representing negative numbers by complements is that this recomplementation is never required, so the operations $\oplus$ and $\ominus$ may be slightly faster. However, it may make multiplication a little more complicated. [See Flores (1963).]

When we use complements to represent negative numbers, the mantissa m of a number in $S(r, p)$ will be represented by a positive number m'. If $m \geq 0$, $m' = m$. If $m < 0$, then $m' = r - |m|$ if we use the r's complement, and $m' = r - r^{-p} - |m|$ if we use the $(r - 1)$'s complement. The number m' can always be expressed with $p + 1$ digits in the radix r, and $0 \leq m' < r$. The leading digit of m', which represents the sign of m, is usually required to be either zero or $r - 1$. But this is wasteful unless $r = 2$. We could avoid using a whole digit to hold the sign by using a mixed radix representation for m' with one bit to the left of the point. Nevertheless, we are more likely to find complements used to represent negative numbers on a binary machine than on other machines.

Many machines have a *minus zero* which appears to be different from a *plus zero*. With the sign and true magnitude representation of numbers,

---

†One way to reduce the frequency of recomplementation is to complement the operand with the smaller exponent when the exponents are unequal. Then recomplementation will be required only if the operands have the same exponent and the machine complemented the number with the larger magnitude. Sweeney (1965) reported the results obtained by tracing several programs and counting the number of times the operands had the same exponent in the subtract magnitude case. He also showed how often this situation would arise on machines with different radices. As we would expect, increasing the radix increases the number of times the operands have the same exponent.

minus zero and plus zero have zero mantissas with different sign bits. We also have two zeros if we use $(r - 1)$'s complements, because the $(r - 1)$'s complement of zero is a number whose digits are all $(r - 1)$s. When the machine has both a plus zero and a minus zero, we would like to be assured that they will both be treated as zero in all contexts. We seldom encounter difficulty in the arithmetic operations, but we must be careful about the branching instructions. For example, if the machine has a BRANCH ON MINUS instruction, will it branch on a minus zero? Similarly, if the machine has a COMPARE instruction and we compare a plus zero with a minus zero, will they be treated as equal? When we write programs in Assembler language we must know the answers to these questions to avoid taking the wrong branches.

Some machines guarantee that a minus zero will never be produced as the result of a floating-point operation. This is usually enough to protect us from encountering them, although we must be careful when floating-point numbers are produced by non-floating-point operations such as those described in Section 4.4. We would also want to be assured that the library programs never produce minus zeros.

We do not have a minus zero when we use r's complements, since the r's complement of zero is zero.† But the number $r/2$ is its own complement, and this may present special problems. For example, on a binary machine we can represent the mantissa -1 but not the mantissa $+1$. This anomaly must be accommodated in some way. We could allow the floating-point numbers to have the mantissa -1, but then changing the sign of x could change its exponent. Alternatively, the arithmetic operations could be designed so that they would never produce a result with $m' = 1$. In either case, we encounter many of the problems presented by minus zero. Either the arithmetic operations must be designed to handle operands whose mantissas are -1, or else they must guarantee that the results they produce will never have this mantissa.

Some machines have special bit patterns that will not be treated as valid floating-point numbers. For example, the CDC 6600 represents INFINITY and INDEFINITE by a zero mantissa with special values for the exponent. The hardware must recognize these bit patterns whenever they appear as operands in floating-point operations and set the result accordingly. (See Section 2.2.) The IBM 7030 had a different way to indicate that a number was abnormal. Each floating-point number had three extra bits called flag bits. The program could set these bits, test them, or use them to cause interrupts.

The CDC 6600 produces the bit pattern INDEFINITE as the result of an indeterminant form. But if we think of this bit pattern as simply meaning that the number is not a valid floating-point number, it has other uses. For example, consider a machine which appends extra bits to the word for error

†This is because we retain only one digit to the left of the radix point in m'.

detection. If we want to proceed after a machine error has been detected in a number, we can set the number to INDEFINITE. This approach can also be used when a subroutine is unable to produce a reasonable answer for certain input data. Similarly, in statistical programs it is desirable to distinguish between missing data and zero. Missing data can be set to INDEFINITE.

Thus, we would like to produce the value INDEFINITE whenever we are unable to compute a reasonably good value for a number. To make this approach feasible, there are two properties the hardware should have. First, the result produced by a floating-point operation should be INDEFINITE whenever one of the operands is INDEFINITE. Then we can let the calculation proceed, and those numbers which have been contaminated by the error will be INDEFINITE, but the other numbers will be printed correctly. The second requirement is that we must be able to test a number to determine whether or not it is INDEFINITE. This may be necessary in order to avoid infinite loops and to force the program to take the proper branches. It is also essential if we want to use INDEFINITE to handle missing data. Even if the hardware satisfies these two requirements, careful programming may be necessary.

Next, consider the scaling of the mantissa. We have assumed that the mantissa m is a p-digit number with the radix point at the left, so $r^{-1} \leq |m|$ unless $m = 0$. But some machines store the mantissa as a p-digit integer with the radix point at the right. Hamming and Mammell (1965) suggest placing the radix point at the right of the leading digit of m. Then we would have $1 \leq |m| < r$ unless $m = 0$. The scaling of the mantissa has no effect on the operations $\oplus$ and $\ominus$, and the only impact it has on the operations $*$ and $\div$ is in the calculation of the exponent. But it may have an effect on the range of the floating-point numbers. For example, consider a binary machine which uses k bits to hold the characteristic. Suppose that the characteristic c is the exponent plus γ and that $0 \leq c \leq 2^k - 1$. The commonest choice for γ is 2^{k-1}, so the exponent e satisfies

$$-2^{k-1} \leq e \leq 2^{k-1} - 1.$$

If the mantissa is an integer, ω is several orders of magnitude larger than $1/\Omega$. Then there are many floating-point numbers in the machine whose reciprocals cannot be represented. Unless the range of the exponents is extremely large, this lack of symmetry can be annoying. We might prefer to use a different value of γ which would make the range of the floating-point numbers symmetric. (See Exercise 9.)

Still another variation in the representation of floating-point numbers in a binary machine was proposed by Goldberg (1967) and McKeeman (1967). Consider a machine which uses the sign and true magnitude representation for negative numbers, and suppose that the characteristic is stored as a nonnegative integer. Since the high-order bit of the mantissa of a normalized, nonzero number is always 1, it need not be stored. Instead, we can store the remaining $p - 1$ bits of the mantissa, and the hardware can supply the leading bit when it is needed. The number zero requires special treatment; we can specify that a floating-point number is zero when all the bits appearing in its characteristic and mantissa are zero. Then every nonzero number that appears in the machine is normalized. This may make operations such as FLOAT TO FIXED and FIXED TO FLOAT more difficult to program unless the machine has special instructions to perform them, but we obtain one more bit of precision without increasing the word length.

12.3. FP(r, p, c) AND FP(r, p, R)

The advantage of the systems FP(r, p, c) and FP(r, p, R) is that they are quite easy for the user to understand. The system FP(r, p, a) in which the machine performs arithmetic often resembles one of these systems, but there are usually some slight differences which the user must consider when he wants to produce high-quality programs. In this section we shall consider the way the hardware can be designed to perform arithmetic in FP(r, p, c) and FP(r, p, R). We shall assume that the operands are normalized.

Arithmetic in FP(r, p, c) can be produced by a slight modification of the arithmetic operations in FP(r, p, clq) which was proposed by Harding (1966a, 1966b). Floating-point division was defined to be the same in FP(r, p, clq) as it is in FP(r, p, c), and we saw in Section 1.8 that floating-point multiplication produces the same result in FP(r, p, clq) as it does in FP(r, p, c) whenever $q \geq 1$. Also, in the add magnitude case, the operations $\oplus$ and $\ominus$ produce the same results in FP(r, p, clq) as they do in FP(r, p, c). Therefore, it suffices to consider $a \ominus b$, where we may assume that $a > b > 0$. Let

$$a = r^e m, \qquad r^{-1} \leq m < 1$$
$$b = r^f n, \qquad r^{-1} \leq n < 1.$$

Since $a > b > 0$, we have $e \geq f$. Now $b = r^e n'$, where $n' = r^{-(e-f)}n$ is obtained by shifting n to the right $e - f$ places. Let $q \geq 1$ and let n'' be the first $p + q$ places to the right of the radix point in n'. Instead of using n'', as we would in FP(r, p, clq), we use n''', where

(12.3.1) $$n''' = n'' \qquad\qquad \text{if } n'' = n'$$
$$n''' = n'' + r^{-(p+q+1)} \qquad \text{if } n'' < n'.$$

Thus, n'' is a $(p + q + 1)$-digit fraction, and its last digit is nonzero if and only if nonzero digits of n' were chopped to produce n''. We now let

$\mu' = m - n'''$, and we note that the first $p + q$ digits to the right of the radix point in μ' are the same as those of $m - n'$. To complete the operation, we normalize $r^e\mu'$ and chop it to $S(r, p)$. The analysis in Section 1.8 shows that this produces the result $\overline{a - b}$ as long as $q \geq 1$.

The function of the $(p + q + 1)$st digit of n''' is to force a borrow from the $(p + q)$th digit of m whenever $n'' < n'$. In place of (12.3.1), we could have used

(12.3.2) $n''' = n''$ if $n'' = n'$

                $n''' = n'' + ir^{-(p+q+1)}$ if $n'' < n'$,

where i is any integer in the range $1 \leq i \leq r - 1$. If $r = 2^k$, we can take $i = 2^{k-1}$ in (12.3.2), so we need only one extra bit instead of an extra digit. Harding calls this extra bit a *sticky bit*, because any nonzero bit shifted through this bit position sticks there. Even if r is not a power of 2, we shall refer to the $(p + q + 1)$st digit of n''' as a *sticky digit*. Thus, to produce arithmetic in FP(r, p, c), we take $q \geq 1$ and modify the arithmetic in FP(r, p, clq) by introducing a sticky digit in the subtract magnitude case of the operations $\oplus$ and $\ominus$.

When r is even, it is easy to modify this approach to produce arithmetic in FP(r, p, R). For division, we develop the first $p + 1$ digits of the quotient and then round the answer to p digits. For the operations $\oplus$, $\ominus$, and $*$, we can simply proceed as we would in FP$(r, p, cl2)$, using a sticky digit in the subtract magnitude case of the operations $\oplus$ and $\ominus$ and then round the result to $S(r, p)$ instead of chopping it. Since the final operation of rounding a number to p digits may require us to perform an extra addition, rounded arithmetic is likely to be a little slower than chopped arithmetic.

12.4. UNNORMALIZED NUMBERS AND UNNORMALIZED ARITHMETIC

We have assumed that a nonzero number x in $S(r, p)$ was always written in the form

$$x = r^e m, \qquad r^{-1} \leq |m| < 1.$$

We shall now consider unnormalized numbers, so the mantissa will be allowed to have an absolute value less than r^{-1}. But then the representation of the floating-point number is not unique, because there are many numbers in $S(r, p)$ which can be written with more than one exponent. Unfortunately, the result produced by the arithmetic operations often depends on the representation of the operands. Therefore, we shall now consider a floating-point number to be an ordered pair (e, m), where e is its exponent and m is its mantissa. The exponent is a signed integer and the mantissa is a signed fraction which can be represented in the radix r with p digits to the right of the radix point. Then (e, m) is a representation for the number $r^e m$ in $S(r, p)$, and $|m| < 1$.

The hardware could be designed so that each arithmetic operation normalized the operands before performing the arithmetic, so the results would be the same regardless of whether the operands were normalized or not. But prenormalizing the operands is extra work, and it makes the arithmetic slower. It may even affect the speed with which arithmetic can be performed on normalized numbers, because the machine must test the numbers to see whether they are normalized. Therefore, the hardware often performs the arithmetic without prenormalizing the operands. Instead, it simply uses the exponent e and mantissa m which appear in the representation (e, m) of the floating-point numbers. But then the result produced by the operation may depend on the representation of the operands. We shall illustrate this by considering the case in which the machine is designed so that the arithmetic will be performed in the system FP(r, p, clq) whenever the operands are normalized.

To compute the product $(e, m) * (f, n)$, let μ' be the first $p + q$ digits to the right of the radix point in the $2p$-digit product mn. Then the result produced by the operation $(e, m) * (f, n)$ is the normalized number (g, μ), where $r^g \mu = \overline{r^{e+f} \mu'}$. That is, we normalize the number $r^{e+f} \mu'$ and then chop it to $S(r, p)$. If $q \geq p$, then $\mu' = mn$, and we shall produce the same result as we would in the system FP(r, p, c). But if $q < p$ and the operands are unnormalized, we may have chopped digits that would have been retained in FP(r, p, c). In fact, we could have $\mu = 0$ even though $mn \neq 0$. The smaller q is, the more severe this problem becomes, and it would be particularly annoying if q were 0 or 1. For this reason, the IBM System/360 normalizes the operands before it performs the multiplication.[†]

Next, we shall consider the quotient $(e, m) \div (f, n)$. If neither m nor n is zero, then

(12.4.1)
$$r^{-1} < \left| \frac{m}{n} \right| < r$$

holds whenever m and n are normalized. But when the operands are allowed to be unnormalized, we have only

$$r^{-p} < \left| \frac{m}{n} \right| < r^p.$$

[†]There are many variations in the way machines compute the product of unnormalized numbers. For example, the IBM 7090 never shifts the result more than one place to postnormalize it. If the operands are normalized, this will produce a normalized result. (See Section 1.8.) Otherwise, the result may be unnormalized, and the product can vanish even though $mn \neq 0$.

To perform the division of normalized numbers, the machine must be able to handle the case in which (12.4.1) holds. But the case in which $|m/n| \geq r$ is often troublesome, and machines differ in the way they treat it. For example, the IBM 7090 produced a *divide check* condition whenever $|m/n| \geq r$, and it did not perform the division. On the other hand, the IBM System/360 avoids the problem by prenormalizing the operands.

Finally, consider the operations $\oplus$ and $\ominus$. We can compute $(e, m) \ominus (f, n)$ by changing the sign of n and adding, so it suffices to consider $(e, m) \oplus (f, n)$. Also, we may assume that the notation is chosen so that $e \geq f$. As in Section 1.8, we form $n' = r^{-(e-f)}n$ by shifting n to the right $e - f$ places. Let n'' be the first $p + q$ digits to the right of the radix point in n', and form $\mu' = m + n''$. The result produced by the operation $(e, m) \oplus (f, n)$ is the normalized number (g, μ), where $r^g \mu = \overline{r^e \mu'}$. Thus if $|\mu'| < 1$, $r^e \mu'$ is normalized and chopped to p digits. On the other hand, if $|\mu'| \geq 1$, $g = e + 1$ and μ is obtained by chopping $r^{-1}\mu'$ to p digits. If (e, m) is normalized, it is easy to see that we shall produce the same result regardless of whether (f, n) is normalized. But if (e, m) is unnormalized and $e - f > q$, then we shall chop more digits of n' than we would have if (e, m) had been prenormalized, and the operation may produce a different result. In fact, when (e, m) is unnormalized, we may shift the number with the larger magnitude. This will happen if $e > f$ but $|r^e m| < |r^f n|$.

The behavior of $(e, 0)$ in addition and subtraction is particularly important. Suppose that

(12.4.2) $(e, 0) \oplus (f, n) = (g, \mu)$.

Clearly

(12.4.3) $r^f n = r^g \mu$

holds whenever $f \geq e$. We recall that in Section 1.4 we defined a normalized zero to be a representation $(e, 0)$ for zero in which e is the smallest allowable exponent, so (12.4.3) always holds if $(e, 0)$ is normalized. Since the operation $\oplus$ always produces a normalized result, the effect of adding a normalized zero to (f, n) is to normalize it—that is, to produce a normalized number (g, μ) satisfying (12.4.3). But we may change the value of a floating-point number if we add an unnormalized zero to it, because n will be chopped in the computation of $(e, 0) \oplus (f, n)$ if $e - f > q$. For this reason, unnormalized zeros can be troublesome. Our definition of floating-point arithmetic in Chapter 1 specified that the result produced by the operations $\oplus$, $\ominus$, $*$, and $\div$ will always be normalized, even if it is zero. When this holds, as it does on many machines, including the IBM System/360, we usually will not encounter any unnormalized zeros in our computation unless we make a special effort to produce them.

There are some situations in which unnormalized zeros can be quite useful. For example, if x has the representation (f, n), then we can chop x to an integer by forming $(f, n) \oplus (p + q, 0)$. This is an easy way to extract the integer part of a floating-point number when we are coding in Assembler language, and the compiler can use this approach to handle the FORTRAN functions AINT and AMOD.

In programming the FIXED TO FLOAT conversion, it is often convenient to begin by constructing an unnormalized number. Let I be an integer with $|I| < r^p$, and suppose that we want to convert the representation of I from a fixed-point to a floating-point number. By converting p to a characteristic and inserting it in the correct portion of the word that holds I, we can form the floating-point representation $(p, r^{-p}I)$ for I. But $(p, r^{-p}I)$ may be unnormalized, so we complete the FIXED TO FLOAT conversion by normalizing it. Some machines have a NORMALIZE instruction which can be used for this purpose; otherwise we normalize the number by adding a normalized zero to it.

We have assumed that the floating-point arithmetic operations always produce normalized results. But many machines have additional instructions to perform *unnormalized arithmetic*, and these instructions can produce unnormalized results. We shall assume that these operations are performed in the same way as the operations in FP(r, p, clq), except that the postnormalization is omitted. That is, we never normalize μ' before chopping it. The result may be unnormalized even though both of the operands are normalized. On some machines, for example, the CDC 6600, all the instructions perform unnormalized arithmetic. To produce a normalized result on such a machine, we first perform the arithmetic operation and then use a NORMALIZE instruction.

Unnormalized arithmetic is often used in programming the FLOAT TO FIXED conversion. We shall illustrate this by considering the coding required for the FORTRAN statement

$$I = X$$

Suppose that the representation for the floating-point number X is (e, m), where $e \leq p$. If we add $(p, 0)$ to (e, m) using unnormalized addition, the result is (p, n), where $r^p n$ is the integer part of X. The integer we want to store in I is comprised of the p digits of n with the radix point at the right. We can usually obtain the result by extracting the p digits and sign of n and storing them in I.† The advantage of using unnormalized addition here is that we know the scaling of the result.

---

†The coding is slightly more complicated on the IBM System/360, because the machine uses the sign and true magnitude representation for floating-point numbers, but it uses the 2's complement representation for negative fixed-point numbers.

Two other sources of unnormalized numbers are the gradual underflow described in Section 2.8 and the significance arithmetic discussed in Section 7.2. With the gradual underflow, a result which underflows may be replaced by an unnormalized number. Significance arithmetic produces unnormalized results, but it uses arithmetic operations which differ somewhat from the unnormalized arithmetic discussed above.

EXERCISES

1. Consider a decimal machine in which each word contains 12 decimal digits and a sign, and suppose that we want to use these words to store the numbers in $S(10, 10)$. We have two decimal digits left to hold the characteristic, which we shall define to be the exponent plus 50. Assume that the machine uses four bits to represent a decimal digit and one bit for the sign, so each word requires 49 bits. We want to compare this machine with a binary machine having a 49-bit word.

   a. Find the largest value of p for which we can represent the numbers in $S(2, p)$ in a 49-bit word if we use enough bits for the characteristic so that we can represent at least as large a range of floating-point numbers as the decimal machine does.

   b. Let p' be the value of p found in part a. Using the criterion (8.1.9) based on the comparison of worst cases, find the largest value of p'' for which we would prefer to use the system $FP(2, p', c)$ instead of $FP(10, p'', c)$.

2. Consider a machine which performs arithmetic in the system $FP(4, p, c)$. The mantissa requires $2p$ bits, so if we use one bit for the sign and k bits for the characteristic, we shall need $2p + k + 1$ bits to hold a floating-point number. We want to compare this machine with a binary machine which uses $2p + k + 1$ bits for each floating-point number and performs arithmetic in $FP(2, p', c)$. Suppose that the binary machine uses k' bits to hold the characteristic, where k' is chosen to produce approximately the same range of floating-point numbers that we had on the machine with radix 4.

   a. How large can p' be?

   b. Which machine provides the better bound for the relative error introduced by chopping a number?

   c. Compare the average relative error introduced by chopping numbers on these machines.

3. Compare the average relative error introduced by chopping numbers to $S(16, 6)$ with that introduced by chopping numbers to $S(2, 22)$, assuming that the mantissas are

   a. Uniformly distributed.

   b. Logarithmically distributed.

4. Suppose that we are using a machine which performs arithmetic in $FP(16, 6, c)$. To compute $x \cdot (\pi/2)$, we can either multiply x by $\pi/2$ or divide x by $2/\pi$. Compare the accuracy with which we can approximate $\pi/2$ and $2/\pi$ by numbers in $S(16, 6)$.

5. Suppose that you were designing the floating-point number system for a binary machine with a 48-bit word. How would you partition the word into the sign, characteristic, and mantissa?

6. Consider a machine which uses complements to represent negative numbers.
   a. If the machine uses $(r - 1)$'s complements, how can you determine whether or not a number is normalized?
   b. If the machine uses $(r - 1)$'s complements, how would you normalize an unnormalized number?
   c. If the machine uses r's complements, how can you determine whether or not a number is normalized?
   d. If the machine uses r's complements, how would you normalize an unnormalized number?

7. Consider the problem of shortening a number from double-precision to single-precision on a binary machine which uses complements to represent negative numbers. Let x be a positive number in $S(2, 2p)$, let $\bar{x}$ be x chopped to $S(2, p)$, and let $\bar{x}^\circ$ be x rounded to $S(2, p)$. Suppose that we are given $-x$, with its mantissa represented as a complement, and that we want to form either $-\bar{x}$ or $-\bar{x}^\circ$. The result is to be stored as a single-precision number with its mantissa represented as a complement. (Note that the 1's complement of the mantissa m of a single-precision number is $2 - 2^{-p} - m$.)
   a. How would you form $-\bar{x}$ if the machine uses 1's complements?
   b. How would you form $-\bar{x}^\circ$ if the machine uses 1's complements?
   c. How would you form $-\bar{x}$ if the machine uses 2's complements?
   d. How would you form $-\bar{x}^\circ$ if the machine uses 2's complements?

8. Consider a machine which has a special bit pattern which is treated as INDEFINITE.
   a. What value would you want stored in I by the FORTRAN statement

$$I = X$$

   when the value of X is INDEFINITE?
   b. Give an example of a situation in which it is essential to know whether or not a number is INDEFINITE.

9. Consider a decimal machine in which the representation of the floating-point numbers consists of a sign, a two-digit characteristic, and an eight-digit mantissa. The characteristic c is the exponent plus γ, and it lies in the interval $0 \leq c \leq 99$. We consider three possible forms for the mantissa:

$$.xxxxxxxx$$
$$x.xxxxxxx$$
$$xxxxxxxx.$$

For each of these forms,
   a. Find ω if $\gamma = 50$.
   b. Find the value of γ that would make the range of the floating-point numbers symmetric. That is, find the value of γ that will yield $\Omega\omega = 1 - 10^{-8}$.

c. Describe the calculation of the characteristic of the product of two normalized floating-point numbers.

10. Let x and y be positive numbers in $S(r, p)$, and let $z = x \oplus y$, where the sum is computed in $FP(r, p, clq)$. Suppose that (e, m) and (f, n) are representations for x and y which are not required to be normalized, and let

$$(g, \mu) = (e, m) \oplus (f, n).$$

Suppose that the machine which we are using is designed so that the arithmetic will be performed in $FP(r, p, clq)$ whenever the operands are normalized, and suppose that it computes the sum $(e, m) \oplus (f, n)$ in the way described in Section 12.4. By how many units in the last place of z can $r^g \mu$ differ from z?

11. Consider a machine which has only unnormalized arithmetic operations. Suppose that it performs arithmetic as it would in $FP(r, p, clq)$ but that it never normalizes the result before chopping it. After each arithmetic operation, we use a NORMALIZE instruction to normalize the result. How does the result differ from the result that would be produced if the arithmetic were performed in the system $FP(r, p, cl0)$ using normalized arithmetic operations?

12. Program the FLOAT TO FIXED and FIXED TO FLOAT conversions in Assembler language for the machine you are using. What special cases must be considered?

13. What would be the effect of using a sticky digit in the add magnitude case of the operations $\oplus$ and $\ominus$ when
a. The operands are normalized?
b. The operands are not required to be normalized?
c. The operands are not required to be normalized and the operation is unnormalized addition?

14. Consider a machine which performs arithmetic in $FP(r, p, c)$ when the operands are normalized. To accomplish this, it performs the arithmetic in $FP(r, p, clq)$, where $q \geq 1$, and it uses a sticky digit in the subtract magnitude case of the operations $\oplus$ and $\ominus$. Assume that it does not prenormalize the operands before performing the arithmetic. Let (g, μ) be the result produced by the operation $(p + q, 0) \oplus (e, m)$. Describe (g, μ) if
a. The operation is normalized addition.
b. The operation is unnormalized addition.

15. How does the unnormalized arithmetic described in Section 12.4 differ from the significance arithmetic discussed in Section 7.2?

13 COMPLEX NUMBERS

13.1. PROGRAMS USING COMPLEX NUMBERS

Although the hardware seldom has operation codes for complex arithmetic, some higher-level languages support the data type COMPLEX. With this data type, the complex number is carried in the form $x + iy$, where x and y are floating-point numbers stored in adjacent storage cells. Another possibility is to represent the complex number in the form $Ae^{i\theta}$, where A and θ are floating-point numbers. While this is more convenient for multiplication and division, it is less convenient for addition and subtraction. We shall assume that the complex numbers are always written in the form $x + iy$.

When the compiler does not support the data type COMPLEX, we encounter many of the same problems we face when we want to use double-precision arithmetic and the compiler does not support the double-precision data type. First, complex numbers require twice as much storage as real numbers do, so we can use one of the techniques decribed in Section 5.7 to allocate the storage for them. We might use different names, say AR and AI, for the real and imaginary parts of A, or we might use an additional subscript to refer to the two parts of the number. Second, the arithmetic must be changed, and in many cases it will be performed by calling subroutines. Finally, we may want to have complex versions of some of the basic library programs.

An annoying problem that arises with complex numbers is that the intermediate results may overflow or underflow even though the answer is within range. For example, if $z = x + iy$, then $|z| = \sqrt{x^2 + y^2}$. Here x^2 and y^2 can overflow even though $|z|$ is much less than Ω. Similarly, they can underflow even though $|z|$ is much larger than ω. Therefore, we would like to have a library program which would avoid these overflows and underflows and compute the absolute value of a complex number z whenever $\omega \leq |z| \leq \Omega$. (See Exercise 8 of Chapter 2.)

13.2. RELATIVE ERROR

If $\tilde{z}$ is an approximation for the complex number $z \neq 0$, we may define the relative error ρ by

$$\rho = \frac{\tilde{z} - z}{z},$$

so $\tilde{z} = (1 + \rho)z$. Here ρ is a complex number, and we are usually interested in a bound for $|\rho|$.

Another approach is to consider the real and imaginary parts of z separately. Let $z = x + iy$ and $\tilde{z} = \tilde{x} + i\tilde{y}$. If neither x nor y vanishes, we may set

$$\sigma_1 = \frac{\tilde{x} - x}{x}$$

and

$$\sigma_2 = \frac{\tilde{y} - y}{y}.$$

Then we shall be interested in the bounds for the $|\sigma_k|$.

Suppose that $|\sigma_k| \leq \sigma$ for $k = 1, 2$. Since

$$\rho = \frac{x + \sigma_1 x + i(y + \sigma_2 y) - (x + iy)}{x + iy} = \frac{\sigma_1 x + i\sigma_2 y}{x + iy},$$

we have

$$|\rho|^2 = \frac{\sigma_1^2 x^2 + \sigma_2^2 y^2}{x^2 + y^2} \leq \frac{\sigma^2 x^2 + \sigma^2 y^2}{x^2 + y^2},$$

so $|\rho| \leq \sigma$. Thus, $|\rho| \leq \max(|\sigma_1|, |\sigma_2|)$, so $|\rho|$ is small if both the $|\sigma_k|$ are small.

But the converse does not hold. Suppose that $z = 1 + 10^{-10}i$ and that $\rho = 10^{-8}i$. Then

$$\tilde{z} = z + \rho z = 1 - 10^{-18} + (10^{-8} + 10^{-10})i,$$

so $|\sigma_2|$ is on the order of 100. Thus, one of the $|\sigma_k|$ may be large even though $|\rho|$ is small.

However, the relative error ρ is still very useful. In many cases we can obtain a bound for $|\rho|$ but not for the $|\sigma_k|$. Moreover, it is often the relative error ρ that is propagated, not the σ_k.†

13.3. COMPLEX ARITHMETIC

We shall now consider the addition, subtraction, multiplication, and division of complex numbers. The complex numbers will be stored in the machine in the form $x + iy$, where x and y are real, floating-point numbers. To simplify the discussion, we shall assume that the machine performs floating-point arithmetic in the system $FP(r, p, cl1)$.

For $k = 1, 2$, let

$$z_k = x_k + iy_k,$$

where x_k and y_k are in $S(r, p)$. If $z = x + iy$ is one of the complex numbers $z_1 + z_2$, $z_1 - z_2$, $z_1 z_2$, or z_1/z_2, we want to compute an approximation $\tilde{z} = \tilde{x} + i\tilde{y}$ for z. Let ρ be the relative error in this approximation, so

$$\tilde{z} = (1 + \rho)z.$$

We want this relative error to be small, and it would be nice if the relative errors σ_1 and σ_2 in the approximations $\tilde{x} \approx x$ and $\tilde{y} \approx y$ were also small. Unfortunately, we cannot always guarantee that the $|\sigma_k|$ will be small, unless we use higher-precision arithmetic. But even if we use only single-precision arithmetic, we can produce a small bound for $|\rho|$.

For the sum $z_1 + z_2$, we set $\tilde{x} = x_1 \oplus x_2$ and $\tilde{y} = y_1 \oplus y_2$. The results of Section 3.3 show that $\tilde{x} = (1 + \sigma_1)x$ and $\tilde{y} = (1 + \sigma_2)y$, where $|\sigma_k| < r^{-(p-1)}$ for $k = 1, 2$. Then, as we saw in Section 13.2, $|\rho| < r^{-(p-1)}$. Subtraction is handled similarly. Since these computations are so simple, many of the compilers that handle the COMPLEX data type compile in-line code instead of calls to subroutines for the addition and subtraction of complex numbers.

Before discussing complex multiplication, we shall prove two theorems which will be used to obtain bounds for the relative error.

THEOREM 13.3.1

Let a, b, c, and d be positive numbers in $S(r, p)$, let $u = ab + cd$, and let $\tilde{u} = (a * b) \oplus (c * d)$, where the floating-point arithmetic is performed in the system $FP(r, p, cl1)$. Suppose that $ab \geq cd$, and let e and f be the integers for which $r^{e-1} \leq ab < r^e$ and $r^{f-1} \leq \tilde{u} < r^f$. Then $|\tilde{u} - u| < 2r^{f-p}$ and $|(\tilde{u} - u)/u| < 2r^{-(p-1)}$. Moreover, if $f > e$, then $|\tilde{u} - u| < (1 + r^{-1})r^{f-p}$.

---

†The same situation arises in matrix problems when we study the error in a vector. We usually consider a norm of the error instead of the relative error in the components of the vector.

Proof. Since $a * b$ and $c * d$ are positive and $cd \leq ab$, it follows that

$$\tilde{u} = \overline{(a * b) + cd}.$$

Then we may write

$$(a * b) + cd = \tilde{u} + \epsilon_1,$$

where $0 \leq \epsilon_1 < r^{f-p}$. We also have $ab = a * b + \epsilon_2$, where $0 \leq \epsilon_2 < r^{e-p}$, so $u = \tilde{u} + \epsilon_1 + \epsilon_2$ and

(13.3.1) $$|\tilde{u} - u| < r^{f-p} + r^{e-p}.$$

Since $f \geq e$, this yields $|\tilde{u} - u| < 2r^{f-p}$ and

$$\left| \frac{\tilde{u} - u}{u} \right| < \frac{2r^{f-p}}{|u|} \leq \frac{2r^{f-p}}{r^{f-1}} = 2r^{-(p-1)}.$$

If $f > e$, then $f = e + 1$, so (13.3.1) reads

$$|\tilde{u} - u| < (1 + r^{-1})r^{f-p}.$$

THEOREM 13.3.2

Let a, b, c, and d be positive numbers in $S(r, p)$, let $v = ab - cd$, and let $\tilde{v} = (a * b) \ominus (c * d)$, where the floating-point arithmetic is performed in the system $FP(r, p, cl1)$. Suppose that $ab \geq cd$, and let e be the integer for which $r^{e-1} \leq ab < r^e$. Then

    1. $|\tilde{v} - v| < (2 - r^{-1})r^{e-p}$.
    2. If $cd \geq r^{e-1}$, then $|\tilde{v} - v| < r^{e-p}$.
    3. If $v \neq 0$ and $|(\tilde{v} - v)/v| \geq 2r^{-(p-1)}$, then
       a. $\frac{1}{2}r^{e-1} < cd < r^e$.
       b. $|\tilde{v} - v| < r^{e-p}$.

Proof. If $ab = cd$, then $a * b = c * d$, so $\tilde{v} = v = 0$ and the theorem holds. Therefore, we may assume that $ab > cd$. Let

$$ab = (a * b) + \delta_1$$
$$cd = (c * d) + \delta_2$$

and

$$(a * b) - (c * d) = [(a * b) \ominus (c * d)] + \delta_3,$$

so

$$v = \tilde{v} + \delta_1 - \delta_2 + \delta_3.$$

Let f be the integer for which $r^{f-1} \leq cd < r^f$. Then $0 \leq \delta_1 < r^{e-p}$ and

$0 \leq \delta_2 < r^{f-p}$. Since the subtraction of $c * d$ from $a * b$ is performed in FP($r, p, cl1$), we have

$$-r^{e-p-1} < \delta_3 \leq (r - 1)r^{e-p-1}.$$

Then

$$|\tilde{v} - v| = |\delta_1 - \delta_2 + \delta_3| < (2 - r^{-1})r^{e-p},$$

so assertion 1 holds. The second assertion follows from the fact that $\delta_3 = 0$ whenever $e = f$.

To prove the third assertion, we assume that $v \neq 0$ and that

(13.3.2)
$$\left| \frac{\tilde{v} - v}{v} \right| \geq 2r^{-(p-1)}.$$

If $e = f$, then both assertions a and b are true, regardless of whether or not (13.3.2) holds. Therefore, we may assume that $e > f$. Then

$$\delta_3 \geq -r^{e-p-1}[1 - r^{-(e-f-1)}],$$

so

$$-\delta_2 + \delta_3 > -r^{e-p-1}.$$

We first show that $v < r^{e-1}$. If $v > r^{e-1}$, we have

$$\left| \frac{\tilde{v} - v}{v} \right| < \frac{(2 - r^{-1})r^{e-p}}{r^{e-1}} < 2r^{-(p-1)},$$

so (13.3.2) implies that $v < r^{e-1}$. If $\tilde{v} \geq r^{e-1} > v$, then $|\tilde{v} - v| < r^{e-p-1}$ and $v > r^{e-1} - r^{e-p-1}$, so

$$\left| \frac{\tilde{v} - v}{v} \right| < \frac{r^{e-p-1}}{r^{e-1} - r^{e-p-1}} < 2r^{-(p-1)}.$$

Therefore, (13.3.2) implies that $\tilde{v} < r^{e-1}$. Then there is a postshift of at least one place in the floating-point subtraction of $c * d$, from $a * b$ and since the arithmetic is performed in the system FP($r, p, cl1$), this implies that $\delta_3 \leq 0$. Then

$$-r^{e-p-1} < \delta_1 - \delta_2 + \delta_3 < r^{e-p},$$

so assertion b holds. If $v \geq \frac{1}{2}r^{e-1}$, this yields

$$\left| \frac{\tilde{v} - v}{v} \right| < \frac{r^{e-p}}{\frac{1}{2}r^{e-1}} = 2r^{-(p-1)}.$$

Therefore, (13.3.2) implies that $v < \frac{1}{2}r^{e-1}$. But

$$v = ab - cd \geq r^{e-1} - cd,$$

so $cd \geq r^{e-1} - v > \frac{1}{2}r^{e-1}$. Thus, assertion a holds, which completes the proof of the theorem.

We shall now consider the product z of the complex numbers z_1 and z_2. The real and imaginary parts of z are given by

$$(13.3.3) \qquad \begin{aligned} x &= x_1 x_2 - y_1 y_2 \\ y &= x_1 y_2 + x_2 y_1, \end{aligned}$$

and we shall use the approximations

$$(13.3.4) \qquad \begin{aligned} \tilde{x} &= (x_1 * x_2) \ominus (y_1 * y_2) \\ \tilde{y} &= (x_1 * y_2) \oplus (x_2 * y_1) \end{aligned}$$

for x and y.

It is easy to see that if the calculation of $\tilde{x}$ involves the add magnitude case, then the calculation of $\tilde{y}$ involves the subtract magnitude case, and vice versa. If, say, the calculation of $\tilde{x}$ involves the add magnitude case, then the approximation $\tilde{x} \approx x$ will have a small relative error. But when the calculation of $\tilde{x}$ involves the subtract magnitude case, we cannot guarantee that the approximation $\tilde{x} \approx x$ will have a small relative error unless we use higher-precision arithmetic. However, even though the relative error in one of the components may be large, the following theorem provides a small bound for $|\rho|$ when the arithmetic is performed in FP($r, p, c/1$).

THEOREM 13.3.3

Let z_1 and z_2 be complex numbers of the form $z_k = x_k + iy_k$, where the x_k and y_k are in $S(r, p)$; let $z = z_1 z_2$; and let $\tilde{z} = \tilde{x} + i\tilde{y}$, where $\tilde{x}$ and $\tilde{y}$ are given by (13.3.4). Assume that the calculations in (13.3.4) are performed in the system FP($r, p, c/1$). Then $\tilde{z} = (1 + \rho)z$, where $|\rho| < 2r^{-(p-1)}$.

Proof. Since

$$(iz_1)z_2 = (-y_1 + ix_1)z_2 = i(z_1 z_2) = -y + ix,$$

the calculation of $(iz_1)z_2$ requires us to perform the same arithmetic operations that we perform in the calculation of $z_1 z_2$, except that the signs of the terms in y and $\tilde{y}$ are changed. Therefore, the calculation of $(iz_1)z_2$ produces the same relative error that the calculation of $z_1 z_2$ does, so we may restrict our attention to the case in which $|x_1| \geq |y_1|$. Similarly, if we replace z_1 by $-z_1$, we merely change the signs of all the terms in (13.3.3) and (13.3.4), so it suffices to consider the case in which $x \geq 0$. If $y_1 = 0$, (13.3.3) reduces to $x = x_1 x_2$ and $y = x_1 y_2$. Then the relative errors in the approximations $x_1 * x_2 \approx x$ and $x_1 * y_2 \approx y$ have absolute values less than $r^{-(p-1)}$, so $|\rho| < r^{-(p-1)}$. Therefore, we may assume that $y_1 \neq 0$. A similar argument applies to z_2, so we may assume that $x_k \geq |y_k| > 0$ holds for $k = 1, 2$. Also, we shall assume that the notation is chosen so that $|x_2 y_1| \geq |x_1 y_2|$. Finally, if we replace z_1 and z_2 by their conjugates $x_1 - iy_1$ and $x_2 - iy_2$, we merely change the signs of the terms in y and $\tilde{y}$. Therefore, we may assume that $y_1 > 0$. Thus, it suffices to consider the case in which

(13.3.5) $x_1 \geq y_1 > 0$

(13.3.6) $x_2 \geq |y_2| > 0$

and

(13.3.7) $x_2 y_1 \geq |x_1 y_2|$.

Let e, f, g, and h be in the integers for which

$$r^{e-1} \leq x_1 x_2 < r^e$$
$$r^{f-1} \leq x < r^f$$
(13.3.8)
$$r^{g-1} \leq x_2 y_1 < r^g$$
$$r^{h-1} \leq y < r^h.$$

Clearly $g \leq e$. If $\tilde{x} = (1 + \sigma_1)x$ and $\tilde{y} = (1 + \sigma_2)y$, where $|\sigma_k| < 2r^{-(p-1)}$ for $k = 1, 2$, the theorem follows from the results of Section 13.2. Therefore, we need consider only the case in which one of the $|\sigma_k|$ is at least $2r^{-(p-1)}$.

First, suppose that y_2 is negative. Then we have the add magnitude case in the calculation of $\tilde{x}$ and the subtract magnitude case in the calculation of $\tilde{y}$. By Theorem 13.3.1, $\tilde{x} = (1 + \sigma_1)x$, where $|\sigma_1| < 2r^{-(p-1)}$. Therefore, we need consider only the case in which $|(\tilde{y} - y)/y| \geq 2r^{-(p-1)}$. Now $\tilde{y} = (x_2 * y_1) \ominus (x_1 * |y_2|)$, so, by Theorem 13.3.2,

$$|\tilde{y} - y| < r^{g-p}$$
(13.3.9)
$$\tfrac{1}{2} r^{g-1} < |x_1 * y_2| < r^g.$$

First, suppose that $f > e$. Then $|z| \geq |x| \geq r^{f-1}$, and, by Theorem 13.3.1,

$$|\tilde{x} - x| < (1 + r^{-1})r^{f-p},$$

so

$$|\rho| < \frac{|(1 + r^{-1})r^{f-p} + ir^{e-p}|}{r^{f-1}} = r^{-(p-1)}|(1 + r^{-1}) + ir^{-1}|.$$

But

(13.3.10) $|(1 + r^{-1}) + ir^{-1}| = \sqrt{1 + 2r^{-1} + 2r^{-2}} < 2,$

so $|\rho| < 2r^{-(p-1)}$.

Thus, we may assume that $f = e$. Then we have

$$\rho < \frac{|2r^{e-p} + ir^{g-p}|}{|z|} \leq 2r^{-(p-1)} \frac{|1 + \frac{1}{2}r^{-(e-g)}i|}{|x/r^{e-1}|},$$

so it suffices to show that

$$|1 + \tfrac{1}{2}r^{-(e-g)}i| \leq \left|\frac{x}{r^{e-1}}\right|$$

or

(13.3.11) $$1 + \tfrac{1}{4}r^{-2(e-g)} \leq \left(\frac{x}{r^{e-1}}\right)^2.$$

From (13.3.8) and (13.3.9), we obtain

$$|x_1 x_2 y_1 y_2| > \tfrac{1}{2}r^{2g-2},$$

so

$$|y_1 y_2| > \frac{\tfrac{1}{2}r^{2g-2}}{x_1 x_2}.$$

Then

$$x = x_1 x_2 + |y_1 y_2| > x_1 x_2 + \frac{\tfrac{1}{2}r^{2g-2}}{x_1 x_2}.$$

Let

$$F(t) = t + \frac{\tfrac{1}{2}r^{2g-2}}{t}.$$

For $t \geq r^{e-1}$, we have

$$F'(t) = 1 - \frac{\tfrac{1}{2}r^{2g-2}}{t^2} \geq 1 - \tfrac{1}{2}r^{-2(e-g)} > 0,$$

so the minimum of $F(t)$ on the interval $r^{e-1} \leq t \leq r^e$ is

$$F(r^{e-1}) = r^{e-1} + \frac{\tfrac{1}{2}r^{2g-2}}{r^{e-1}} = r^{e-1}[1 + \tfrac{1}{2}r^{-2(e-g)}].$$

Then $x > F(r^{e-1})$, so

$$\left(\frac{x}{r^{e-1}}\right)^2 > 1 + r^{-2(e-g)},$$

and (13.3.11) holds. Thus, $|\rho| < 2r^{-(p-1)}$ if $y_2 < 0$.

Now suppose that y_2 is positive, so we have the subtract magnitude case in the calculation of $\tilde{x}$ and the add magnitude case in the calculation of $\tilde{y}$. As above, we need consider only the case in which $|\sigma_1| \geq 2r^{-(p-1)}$, so by

Theorem 13.3.2,

(13.3.12) $$\tfrac{1}{2}r^{e-1} < y_1 y_2 < r^e$$
$$|\tilde{x} - x| < r^{e-p}.$$

Then

(13.3.13) $$\tfrac{1}{2}r^{e-1} < y_1 y_2 \leq x_1 y_2 \leq x_2 y_1 \leq x_1 x_2 < r^e,$$

so

$$r^{e-1} \leq \tilde{y} \leq y < 2r^e.$$

Thus, h is either e or $e + 1$. If $h = e + 1$, we have

$$|\tilde{y} - y| < (1 + r^{-1})r^{h-p}$$

by Theorem 13.3.1, so

$$|\rho| < \frac{|r^{e-p} + i(1 + r^{-1})r^{e+1-p}|}{r^e} = r^{-(p-1)}|r^{-1} + i(1 + r^{-1})|.$$

Using (13.3.10), this yields $|\rho| < 2r^{-(p-1)}$. Therefore, we may assume that $h = e$. If $x_2 y_1 \geq r^{e-1}$, then (13.3.13) implies that $y \geq \tfrac{3}{2}r^{e-1}$, so, using (13.3.12), we have

$$|\rho| < \frac{|r^{e-p} + 2r^{e-p}i|}{\tfrac{3}{2}r^{e-1}} = \tfrac{2}{3}r^{-(p-1)}|1 + 2i| < 2r^{-(p-1)}.$$

Therefore, we may assume that

$$r^{e-1} > x_2 y_1 \geq x_1 y_2 \geq \tfrac{1}{2}r^{e-1}.$$

But then $h = g + 1$, so, by Theorem 13.3.1, $|\tilde{y} - y| < (1 + r^{-1})r^{e-p}$, and this yields

$$|\rho| < \frac{|r^{e-p} + i(1 + r^{-1})r^{e-p}|}{r^{e-1}} = r^{-(p-1)}|1 + i(1 + r^{-1})| < 2r^{-(p-1)}.$$

Thus, $|\rho| < 2r^{-(p-1)}$ also holds for $y_2 > 0$, which completes the proof of the theorem.

We may have some problems with overflow and underflow in the calculations in (13.3.4). We can encounter exponent spill in the intermediate results, even though the numbers we are trying to compute are within range. For example, $|x_1 * x_2|$ can be greater than Ω even though $|x|$ is less than Ω. Similarly, $x_1 * x_2$ and $y_1 * y_2$ can underflow even though $x > \omega$. Now if one of the intermediate results, say $x_1 * x_2$, overflows, we have

$$|z| = |z_1| \cdot |z_2| \geq |x_1| \cdot |x_2|,$$

so $|z| > \Omega$. We might be willing to accept the Ω-zero fixup in this case, although it is not completely satisfactory. (See Exercise 6.) But underflows may be more annoying. Suppose that $y_1 * y_2$ underflows and that we use the Ω-zero fixup. If $|x_1 * x_2| < r^{p-1}\omega$, replacing $y_1 * y_2$ by zero can produce an abnormally large relative error in x. If $|y|$ is also small, this can produce an abnormally large value for $|p|$.

There are two ways we can avoid this problem. One is to use gradual underflow, described in Section 2.8. Another approach is to scale z_1 and z_2 before performing the calculations in (13.3.4) and then adjust the final answer. If we use scale factors which are powers of the radix, scaling will not introduce any error. The scale factors can be chosen in various ways. (See Exercise 7.) For example, we might scale z_1 and z_2 so that the exponent of the larger component of each z_k is zero. Then we shall never encounter overflow in the calculations in (13.3.6), and numbers which underflow can be replaced by zero with negligible effect on $|p|$. Therefore, we can use the Ω-zero fixup for underflow.† If z_1 and z_2 were multiplied by the scale factors s_1 and s_2, we complete the operation by dividing $\tilde{x}$ and $\tilde{y}$ by $s_1 s_2$. If $s_1 s_2$ is a power of the radix, we can do this by adjusting the exponents of $\tilde{x}$ and $\tilde{y}$, and we can provide whatever treatment we want when the resulting exponents are out of range.

Finally, consider division. Assume that $z_2 \neq 0$, and let $z = z_1/z_2$. Then

$$(13.3.14) \qquad z = \frac{(x_1 + iy_1)(x_2 - iy_2)}{x_2^2 + y_2^2},$$

so the real and imaginary parts of z are

$$(13.3.15) \qquad \begin{aligned} x &= \frac{x_1 x_2 + y_1 y_2}{x_2^2 + y_2^2} \\ y &= \frac{x_2 y_1 - x_1 y_2}{x_2^2 + y_2^2}. \end{aligned}$$

The natural approximation to use is $\tilde{z} = \tilde{x} + i\tilde{y}$, where $\tilde{x}$ and $\tilde{y}$ are computed by replacing the arithmetic operations in (13.3.15) by the corresponding floating-point operations. That is, we set

$$(13.3.16) \qquad \tilde{D} = (x_2 * x_2) \oplus (y_2 * y_2)$$

and compute $\tilde{x}$ and $\tilde{y}$ from

$$(13.3.17) \qquad \begin{aligned} \tilde{x} &= [(x_1 * x_2) \oplus (y_1 * y_2)] \div \tilde{D} \\ \tilde{y} &= [(x_2 * y_1) \ominus (x_1 * y_2)] \div \tilde{D}. \end{aligned}$$

†Since these underflows have negligible effect on $|p|$, we might want to avoid printing underflow messages for them.

Now $z = w/D$, where

$$w = (x_1 + iy_1)(x_2 - iy_2)$$

and

$$D = x_2^2 + y_2^2.$$

Let $\tilde{w} = \tilde{u} + i\tilde{v}$, where

$$\tilde{u} = (x_1 * x_2) \oplus (y_1 * y_2)$$
$$\tilde{v} = (x_2 * y_1) \ominus (x_1 * y_2).$$

Then $\tilde{x} = \tilde{u} \div \tilde{D}$ and $\tilde{y} = \tilde{v} \div \tilde{D}$. By Theorems 13.3.3 and 13.3.1, we may write $\tilde{w} = (1 + \sigma)w$ and $\tilde{D} = (1 + \tau)D$, where $|\sigma|$ and $|\tau|$ are less than $2r^{-(p-1)}$. Also,

$$\tilde{u} \div \tilde{D} = (1 - \delta_1)\frac{\tilde{u}}{\tilde{D}}$$

$$\tilde{v} \div \tilde{D} = (1 - \delta_2)\frac{\tilde{v}}{\tilde{D}},$$

where the δ_k are real and $0 \le \delta_k < r^{-(p-1)}$. Then $\tilde{z} = (1 + \delta)\tilde{w}/\tilde{D}$, with $|\delta| < r^{-(p-1)}$. Thus

$$\tilde{z} = \frac{(1 + \delta)(1 + \sigma)}{1 + \tau}z.$$

We may write $\tilde{z} = (1 + \rho)z$, where

$$1 + \rho = \frac{(1 + \delta)(1 + \sigma)}{1 + \tau},$$

which yields

(13.3.18) $$\rho = \frac{\delta + \sigma - \tau + \delta\sigma}{1 + \tau}.$$

Then $\rho \approx \sigma + \delta - \tau$, which shows that the calculations in (13.3.16) and (13.3.17) yield a small relative error ρ. Using (13.3.18), we can obtain a bound for $|\rho|$ on the order of $5r^{-(p-1)}$, but this is not a very good bound. If we exclude trivial combinations of r and p, it can be shown that $|\rho| < 3r^{-(p-1)}$.

Overflow and underflow present a much more serious problem in complex division than they do in complex multiplication. If one of the products in (13.3.4) overflows, $|z_1 z_2|$ must be greater than Ω. But the intermediate results in (13.3.16) and (13.3.17) can overflow even when $|z_1/z_2|$ is on the order of 1. For example, if $|z_1| \approx |z_2| > \sqrt{\Omega}$, we have $|z_1/z_2| \approx 1$, but $D > \Omega$. Similarly, we can have $|z_1/z_2| \approx 1$ and $D < \omega$. Thus, we can have exponent spill in (13.3.16) and (13.3.17) even though $|z_1|, |z_2|$, and $|z_1/z_2|$ all differ from

ω and Ω by many orders of magnitude. Unfortunately, if we use the Ω-zero fixup for exponent spill in (13.3.16) and (13.3.17), we can produce some very misleading results. (See Exercise 9.) Therefore, most programs for complex division try to avoid exponent spill, either by rearranging the calculations in (13.3.16) and (13.3.17) or by scaling z_1 and z_2.

One approach, suggested by Smith (1962), is to divide the factors $(x_2 - iy_2)$ and $(x_2^2 + y_2^2)$ in (13.3.14) by either x_2 or y_2, whichever has the larger absolute value. That is, if $|x_2| \geq |y_2|$, we use

$$(13.3.19) \qquad \tilde{Q} = y_2 \div x_2$$
$$\tilde{D} = x_2 \oplus (y_2 * \tilde{Q})$$
$$\tilde{x} = [x_1 \oplus (y_1 * \tilde{Q})] \div \tilde{D}$$
$$\tilde{y} = [y_1 \ominus (x_1 * \tilde{Q})] \div \tilde{D},$$

and if $|x_2| < |y_2|$, we use

$$(13.3.20) \qquad \tilde{Q} = x_2 \div y_2$$
$$\tilde{D} = (x_2 * \tilde{Q}) \oplus y_2$$
$$\tilde{x} = [(x_1 * \tilde{Q}) \oplus y_1] \div \tilde{D}$$
$$\tilde{y} = [(y_1 * \tilde{Q}) \ominus x_1] \div \tilde{D}.$$

This approach avoids many of the most disastrous cases of exponent spill, but we can still encounter some problems if one of the $|z_k|$ is on the order of ω.

Another approach is to scale z_1 and z_2. We could select a scale factor S and compute z_1'/z_2', where $z_k' = z_k/S$ for $k = 1, 2$. For example, we might use $S = \max(|x_2|, |y_2|)$. Then

$$(13.3.21) \qquad\qquad 1 \leq |z_2'| \leq \sqrt{2}$$

so

$$(13.3.22) \qquad\qquad \left|\frac{z_1}{z_2}\right| \leq |z_1'| \leq \left|\frac{z_1}{z_2}\right| \cdot \sqrt{2}.$$

Consequently, we shall not encounter overflow unless $|z_1/z_2|$ is either out of range or almost out of range. Also, underflows are not a problem unless $|z_1/z_2|$ is quite small, so we have eliminated the most disastrous cases of exponent spill. However, we will still have the same sort of problems with overflow and underflow that we had in the multiplication of complex numbers. Instead of using the scale factor $\max(|x_2|, |y_2|)$, we might prefer to make S a power of the radix, so that the scaling will not introduce any rounding errors. Suppose that we choose the scale factor S to be r^e, where e is the exponent of $\max(|x_2|, |y_2|)$. Then in place of (13.3.21) and (13.3.22), we have

$$r^{-1} \leq |z_2'| < \sqrt{2}$$

and

$$\left|\frac{z_1}{z_2}\right| \cdot r^{-1} \leq |z_1'| < \left|\frac{z_1}{z_2}\right| \cdot \sqrt{2},$$

so the scale factor r^e is almost as effective in eliminating exponent spills as $\max(|x_2|, |y_2|)$.

Probably the best approach is to scale z_1 and z_2 independently. For $k = 1, 2$, let e_k be the exponent of $\max(|x_k|, |y_k|)$, and let $z_k' = z_k/r^{e_k}$. We can divide z_1' by z_2' and then add $e_1 - e_2$ to the exponents of the real and imaginary parts of the quotient. Now $r^{-1} \leq |z_k'| < \sqrt{2}$, and

$$\frac{1}{r\sqrt{2}} < \left|\frac{z_1'}{z_2'}\right| < r\sqrt{2}.$$

Therefore, we shall not encounter overflow in the division of z_1' by z_2', and numbers which underflow are so small with respect to $|z_1'/z_2'|$ that replacing them by zero has a negligible effect on the complex relative error $\rho = (\tilde{z} - z)/z$. We can provide any treatment we desire for the cases in which the addition of $e_1 - e_2$ to the exponents of $\tilde{x}$ and $\tilde{y}$ produces numbers that are out of range.

EXERCISES

1. For $k = 1, 2$, let $\tilde{z}_k = \tilde{x}_k + i\tilde{y}_k$ be an approximation for $z_k = x_k + iy_k$, and suppose that $\tilde{x}_k = (1 + \sigma_k)x_k$, $\tilde{y}_k = (1 + \tau_k)y_k$, and $\tilde{z}_k = (1 + \rho_k)z_k$. Let $\tilde{z}_1\tilde{z}_2 = \tilde{x} + i\tilde{y}$ and $z_1z_2 = x + iy$, and write $\tilde{x} = (1 + \sigma)x$, $\tilde{y} = (1 + \tau)y$, and $\tilde{z}_1\tilde{z}_2 = (1 + \rho)z_1z_2$. Here σ, τ, and the σ_k and τ_k are real, but ρ, ρ_1, and ρ_2 may be complex.
   a. Find an example which shows that even if the $|\sigma_k|$ and $|\tau_k|$ are small, we cannot guarantee that both $|\sigma|$ and $|\tau|$ will be small.
   b. Find a bound for $|\rho|$ in terms of $|\rho_1|$ and $|\rho_2|$.

2. Let $\tilde{z} = \tilde{A}e^{i\tilde{\theta}}$ be an approximation for $z = Ae^{i\theta}$, where $0 \leq \theta < 2\pi$. If $\tilde{A} = (1 + \sigma)A$, $\tilde{\theta} = (1 + \tau)\theta$, and $\tilde{z} = (1 + \rho)z$, find a bound for $|\rho|$ in terms of $|\sigma|$ and $|\tau|$.

3. Suppose that $z_k = A_k e^{i\theta_k}$, where A_k and θ_k are in $S(r, p)$ for $k = 1, 2$. Let $\tilde{A} = A_1 * A_2$ and $\tilde{\theta} = \theta_1 \oplus \theta_2$, where the arithmetic is performed in $FP(r, p, c)$, and consider the approximation $\tilde{z} = \tilde{A}e^{i\tilde{\theta}}$ for $z = z_1z_2$. Write $\tilde{z} = \tilde{x} + i\tilde{y}$ and $z = x + iy$, and let $\tilde{x} = (1 + \sigma)x$, $\tilde{y} = (1 + \tau)y$, and $\tilde{z} = (1 + \rho)z$. What can you say about $|\sigma|, |\tau|,$ and $|\rho|$?

4. Find an example which shows that the relative error p in Theorem 13.3.3 can be close to 2.

5. For $k = 1, 2$, let $z_k = x_k + iy_k$, where x_k and y_k are in $S(r, p)$ and $x_k \geq |y_k|$. Let $z = z_1 z_2$, and write $z = x + iy$.
    a. If $x < x_1 x_2$, what is the smallest $|z|$ can be?
    b. Find an example in which $x_1 x_2 > \Omega$ but $|x|$ and $|y|$ are both less than Ω.

6. For $k = 1, 2$, let $z_k = x_k + iy_k$, where the x_k and y_k are in $S(r, p)$ and $x_k \geq |y_k|$. Let $z_1 z_2 = x + iy$, and let $\tilde{z} = \tilde{x} + i\tilde{y}$ be the approximation for $z_1 z_2$ computed by using (13.3.4). Suppose that we perform the floating-point arithmetic in FP$(r, p, cl1)$ and use the Ω-zero fixup for overflow and underflow.
    a. If $x_1 x_2 > \Omega$, what is the smallest $|\tilde{z}|$ can be?
    b. Find an example in which x and y are greater than Ω but $\tilde{x} = 0$ and $\tilde{y} = \Omega$.
    c. Suppose that x and y are greater than Ω but $\tilde{x} = 0$ and $\tilde{y} = \Omega$. How large can x/y be?
    d. Find an example in which $x > y$ but $0 < \tilde{x} < \tilde{y}$.
    e. If $x > y$ but $0 < \tilde{x} < \tilde{y}$, how large can $\tilde{y}/\tilde{x}$ be?
    f. If $\tilde{x}$ and $\tilde{y}$ are zero, how large can $|z_1 z_2|$ be?

7. For $k = 1, 2$, let $z_k = x_k + iy_k$, where the x_k and y_k are in $S(r, p)$. Let $z_1 z_2 = x + iy$ and let $\tilde{z} = \tilde{x} + i\tilde{y}$ be the approximation for $z_1 z_2$ computed by using (13.3.4). Assume that we perform the floating-point arithmetic in FP$(r, p, cl1)$ and use the Ω-zero fixup for overflow and underflow. Suppose that we scale the z_k by forming $z_k' = z_k/S_k$, and let $z_k' = x_k' + iy_k'$.
    a. Let $S_k = r^{e_k}$, where e_k is the exponent of max$(|x_k|, |y_k|)$, and suppose that both $|x_1' y_2'|$ and $|x_2' y_1'|$ are less than ω. Then $\tilde{y} = 0$. How large can $|y|$ be? How large can $|y/x|$ be?
    b. Assume that we want the scale factors S_k to be powers of the radix, so $S_k = r^{e_k}$. Is there a better choice for the e_k than the exponent of max$(|x_k|, |y_k|)$?

8. Let z_1 and z_2 be complex numbers of the form $x_k + y_k$, where the x_k and y_k are in $S(r, p)$. Let $z = z_1/z_2$, and let $\tilde{z} = \tilde{x} + i\tilde{y}$ be the approximation for z produced by the calculations in (13.3.16) and (13.3.17). Find an example in which $\tilde{z} = (1 + p)z$, where $|p|$ is close to 3.

9. Let z_1 and z_2 be complex numbers of the form $x_k + iy_k$, where the x_k and y_k are in $S(r, p)$. Let $z = z_1/z_2$, and let $\tilde{z} = \tilde{x} + i\tilde{y}$ be the approximation for z produced by the calculations in (13.3.16) and (13.3.17). Suppose that we do not scale the z_k and that we use the Ω-zero fixup for overflow and underflow.
    a. What is the range of $|z_1/z_2|$ if $\tilde{D} = 0$ but $D \neq 0$?
    b. Suppose that $|x|$ and $|y|$ are greater than Ω. How small can $|\tilde{z}|$ be?
    c. Suppose that $|x_1, x_2|, |y_1 y_2|, |x_1 y_2|, |x_2 y_1|, |x_2^2|$, and $|y_2^2|$ are all greater than Ω. What numbers will be produced for $\tilde{x}$ and $\tilde{y}$?

10. Let z_1 and z_2 be complex numbers of the form $x_k + iy_k$, where the x_k and y_k are in $S(r, p)$ and $|x_k| \geq |y_k|$. Let $z = z_1/z_2$, and let $\tilde{z} = \tilde{x} + i\tilde{y}$ be the approximation for z produced by the calculations in (13.3.19). Assume that we use the Ω-zero fixup for exponent spill.

a. Suppose that $D > \Omega$ and that neither $\tilde{x}$ nor $\tilde{y}$ is zero. Find a lower bound for $|z_1/z_2|$.

b. Suppose that $|y_2^2/x_2| < \omega$, so $y_2 * \tilde{Q}$ will be replaced by zero in the calculation of $\tilde{D}$. What is the maximum relative error this can introduce in $\tilde{D}$?

c. Find an example in which underflows in the calculation of $\tilde{x}$ and $\tilde{y}$ produce a bad relative error even though $|z_1/z_2|$ is on the order of 1, $\tilde{D}$ does not underflow, and neither $\tilde{x}$ nor $\tilde{y}$ is zero.

11. Let z_1 and z_2 be complex numbers of the form $x_k + iy_k$, where the x_k and y_k are in $S(r, p)$ and $|x_k| \geq |y_k|$. Let e be the exponent of x_2. To avoid the most disastrous cases of exponent spill in the calculation of z_1/z_2, we compute z_1'/z_2', where $z_k' = z_k/r^e$.

a. Find an example in which the calculation of z_1' overflows, even though the calculation of z_1/z_2 would not overflow. How small can $|z_1/z_2|$ be if this happens?

b. Suppose that we are using the Ω-zero fixup for exponent spill and that we encounter underflow, either in the calculation of z_1' or in the division of z_1' by z_2'. How large can the relative error be if the final answer is not zero?

12. Let z_1 and z_2 be complex numbers of the form $x_k + iy_k$, where the x_k and y_k are in $S(r, p)$. Let e_k be the exponent of $\max(|x_k|, |y_k|)$. To avoid exponent spill in the division of z_1 by z_2, we scale z_1 and z_2 individually by forming $z_k' = z_k/r^{e_k}$. Then we adjust the exponents of the answer accordingly.

a. Show that any numbers which underflow may be replaced by zero with negligible effect on the relative error ρ in the answer.

b. Let $z_1/z_2 = x + iy$, and suppose that we produce the answer $\tilde{x} + i\tilde{y}$. If we use the Ω-zero fixup for exponent spill, underflows may cause us to produce $\tilde{y} = 0$, even though $y \neq 0$. Find a better choice for the e_k that would help avoid this situation.

BIBLIOGRAPHY

AMBLE, O. (1961), "On the Accuracy of Floating-Point Computers," *BIT*, 1, 220–221.

American Standards Association (1964), "A Programming Language for Information Processing on Automatic Data Processing Systems; FORTRAN vs. Basic FORTRAN," *Comm. Assoc. Comput. Mach.*, 7, 591–625.

ASHENHURST, R. L. (1962), "The Maniac III Arithmetic System," *Proc. Spring Joint Comput. Conf.*, AFIPS, 21, 195–202.

ASHENHURST, R. L. (1964), "Function Evaluation in Unnormalized Arithmetic," *J. Assoc. Comput. Mach.*, 11, No. 2, 168–187.

ASHENHURST, R. L. (1965a), "Techniques for Automatic Error Monitoring and Control," in *Error in Digital Computation*, Vol. 1 (ed. by L. B. Rall), Wiley, New York, pp. 43–59.

ASHENHURST, R. L. (1965b), "Experimental Investigation of Unnormalized Arithmetic," *Error in Digital Computation*, Vol. 2 (ed. by L. B. Rall), Wiley, New York, pp. 3–37.

ASHENHURST, R. L., and N. METROPOLIS (1959), "Unnormalized Floating-Point Arithmetic," *J. Assoc. Comput. Mach.*, 6, 415–428.

ASHENHURST, R. L., and N. METROPOLIS (1965), "Error Estimation in Computer Calculation" *Amer. Math. Monthly*, 72, No. 2 (part II: Computers and Computing), 47–58.

AZEN, S., and J. DERR (1968), "On the Distribution of the Most Significant Hexadecimal Digit," *Rand Report R. M. 5496 PR*.

BROWN, W. S., and P. L. RICHMAN (1969), "The Choice of Base," *Comm. Assoc. Comput. Mach.*, 12, 560–561.

CAMPBELL, S. G. (1962), "Floating-Point Operation," in *Planning a Computer System* (ed. by W. Buchholz), McGraw-Hill, New York, pp. 92–121.

301

CARR, J. W., III (1959a), "Error Analysis in Floating-Point Arithmetic," *Comm. Assoc. Comput. Mach.*, 2, No. 5, 10–15.

CARR, J. W., III (1959b), "Programming and Coding," in *Handbook of Automation, Computation, and Control*, Vol. 2 (ed. by E. M. Graffe, S. Ramo, and D. E. Wooldridge), Wiley, New York, Chap. 2.

CHARTRES, B. A. (1966), "Automatic Controlled Precision Calculations," *J. Assoc. Comput. Mach.*, 13, 386–403.

CLARK, N. A., W. J. CODY, K. E. HILLSTROM, and E. A. THIELEKER (1967), "Performance Statistics of the FORTRAN IV (H) Library for the IBM System/360," *Argonne National Lab. Report ANL-7231*, reprinted in SHARE Secretary Distribution, *SDD 169*, C4473, pp. 12–46.

CLARK, N. A., W. J. CODY, and H. KUKI (1971), "Self-Contained Power Routines," in *Mathematical Software* (ed. J. R. Rice), Academic Press, New York pp. 399–415.

CODY, W. J. (1967a), "The Influence of Machine Design on Numerical Algorithms," *Proc. Spring Joint Comput. Conf.*, AFIPS, 30, 305–309.

CODY, W. J. (1967b), "Critique of the FORTRAN IV (H) Library for the IBM System/360," SHARE Secretary Distribution, *SSD 169*, C4473, pp. 4–11.

CODY, W. J. (1971a), "Desirable Hardware Characteristics for Scientific Computation," *SIGNUM Newsletter*, 6, No. 1, 16–31.

CODY, W. J. (1971b), "Software for Elementary Functions," in *Mathematical Software* (ed. by J. R. Rice), Academic Press, New York, pp. 171–186.

CROCKETT, J. B., and H. CHERNOFF (1955), "Gradient Methods of Maximization," *Pac. J. Math.*, 5, 33–50.

DESCLOUX, J. (1963), "Note on the Round-Off Errors in Iterative Processes," *Math. Comp.*, 17, 18–27.

ECKERT, W. J., and R. JONES (1955), *Faster, Faster*, McGraw-Hill, New York.

EHRMAN, J. R. (1969), "A Study of Floating-Point Conversions in Some OS/360 Components," SHARE Secretary Distribution, *SDD 196*, C5207, 1–6.

FIKE, C. T. (1968), *Computer Evaluation of Mathematical Functions*, Prentice-Hall, Englewood Cliffs, N.J.

FLEHINGER, B. J. (1966), "On the Probability that a Random Integer Has Initial Digit A," *Amer. Math. Monthly*, 73, 1056–1061.

FLORES, I. (1963), *The Logic of Computer Arithmetic*, Prentice-Hall, Englewood Cliffs, N.J.

FORSYTHE, G. E. (1959), "Reprint of a Note on Rounding-off Errors," *SIAM Rev.*, 1, 66–67.

FORSYTHE, G. E. (1970), "Pitfalls in Computation, or Why a Math Book Isn't Enough," *Amer. Math. Monthly*, 77, 931–955.

FORSYTHE, G. E., and C. B. MOLER (1967), *Computer Solution of Linear Algebraic Systems*, Prentice-Hall, Englewood Cliffs, N.J.

FROBERG, C.E. (1965), *Introduction to Numerical Analysis*, Addison-Wesley, Reading, Mass.

GARNER, H. L. (1965), "Number Systems and Arithmetic," in *Advances in Computers*, Vol. 6 (ed. by F. L. Alt and M. Rubinoff), Academic Press, New York, pp. 131–194.

GARWICK, J. V. (1961), "The Accuracy of Floating-Point Computers," *BIT*, 1, 87–88.

GOLDBERG, I. B. (1967), "27 Bits Are Not Enough for 8-Digit Accuracy," *Comm. Assoc. Comput. Mach.*, 10, 105–106.

GOLDSTEIN, M. (1963), "Significance Arithmetic on a Digital Computer," *Comm. Assoc. Comput. Mach.*, 6, 111–117.

GOOD, D. I., and R. L. LONDON (1970), "Computer Interval Arithmetic: Definition and Proof of Correct Implementation," *J. Assoc. Comput. Mach.*, 17, 603–612.

GORN, S. (1954), "The Automatic Analysis and Control of Computing Errors," *SIAM J.*, 2, 69–81.

GRAM, C. (1964), "On the Representation of Zero in Floating-Point Arithmetic," *BIT*, 4, 156–161.

GRAY, L. H., and J. C. HARRISON (1959), "Normalized Floating-Point Arithmetic with an Index of Significance," *Proc. Eastern Joint Comput. Conf.*, AFIPS, 16, 244–248.

GREGORY, R. T. (1966), "On the Design of the Arithmetic Unit of a Fixed-Word-Length Computer from the Standpoint of Computational Accuracy," *IEEE Trans. Electronic Comput.*, 15, 255–257.

HAMMING, R. W. (1962), *Numerical Methods for Scientists and Engineers*, McGraw-Hill, New York.

HAMMING, R. W., and W. L. MAMMELL (1965), "A Note on the Location of the Binary Point in the Computing Machine," *IEEE Trans. Electronic Comput.*, 14, 260–261.

HANSON, R. J. (1969), "Certifying Linear Equation Solvers," *SIGNUM Newsletter*, 4, No. 3, 21–27.

HARDING, L. J., JR. (1966a), "Idiosyncrasies of System/360 Floating-Point," presented at SHARE XXVII, Toronto, Canada, Aug. 1966.

HARDING, L. J., JR. (1966b), "Modifications of System/360 Floating-Point," SHARE Secretary Distribution, *SSD 157*, C4470 pp. 11–27.

HARTREE, D. R. (1949), "Note on Systematic Roundoff Errors in Numerical Integration," *J. Res. Nat. Bur. Stand.*, 42, 62.

HENRICI, P. (1959), "Theoretical and Experimental Studies on the Accumulation of Error in the Numerical Solution of Initial Value Problems for Systems of Ordinary Differential Equations," *Proc. Int. Conf. Information Processing*, UNESCO, 36–43.

HENRICI, P. (1964), *Elements of Numerical Analysis*, Wiley, New York.

HENRICI, P. (1966), "Tests of Probabilistic Models for Propagation of Roundoff Errors," *Comm. Assoc. Comput. Mach.*, 9, 409–410.

HILDEBRAND, F. B. (1956), *Introduction to Numerical Analysis*, McGraw-Hill, New York.

HILLSTROM, K. E. (1970), "Performance Statistics for the FORTRAN IV (H) and PL/I (Version 5) Libraries in the IBM OS/360 Release 18," *Argonne National Lab. Report ANL-7666*.

HOUSEHOLDER, A. S. (1953), *Principles of Numerical Analysis*, McGraw-Hill, New York.

HOWELL, K. M. (1967), "Multiple Precision Arithmetic Techniques," *Comp. J.*, 9, 383–387.

HULL, T. E., and J. L. SWENSON (1966), "Tests of Probabilistic Models for the Propagation of Roundoff Error," *Comm. Assoc. Comput. Mach.*, 9, 108–111.

HUSKEY, H. D. (1949), "On the Precision of a Certain Procedure of Numerical Integration," *J. Res. Nat. Bur. Stand.*, 42, 57–62.

International Business Machines (1966), *IBM System/360 Model 91; Functional Characteristics*, IBM Systems Reference Library, File No. S360-01, Form A22-6907-2, IBM, White Plains, N.Y.

International Business Machines (1969), *IBM System/360 Model 195; Functional Characteristics*, IBM Systems Reference Library, File No. S360-01, Form A22-6943-0, IBM, White Plains, N.Y.

ISAACSON, E., and H. B. KELLER (1966), *Analysis of Numerical Methods*, Wiley, New York.

KAHAN, W. (1965a), "The Floating-Point Over/Underflow Trap Routine FPTRP," *Programmers' Reference Manual*, Sect. 4.1, Institute of Computer Science, University of Toronto, Toronto, Canada.

KAHAN, W. (1965b), "Further Remarks on Reducing Truncation Errors," *Comm. Assoc. Comput. Mach.*, 8, 40.

KAHAN, W. (1966), "7094 II System Support for Numerical Analysis," SHARE Secretary Distribution, *SSD 159*, C4537, 1–54.

KANNER, H. (1965), "Number Base Conversion in a Significant Arithmetic," *J. Assoc. Comput. Mach.*, 12, 242–246.

KNUTH, D. E. (1969), *The Art of Computer Programming*, Vol. II, Addison-Wesley, Reading, Mass.

KUKI, H. (1967), "Comments on the ANL Evaluation of the OS/360 FORTRAN Math Function Library," SHARE Secretary Distribution, *SSD 169*, C4773, pp. 47–53.

KUKI, H. (1971), "Mathematical Function Subprograms for Basic System Libraries —Objectives, Constraints, and Trade-Off," in *Mathematical Software* (ed. by J. R. Rice), Academic Press, New York, pp. 187–200.

KUKI, H., and J. ASCOLY (1971), "FORTRAN Extended Precision Library," *IBM Sys. J.*, 10, 39–61.

KUKI, H., E. HANSON, J. J. ORTEGA, J. C. BUTCHER, and P. G. ANDERSON (1966), "Evaluation Guidelines SHARE Numercial Analysis Project (N.A.P.)," SHARE Secretary Distribution, *SSD 150*, part II, C4304, 1–42.

LINZ, P. (1970), "Accurate Floating-Point Summation," *Comm. Assoc. Comput. Mach.*, 13, 361–362.

MATULA, D. W. (1967), "Base Conversion Mappings," *Proc. AFIPS Spring Joint Comput. Conf.*, 30, 311–318.

MATULA, D. W. (1968a), "In and Out Conversions," *Comm. Assoc. Comput. Mach.*, 11, 47–50.

MATULA, D. W. (1968b), "The Base Conversion Theorem," *Proc. Amer. Math. Soc.*, 19, 716–723.

MCCALLA, T. R. (1967), *Introduction to Numerical Methods and FORTRAN Programming*, Wiley, New York.

MCCRACKEN, D. D., and W. S. DORN (1964), *Numerical Analysis and FORTRAN Programming*, Wiley, New York.

MCKEEMAN, W. M. (1967), "Representation Error for Real Numbers in Binary Computer Arithmetic," *IEEE Trans. Electronic Comput.*, EC16, 682–683.

METROPOLIS, N. (1965), "Analysis of Inherent Errors in Matrix Decomposition Using Unnormalized Arithmetic," *Proc. IFIP Congress*, 2, 441–442.

METROPOLIS, N., and R. L. ASHENHURST (1958), "Significant Digit Computer Arithmetic," *IRE Trans. Electronic Comput.*, EC7, 265–267.

METROPOLIS, N., and R. L. ASHENHURST (1963), "Basic Operations in an Unnormalized Arithmetic System," *IEEE Trans. Electronic Comput.*, EC12, 896–904.

METROPOLIS, N., and R. L. ASHENHURST (1965), "Radix Conversion in an Unnormalized Arithmetic System," *Math. Comp.*, 19, 435–441.

MILLER, R. H. (1964), "An Example in 'Significant Digit' Arithmetic," *Comm. Assoc. Comput. Mach.*, 7, 21.

MOLLER, R. H. (1965), "Quasi Double-Precision in Floating-Point Addition," *BIT*, 5, 37–50.

MOOD, A. M., and F. A. GRAYBILL (1963), *Introduction to the Theory of Statistics*, McGraw-Hill, New York.

MOORE, R. E. (1965a), "The Automatic Analysis and Control of Error in Digital Computation Based on the Use of Interval Numbers," in *Error in Digital Computation*, Vol. I (ed. by L. B. Rall), Wiley, New York, pp. 61–130.

MOORE, R. E. (1965b), "Automatic Local Coordinate Transformations To Reduce the Growth of Error Bounds in Interval Computation of Solutions of Ordinary Differential Equations," in *Error in Digital Computation*, Vol. II (ed. by L. B. Rall), Wiley, New York, pp. 103–140.

Moore, R. E. (1966), *Interval Analysis*, Prentice-Hall, Englewood Cliffs, N.J.

Muller, M. E. (1959), "Generation of Normal Deviates," *J. Assoc. Comput. Mach.*, 6, No. 3, 376–383.

National Physical Laboratory (1961), *Modern Computing Methods*, 2nd ed., Notes on Applied Science No. 16, Her Majesty's Stationery Office, London.

Nickel, K. (1966), "Uber die Notwendigkeit einer Fehlerschranken-Arithmetik fur Rechenautomaten," *Numer. Math.*, 9, 69–79.

Nickel, K. (1968), "Error Bounds and Computer Arithmetic," *Proc. IFIP Congress 1968*, North-Holland, Amsterdam, pp. 54–60.

Padegs, A. (1968), "Structural Aspects of the System/360 Model 85, III Extension to Floating-Point Architecture," *IBM Sys. J.*, 7, 22–29.

Pinkham, R. S. (1961), "On the Distribution of First Significant Digits," *Ann. Math. Stat.*, 32, 1223–1230.

Ralston, A. (1965), *A First Course in Numerical Analysis*, McGraw-Hill, New York.

Rice, J. R. (1971), *Mathematical Software*, Academic Press, New York.

Smith, R. L. (1962), "Complex Division," *Comm. Assoc. Comput. Mach.*, 5, No. 8, 435.

Stifler, W. W., Jr. (1950), *High Speed Computing Devices*, McGraw-Hill, New York.

Stegun, I. A., and M. Abramowitz (1956), "Pitfalls in Computation," *SIAM J.*, 4, 207–219.

Stuart, F. (1969), *FORTRAN Programming*, Wiley, New York.

Sweeney, D. W. (1965), "An Analysis of Floating-point Addition," *IBM Sys. J.*, 4, 31–42.

Turner, L. R. (1969a), "Input-Output Conversion in System/360," SHARE Secretary Distribution, *SSD 194*, C5173, pp. 1–8.

Turner, L. R. (1969b), "Difficulty in SIN/COS Routine," *SIGNUM Newsletter*, 4, No. 3, 13.

Turner, L. R. (1969c), "Comment on Some IBM Software," SHARE Secretary Distribution, *SSD 199*, C5279, pp. 40–43.

Urabe, M. (1968), "Roundoff Error Distribution in Fixed Point Multiplication and a Remark About the Rounding Rule," *SIAM J. Num. Anal.*, 5, 202–210.

Usow, K. H. (1969), "SIGNUM Subroutine Certification Committee Report," *SIGNUM Newsletter*, 4, No. 3, 15–18.

Usow, K. H. (1970), "Certification Bibliography," *SIGNUM Newsletter*, 5, No. 2, 14–15.

van Wijngaarden, A. (1966), "Numerical Analysis as an Independent Science," *BIT*, 6, 66–81.

von Neuman, J., and H. H. Goldstein (1947), "Numerical Inverting of Matrices of High Order," *Bull. Amer. Math. Soc.*, 53, 1021–1099.

WADEY, W. G. (1960), "Floating-Point Arithmetics," *J. Assoc. Comput. Mach.*, 7, 129–139.

WILKINSON, J. H. (1963), *Rounding Errors in Algebraic Processes*, Prentice-Hall, Englewood Cliffs, N.J.

WYNN, P. (1962), "An Arsenal of Algol Procedures for Complex Arithmetic," *BIT*, 2, 232–255.

YARBOROUGH, L. (1967), "Precision Calculation of e and π Constants," *Comm. Assoc. Comput. Mach.*, 10, 537.

YOUNG, D. M., and A. E. McDONALD (1968), "On the Surveillance and Control of Number Range and Accuracy in Numerical Integration," *Proc. IFIP Congress 1968*, North-Holland, Amsterdam, pp. 145–152.

GLOSSARY OF SYMBOLS

Symbol	Definition	Page
$\oplus$	Floating-point addition	10
$\ominus$	Floating-point subtraction	10
$*$	Floating-point multiplication	10
$\div$	Floating-point division	10
r	Radix (number base) in which the floating-point numbers are written	9
$S(r,p)$	Set of floating-point numbers having p digits of precision in the radix r	10
$FP(r, p, a)$	Floating-point number system using p-digit numbers in the radix r; the symbol substituted for a designates the type of arithmetic used	9
$FP(r, p, c)$	Floating-point number system using p-digit numbers in the radix r; the symbol c specifies that the system uses chopped arithmetic	12
$FP(r, p, R)$	Floating-point number system using p-digit numbers in the radix r; the symbol R specifies that the system uses rounded arithmetic	12
$FP(r, p, clq)$	Floating-point number system using p-digit numbers in the radix r; the symbol clq specifies that the system uses chopped arithmetic with a low-order register that is q digits long	22
$\bar{x}$	Number produced by chopping x to p digits in the radix r	13
$\overset{\circ}{x}$	Number produced by rounding x to p digits in the radix r	13
x_L	Largest number in $S(r, p)$ which is $\leq x$	197
x_R	Smallest number in $S(r, p)$ which is $\geq x$	197
$B(x)$	Number obtained from x by performing the bias removal operation	195
$B'(x)$	Number obtained from x by performing the modified bias removal operation	196

Symbol	Definition	Page
D, H, O, B used as subscripts in numbers	These subscripts specify the radix in which the number is written; D stands for decimal, H for hexadecimal, O for octal, and B for binary	9
A, B, C, D, E, F used as digits in numbers	When a number is written in hexadecimal, the symbols A, B, C, D, E, F represent the hexadecimal digits ten through fifteen, respectively.	10
e^*	Maximum exponent that can be used in the machine representation of floating-point numbers	39
e_*	Minimum exponent that can be used in the machine representation of floating-point numbers	39
γ	Number which must be added to the exponent of a floating-point number to produce its characteristic	11
Ω	Largest positive floating-point number whose exponent is $\leq e^*$	40
ω	Smallest positive floating-point number whose exponent is $\geq e_*$.	40

INDEX